19/332
£136.48

7384R.   BM2

BM 2

7384R

Automated Microbial Identification and Quantitation

Technologies for the 2000s

Edited by
Wayne P. Olson

Interpharm Press, Inc.
Buffalo Grove, IL

Library of Congress Cataloging-in-Publication Data

Automated microbial identification and quantitation:
    technologies for the 2000s / edited by Wayne P. Olson
        p.   cm.
    Includes bibliographical references and index.
    ISBN 0-935184-82-1 (hard)
    1. Microbiology—Automation.   2. Microbiological assay—Automation
  I. Olson, Wayne P.
  QR69.A88A87   1996
  576'.0285—dc20
                                         95-43354
                                           CIP

10 9 8 7 6 5 4 3 2

Interpharm Press, Inc.
1358 Busch Parkway
Buffalo Grove, IL 60089, USA
Phone: + 1 + 847 + 459-8480
Fax: + 1 + 847 + 459-6644

Contents

Types of Systems Available   59

Data Management and Interface Options   68

Applications   69

System Evaluation—A User's Perspective   71

Comparison to Other Automated Systems   73

Quality Control   74

System Validation   76

References   79

Gerald K. Masover
Frances A. Becker

9.   CELLULAR FATTY ACID ANALYSIS FOR THE CLASSFICATION AND IDENTIFICATION OF BACTERIA   179

A. Ray Smith
Joel P. Siegel

10.   ENUMERATION AND CHARACTERIZATION OF MICROBIAL POPULATIONS USING OPTICAL AND ELECTRICAL ZONE-SENSING PARTICLE COUNTERS   223

Thomas A. Barber

Preface

Automated Microbial Identification and Quantitation: Technologies for the 2000s is addressed primarily to two markets: pharmaceutical manufacturers that produce parenteral drugs, including those that employ biotechnology, and clinical microbiologic laboratories, whether independent or in a hospital. The needs of these disparate groups for rapid, reliable microbial assessment of samples overlap.

Most injectable drugs are sterilized by filtration and filled aseptically. The reduction of microbial contamination in these preparations requires that discovered contaminants be identified and their sources detected in a timely manner. Similarly, the detection, quantitation, and identification of aerosolized microbes in filling and test areas is of great importance. It is crucial that the data are available for interpretation as quickly as possible, hence the need for, and widespread use of, automated systems. An increasing number of injectable drugs are biologics, often produced in recombinant microbes grown in axenic cultures. This provides interesting challenges in the detection and quantitation of contaminants. This book address many of these needs and identifies state of the art as well as trends in the industry.

The requirement of pharmaceutical manufacturers for rapid microbial identification is less urgent than the need to identify and characterize the organism(s) infecting septic patients. The sense of urgency in the clinical microbiological laboratory relates to patient treatment and survival. This urgency was the initial driving force for the marketing of machines such as the Microbial Identification System (MIS) and VITEK® (bioMérieux, Inc.)

Automated Microbial Identification and Quantitation is not complete in and of itself. Many important issues have been ignored in this book (e.g., dysgenic, starved, and slow-growing organisms, which presently give rise to issues in pharmaceutical manufacturing and detection/identification in the clinical laboratory).

In another example, although the emergence of DNA/RNA probes is noted and briefly described, the growing role of these probes has not received appropriate attention in this book.

It is proper to note the interplay of texts related to aseptic manufacturing. *Aseptic Pharmaceutical Manufacturing: Technology for the 1990s* (edited by W. P. Olson and M. J. Groves) and *Sterile Pharmaceutical Manufacturing: Applications for the 1990s* (Volumes I and II, edited by M. J. Groves, W. P. Olson, and M. H. Anisfeld) relate to manufacturing methods and are predecessors to the present text. A companion text to *Aseptic* was recently published, *Aseptic Pharmaceutical Manufacturing II: Applications for the 1990s* (edited by M. J. Groves and R. Murty). Also, *Isolator Technology: Applications in the Pharmaceutical and Biotechnology Industries* (edited by C. M. Wagner and J. E. Akers) is equally and directly applicable. All are published by Interpharm Press, Buffalo Grove, Illinois.

It is also appropriate to note that Michael Anisfeld, Amy Davis, and all at Interpharm Press have been crucial members of the team who have conceived, written, edited, produced, and marketed this group of texts that will contribute to improvements in our methods of health care.

Finally, as a memorial to the lasting contributions that Michael Faith made to Organon Teknika, the new BCG building in Durham was dedicated to him in July 1993. The dedication of this text is my tribute to a fine friend.

Wayne P. Olson
January 1996

1

Gas Chromatography, Substrate Utilization, and Systems for Microbial Identification

Wayne P. Olson

Oldevco

Microbes recovered from sterility test failures must be identified if the source is to be determined (Olson 1987). Increasingly, FDA inspectors press for the characterization of microbial flora in the filling rooms and elsewhere in the facilities of pharmaceutical manufacturing firms that perform aseptic fills. In order to control labor costs and to maintain reproducibility in the identification of microbes, quality control (QC) departments in pharmaceutical manufacturing facilities and several FDA sites have purchased systems for automated microbial identification (AMI). The AMI systems most commonly used are based on either substrate utilization (SU) for microbial growth, including reaction to dyes and antibiotics, or gas chromatography (GC) of fatty acid methyl esters. Table 1.1 lists systems currently in the marketplace.

TWO VERY DIFFERENT SYSTEMS

At this writing, the AMI systems in most widespread use appear to be the VITEK® instrument, an SU marketed by bioMérieux Vitek (Hazelwood, MO), and the MIS (microbial identification system), a GC handled by MIDI (Microbial ID, Inc.,

Table 1.1. Automated Systems for Identification of Microorganisms

System	Basis[a]	In this text	% GN correct[b]
ALADIN	SU	No	—[c]
AutoReader	SU	No	—[d]
autoSCAN	SU	No	60–95
AutoSceptor	SU	No	95
Biolog	SU	Yes	56–60
MIS	GC	Yes	81–93
Sensititre	SU	No	35–93
VITEK	SU	Yes	75–97
WalkAway-40 or -96	SU	No	41–92

[a] GC = gas chromatography of fatty acid methyl esters; SU = substrate utilization, and various inhibitors

[b] % correct identification of gram-negative organisms (enteric and nonenteric), adapted from Tables 1 and 2 and text of Stager and Davis (1992).

[c] Limited testing by one group, details missing, unusual protocol comparing machine data with visual data on the same test (Shulman et al. 1987). Percentage correlation of visual observation with automated readouts is 99 percent.

[d] O'Hara et al. (1990). Ibid.

Newark, DE). These systems have been marketed effectively and have been around for many years. The MIS, VITEK, and a variety of other units are used in pharmaceutical companies, clinical microbiological laboratories, environmental monitoring departments and firms and, to a lesser degree, in the food/beverage industries. These and other systems have been reviewed by Stager and Davis (1992).

The systems listed in Table 1.1 are utilitarian devices for organism identification. The microbial taxa outlined in *Bergey's Manual* (Tanner and Socransky 1984) and journals of microbial systematics are based on multiple characteristics, including colony appearance, cell morphology, motility, pathogenesis, G+C ratios, and other attributes not measured by these instruments. The SU machines are capable of indicating antibiotic susceptibility and some growth characteristics, which MIS cannot do.

Table 1.1 presents percentage identification of gram-negative enterics and nonenterics, together with some data on ALADIN and AutoReader that relate only to the visual correlation with machine test information (i.e., the correlation of the test with itself), which is of questionable value.

The information of Master et al. (1991) is extremely interesting. It would appear that a number of *Methylobacterium* spp. and *Pseudomonas vesicularis* were identified correctly by the MIS and but not by any of the SU systems (API 20E; Rapid NFT, BBL Sceptor; Corning Uni-NF Tek; MicroScan, Pasco; Vitek GNI). Also, the MIS is suitable for strain designation (see, for example, Mukwaya and Welch 1989).

Tanner et al. (1981) state that *Wolinella recta* differs from *Wolinella succinogenes* in cellular and ultrastructural morphology, serological reactions, and susceptibility to dyes and antibiotics. On the basis of SU characteristics, they are difficult or impossible to separate (although some dye-susceptibility tests are possible in an SU system like VITEK). *Wolinella recta* and *Bacteroides gracilis* are metabolically identical, but differ in cell morphology and motility (Tanner et al. 1984).

This text is not about microbial taxonomy sensu stricto, but about the practical aspects of routine microbial identification and quantitation, with the emphasis on automation.

LIMITATIONS TO THE VITEK AND MIS SYSTEMS

The commercial identification systems are programmed with analyses of large numbers of species and strains obtained from well-described culture collections such as American Type Culture Collection (ATCC) and National Type Culture Collection (NTCC). Machine identifications tend to be highly reliable to the genus level, but each makes some errors at the species level. The errors are consistent with the information available in the machine library and the attributes that the machine measures. For example, both the MIS (Olson et al. 1989) and any SU instrument (including VITEK) should misidentify the BCG strain of *Mycobacterium bovis* as *Mycobacterium tuberculosis*, where inoculum size is not a consideration. *Mycobacterium tuberculosis* is pathogenic and *Mycobacterium bovis* BCG is a nonpathogenic attenuated vaccine for the treatment of primary bladder cancer and tuberculosis in man. Differentiation is based on pathogenesis in the guinea pig: *Mycobacterium tuberculosis* kills, *Mycobacterium bovis* strain BCG does not. Differentiation based on the structure of mycolic acids (cord factors) *may* be possible, but is not done by either machine.

There are instances in which a given machine will not perform an identification. For example, if analysis to genus and species is based on the utilization of various substrates, the standard VITEK system may not perform adequately if a small inoculum of a dysgenic or a slow-growth organism is analyzed. The semiautomatic Automated Test(er) for Bacteriology (ATB), which is only a card reader, might be suitable for these since incubation is off-line. Where a variety of organisms thrive in a particular niche (e.g., enterics) and have similar fatty acid compositions, identification to species may be difficult in the MIS based on fatty acids containing 9–20 carbons (Welch 1991).

Prior to SU analysis in the VITEK, a Gram stain must be performed on the isolate. If the stain is done improperly or is misread, the gram-positive VITEK card may be selected for the analysis, resulting in gross error. In the MIS system, because there are significant chemical differences between gram-positive and gram-negative organisms, a Gram stain is unnecessary prior to machine analysis. However, consistent results in GC analysis for microbial identification requires uniformity of media (Farshtchi and McClung 1970), temperature of incubation (Marr and Ingraham 1962), injection port temperature (Guerrant et al. 1981), and the nutritional state of the organism (Guckert et al. 1986).

The isolate must be cultured overnight (e.g., 18 hr) for MIS analysis since 20 mg or more (wet weight) of cells is required. This usually is a disadvantage, but can also be an attribute. One prefers results more rapidly than after an overnight delay, but overnight growth establishes whether the cells are nonviable or slow-growers. Overnight culturing also provides adequate material for several other analytical procedures, including the use of labeled antibodies, DNA probes, or RNA probes, if they are indicated. There is a possibility of fatty acid analysis from single colonies when the methyl esters are generated by means of trimethyl-sulfonium hydroxide (Nalik et al. 1992).

Once a satisfactory amount of inoculum has been prepared, the MIS material can be analyzed to determine its viability. This is an advantage in the shipment of samples for analysis. So far as the author is aware, autoclaved samples can be sent by parcel post without any warning labels or special considerations since the material is noninfectious and nonviable.

A superficial comparison of the performances of the MIS and VITEK on the same isolates is given in Table 1.2. Note the consistent tendency of the two systems to assign different species names to the same isolate. Some of the characteristics of the two instruments, based on direct comparisons and information from users and the manufacturers, are summarized in Table 1.3.

OTHER MACHINES

Biolog (chapter 2) is a 96-well system that operates much as does an enzyme-linked immunosorbent assay (ELISA) system. Different substrates or reagents are placed in the various wells that are inoculated. The system is incubated (aerobically, anaerobically, or under microaerophilic conditions) and read in an ELISA-type plate reader programmed for microbial identification. Biolog, unlike most of the other systems, has the immediate potential for a variety of nontaxonomic applications, such as the screening of bacteria capable of degrading toxic substances (Gorden et al. 1993). However, this text is about identification and quantitation, and not directly about bioremediation.

Parties interested in the identification performance of the various systems are urged to consult Table 6 in Stager and Davis (1992) for data on the machines listed in Table 1.1. The referee manual system to which Biolog, VITEK, and other SU-type system results are compared is Analytical Product Identification (API, Analytab Products distributed by bioMérieux, St. Louis, MO). See, for example, Karachewski et al. (1985).

Systems That Are Unlikely to Be Available

Irwin (1982) opined that mass pyrolysis of the intact organism is a suitable method for the identification of organisms. However, he pointed out that the

Table 1.2. Identifications Performed with a GC System (MIS) and an SU System (VITEK) on a Small Number of Isolates

Isolate	MIS identification	VITEK identification
1	*Staphylococcus haemolyticus*	*Staphylococcus epidermidis*
2	*Bacillus pumilus*	*Bacillus* sp.
3	*Staphylococcus hominus*	*Staphylococcus epidermidis*
4	*Penicillium* sp.	Mold[a]
5	*Corynebacterium* sp.	Nonviable[b]
6	*Staphylococcus hominus*	*Staphylococcus cohnii*
7	*Staphylococcus hominus*	*Staphylococcus auricularis*
8	*Staphylococcus haemolyticus*	*Staphylococcus cohnii*
9	*Staphylococcus aureus*	*Staphylococcus aureus*
10	*Alternaria* sp.	Mold[a]
11	*Micrococcus roseus*	Gram-positive coccus
12	*Bacillus thuringiensis*	*Bacillus* sp.

[a] Identified by microscopic observation

[b] Slow grower

mass pyrograms of *Pseudomonas putida* and of *Bacillus subtilis* var. *niger* are remarkably similar although the former is gram negative and the latter is gram positive. Gas chromatography–mass spectrometry of microbial pyrolysis products has been examined critically (Snyder et al. 1990). In this author's opinion the Py/GC–MS approach appears to offer a significant additional investment in hardware (the mass spectrometer)—an approach that is not cost-effective on the basis of the information currently available.

Analysis of growth medium by high performance liquid chromatography (HPLC), before and after incubation with pure cultures, has been attempted with some success in the identification of several bacterial species (Harpold and Wasilauskas 1987; Radin et al. 1988). The number of studies required to bring either the Py/GC–MS or the HPLC systems to market is considerable, and from the literature one concludes that they have not been done.

Table 1.3. Characteristics of the MIS and VITEK Systems for the Automated Identification of Microbes

Characteristic[a]	MIS	VITEK
Gram stain required	No	Yes
Identifies slow growers	Yes[b]	No/Yes[b]
Identifies microaerophiles	Yes	Yes
Identifies anaerobes	Yes	Yes
Differentiates *Bacillus* spp. well	Yes	Variable[c]
Number of yeasts identified	195	36
Inoculum required	20–60 mg (wet)	Loopful
Processes multiple isolates[d]	Yes	Yes
Differentiates Enterobacteriaceae	Variable	Yes
Differentiates strains	Many	No
Libraries are updated	Yes	Yes
User can add to library	Yes	Yes
Sample preparation required	Yes	No/Yes[b]
Average hours, isolate to ID	30	3–24
Price of hardware $\times 10^{-3}$[e,f]	$46+	$39.5–70.5
Cost per sample[f]	$1.30–1.50[h]	$3.00–5.00[g]
Yearly maintenance[f]	$1,000	$1,000

[a] Characteristics listed in approximate priority order; the requirement for a Gram stain is most important.

[b] Growth of the organisms *must* be external to the system. For slow growers, if they are not cultured external to the system, the VITEK will indicate that the isolate is nonviable. Whether external culturing is required affects sample preparation and the time from isolation to identification.

[c] A *Bacillus* identification system is available but was not used in this study.

[d] 100 or more samples can be run by either system.

[e] Fully automated systems range from $39,500 for the VITEK Jr., which handles 30 tests simultaneously, to VITEK 60 (60 tests, $60,500) and the VITEK 120 (120 tests, $70,500). A VITEK 240 exists but is not used by pharmaceutical firms. The semiautomatic ATB incubates off-line but cannot identify *Bacillus* spp. ($19,500).

[f] Data provided by the manufacturer.

[g] Gram-negative expendables, $3.00–5.00; gram-positive expendables, $3.50; yeast, $4.50; *Bacillus* spp. system is priced at $19,500. Discounts available.

[h] Growth media, reagents, and solvents.

METHODS WITH GREAT PROMISE

Phenotypic characterization of closely related organisms is not always differential. All of the systems cited thus far examine phenotypic characters. Guanine + cytosine (G+C) ratios (relative to adenine + thymine [A + T]) are a move in the direction of genotypic characterization, the ultimate method for the identification of an organism. The base sequence of the chromosome should be definitive. Forbes and Hicks (1993) and Luk (1994) devised detection methods for *Mycobacterium tuberculosis* and *Salmonella typhi*, respectively; both tests are based on polymerase chain reaction (PCR) amplifications of the genes of the organisms.

Since it is derivative of the DNA sequence, the sequence of the 16S rRNA is useful taxonomically (Gutell et al. 1994; Head et al. 1993; Lau et al. 1987; Olsen and Woese 1993; Paster and Dewhirst 1988; Teske et al. 1994; Thompson et al. 1988). The phenol-extracted RNA is purified and converted to DNA by the reverse transcriptase method, the DNA is multiplied manyfold by means of the PCR (Lane et al. 1985; Stahl et al. 1988). DNA sequencing is done by the method of Lane et al. Aside from DNA sequencing, these steps are done by hand, are tedious, and require considerable expertise.

Conserved regions of the rRNA are common to groups such as the methanogens (Raskin et al. 1994a, 1994b) and sulfate-reducing bacteria (Risatti et al. 1994). Radioactive- or fluorescent-labeled probes of conserved areas are useful for the detection and the quantitation of such microbial groups. Probes of this type are powerful tools (Amann et al. 1990b) and target cells can be quantified in an excess of nontarget cells by means of flow cytometry (Amann et al. 1990a). Probes labeled with ^{32}P make possible the identification and quantitation of cells even in a microbial mat (Risatti et al. 1994).

Avaniss-Aghajani et al. (1994) employed PCR amplification of what they term small subunit rRNA and digested the amplification products with restriction endonucleases. They then analyzed the restriction enzyme(s) products by capillary electrophoresis rather than by conventional polyacrylamide gel electrophoresis (PAGE). With the exception of the analysis by capillary electrophoresis, this seems to be an extraordinary amount of work. However, Chen et al. (1994) applied a similar approach to the detection and typing of papillomaviruses. The use of probes should be simpler, less tedious, and less expensive.

Gen-Probe, Inc. (San Diego, CA) currently sells probes used in the identification of chlamydia and gonococcus. They now are marketing a chemiluminiscent-labeled DNA probe that targets the rRNA of *Streptococcus* Group A, notably *Streptococcus pyogenes*, that diagnoses Group A strep pharyngitis within 1 hour. Gen-Probe currently has a DNA probe assay for *Mycobacterium tuberculosis* in culture and has submitted it to the FDA for review data on a direct, amplified test for *Mycobacterium tuberculosis* in small numbers in sputum. Microprobe (Bothell, WA) has developed DNA probes (for targeting rRNA), but the probes are attached to nylon beads. Microprobe's Affirm VPIII is a probe system used in the detection of the three pathogens (a bacterium, a protozoan, and a yeast) associated with vaginitis.

At this writing, Gen-Probe claims patent rights to rRNA as a target molecule and is in litigation with Microprobe, but not with Gene-Track Systems (a division of Amoco Oil, Framingham, MA), which makes a diagnostic test for food pathogens that targets rRNA. A database on probes and applications is in preparation

(Larsen et al. 1993). Aprogenex (Houston, TX) produces DNA probes (trade-named GenSite) of mRNA that determine the number and identity of chromosomes within fetal-nucleated erythrocytes (Danheiser 1994).

At this writing and within the immediate future, it would appear that such probes are not competitive with MIS, VITEK and other well-established systems. However, it is clear that probe technology is emerging rapidly. If the use of probes is automated and made as user-friendly as the GC and SU systems, it will see routine use in microbiological laboratories everywhere.

ACKNOWLEDGMENTS

Charles H. Moore, Rhone-Poulenc Rorer Pharmaceutical Co., Kankakee, IL, provided the VITEK analyses; Prof. A. R. Smith, Dept. of Veterinary Pathobiology, Univ. of Illinois-Urbana Champaign, provided the MIS analyses. I thank them both for their expertise, data, and occasional advice.

REFERENCES

Amann, R. I., B. J. Binder, R. J. Olson, S. W. Chisholm, R. Devereux, and D. A. Stahl. 1990a. Combination of 16S rRNA-targeted oligonucleotide probes with flow cytometry for analyzing mixed microbial populations. *Appl. Environ. Microbiol.* 56:1919–1925.

Amann, R. I., L. Krumholz, and D. A. Stahl. 1990b. Fluorescent-oligonucleotide probing of whole cells for determinative, phylogenetic, and environmental studies in microbiology. *J. Bacteriol.* 172:762–770.

Avaniss-Aghajani, E., K. Jones, D. Chapman, and C. Brunk. 1994. A molecular technique for identification of bacteria using small subunit ribosomal RNA sequences. *BioTechniques* 17:144–149.

Chen, S., S. N. Tabriz, C. K. Fairley, A. J. Borg, and S. M. Garland. 1994. Simultaneous detection and typing strategy for human papillomaviruses based on PCR and restriction endonuclease mapping. *BioTechniques* 17:138–143.

Danheiser, S. L. 1994. Advances in DNA probe-based assays lead to second generation products. *Gen. Engr. News* 14 (12):6–7.

Farshtchi, D., and N. M. McClung. 1970. Effect of substrate on fatty acid production in *Nocardi esteroides*. *Can. J. Microbiol.* 16:213–217.

Forbes, B. A., and K. E. S. Hicks. 1993. Direct detection of *Mycobacterium tuberculosis* in respiratory specimens in a clinical laboratory by polymerase chain reaction. *J. Clin. Microbiol.* 31:1688–1694.

Gorden, R. W., T. C. Hazen, and C. B. Fliermans. 1993. Rapid screening for bacteria capable of degrading toxic organic compounds. *J. Microbiol. Methods* 18:339–347.

Guckert, J. B., M. A. Hood, and D. C. White. 1986. Phospholipid ester-linked fatty acid profile changes during nutrient deprivation of *Vibrio cholerae*: Increases in the *trans/cis* ratio and proportions of cyclopropyl fatty acids. *Appl. Environ. Microbiol.* 52:794–801.

Guerrant, G. O., M. A. Lambert, and C. W. Moss. 1981. Gas-chromatographic analysis of mycolic acid cleavage products in mycobacteria. *J. Clin. Microbiol.* 13:899–907.

Gutell, R. R., N. Larsen, and C. R. Woese. 1994. Lessons from an evolving rRNA: 16S and 23S rRNA structures from a comparative perspective. *Microbiol. Rev.* 58:10–26.

Harpold, D. J., and B. L. Wasilauskas. 1987. Rapid identification of obligately anaerobic gram-positive cocci using high-performance liquid chromatography. *J. Clin. Microbiol.* 25:996–999.

Head, I. M., W. D. Hiorns, T. M. Embley, A. J. McCarthy, and J. R. Saunders. 1993. The phylogency of autotrophic ammonia-oxidizing bacteria as determined by analysis of 16S ribosomal RNA gene sequences. *J. Gen. Microbiol.* 139:1147–1153.

Irwin, W. J. 1982. *Analytical pyrolysis.* New York: Marcel Dekker.

Karachewski, N. O., E. L. Busch, and C. L. Wells. 1985. Comparison of PRAS II, RapID ANA, and API 20A systems for identification of anaerobic bacteria. *J. Clin. Microbiol.* 21:122–126.

Lane, D. J., B. Pace, G. J. Olsen, D. A. Stahl, M. L. Sogin, and N. R. Pace. 1985. Rapid determination of 16S ribosomal RNA sequences for phylogenetic analyses. *Proc. Natl Acad. Sci. USA* 82:6955–6959.

Larsen, N., G. J. Olsen, B. L. Maidak, M. J. McCaughery, R. Overbeek, T. J. Macke, T. L. Marsh, and C. R. Woese. 1993. The ribosomal RNA data base project. *Nucleic Acids Res.* 21:3021–3023.

Lau, P. P., B. DeBrunner-Vossbrinck, B. Dunn, K. Miotto, M. T. MacDonnell, D. M. Rollins, C. J. Pillidge, R. B. Hespell, R. R. Colwell, N. L. Sogink, and G. E. Fox. 1987 Phylogenetic diversity and position of the genus *Campylobacter. Syst. Appl. Microbiol.* 9:231–238.

Luk, J. M. C. 1994. A PCR enzyme immunoassay for detection of *Salmonella typhi.* *BioTechniques* 17:1038–1042.

Marr, A. G., and J. L. Ingraham. 1962. Effect of temperature on the composition of fatty acids in *Escherichia coli. J. Bacteriol.* 84:1260–1267.

Master, R. N., R. L. Sautter, W. J. Brown, J. W. Hong, and A. E. Smith, Jr. 1991. Comparison of the Microbial Identification System with conventional and commercially available systems for the identification of pink-pigmented, oxidase-positive bacteria. Abstract C-218, 91st Gen. Mtg. Washington, DC: American Society for Microbiology.

Mukwaya, G. M., and D. F. Welch. 1989. Subgrouping of *Pseudomonas cepacia* by cellular fatty acid composition. *J. Clin. Microbiol.* 27:2640–2646.

Nalik, H. P., K. D. Muller, and R. Ansorg. 1992. Rapid identification of *Legionella* species from a single colony by gas-liquid chromatography with trimethysulphonium hydroxide for transesterification. *J. Med. Microbiol.* 36:371–376.

O'Hara, C. M., D. L. Rhoden, and P. B. Smith. 1990. Agreement between visual and automated UniScept API readings. *J. Clin. Microbiol.* 28:452–454.

Olsen, G. J., and C. R. Woese. 1993. Ribosomal RNA: A key to phylogeny. *FASEB J.* 7:113–123.

Olson, W. P. 1987. Sterility testing. In *Aseptic pharmaceutical manufacturing: Technology for the 1990s,* edited by W. P. Olson, and M. J. Groves. Buffalo Grove, IL: Interpharm Press.

Olson, W. P., M. J. Groves, and M. E. Klegerman. 1989. Fatty acids of *Mycobacterium bovis* BCG. *Microbios* 57:151–155.

Paster, B. J., and F. E. Dewhirst. 1988. Phylogeny of campylobacters, wolinellas, *Bacteroides gracilis* and *Bacteroides ureolyticus* by 16S ribosomal ribonucleic acid sequencing. *Int. J. Syst. Bacteriol.* 38:56–62.

Radin, L., A. Arzese, C. Lucarelli, and G. A. Botta. 1988. Rapid identification of *Bacteroides* species by high-performance liquid chromatography. *J. Chromatog.* 459:331–335.

Raskin, L., L. K. Poulsen, D. R. Noguera, B. E. Rittmann, and D. A. Stahl. 1994a. Quantification of methanogenic groups in anaerobic biological reactors by oligonucleotide probe hybridization. *Appl. Environ. Microbiol.* 60:1241–1248.

Raskin, L., J. M. Stromley, B. E. Rittmann, and D. A. Stahl. 1994b. Group-specific 16S rRNA hybridization probes to describe natural communities of methanogens. *Appl. Environ. Microbiol.* 60:1232–1240.

Risatti, J. B., W. C. Capman, and D. A. Stahl. 1994. Community structure of a microbial mat: The phylogenetic dimension. *Proc. Natl Acad. Sci. USA* 91:10173–10177.

Shulman, M. A., M. C. Navarro, J. A. Bahrenburg, S. Jahnke, and D. Schreier. 1987. Comparison of the ALADIN video image processing to manual interpretation of UniScept 20E. Abstract C-94, 37th Annu. Mtg. Washington, DC: American Society for Microbiology.

Snyder, A. P., W. H. McClennen, J. W. Dworzanksi, and H. L. C. Meuzelaar. 1990. Characterization of underivatized lipid biomarkers from microorganisms with pyrolysis short-column gas chromatography/ion-trap mass spectrometry. *Anal. Chem.* 62:2565–2568.

Stager, C. E., and J. R. Davis. 1992. Automated systems for identification of microorganisms. *Clin. Microbiol. Rev.* 5:302–327.

Stahl, D. A., B. Flesher, H. R. Mansfield, and L. Montgomery. 1988. Use of phylogenetically based hybridization probes for studies of ruminal microbial ecology. *Appl. Environ. Microbiol.* 54:1079–1084.

Tanner, A. C. R., and S. S. Socransky. 1984. Genus *Wolinella*. In *Bergey's Manual of Systemic Bacteriology*, vol 1, edited by N. R. Krieg, and J. G. Holt. Baltimore: The Williams and Wilkins Co.

Tanner, A. C. R., S. Badger, C.-H. Lai, M. A. Listgarten, R. A. Visconti, and S. S. Socransky. 1981. *Wolinella* gen. nov., *Wolinella succinogenes* (*Vibrio succinogenes* Wolin *et al.*) comb. nov., and description of *Bacteroides gracilis* sp. nov., and *Eikenella corrodens* from humans with periodontal disease. *Int. J. Syst. Bacteriol.* 31:432–445.

Teske, A., E. Alm, J. M. Regan, S. Toze, B. E. Rittmann, and D. A. Stahl. 1994. Evolutionary relationships among ammonia- and nitrite-oxidizing bacteria. *J. Bacteriol.* 176:6623–6630.

Thompson, III, L. M., R. N. Smibert, J. L. Johnson, and N. R. Krieg. 1988. Phylogenetic study of the genus *Campylobacter*. *Int. J. Syst. Bacteriol.* 38:190–200.

Welch, D. F. 1991. Applications of cellular fatty acid analysis. *Clin. Microbiol. Rev.* 4:422–438.

2

The Biolog MicroStation System and General Procedures for Identifying Environmental Bacteria and Yeast

Barry R. Bochner

Biolog, Inc.

IDENTIFICATION OF MICROORGANISMS BY SUBSTRATE UTILIZATION TESTING IN 96-WELL MICROPLATES

The concept of identifying microorganisms based on their substrate utilization (SU) patterns originated with the doctoral dissertation of L. E. den Dooren de Jong in 1926. Although it is generally viewed as an old methodology, it remains the most popular and commonly used method of microbial identification. The principal reason for the popularity of SU testing is that it is a very simple and low-cost approach. However, the methodology has never reached its full potential due to certain technical limitations. The two most important limitations have been (1) the lack of a sufficient number of tests, and (2) a lack of standardization and automation of both the testing procedures and the interpretation of test results.

This chapter describes the Biolog MicroStation System, which uses standardized 96-well microplates in which comprehensive SU testing is performed. Species and strains can be identified by their characteristic "metabolic fingerprints" in the 96-well test panels. Interpretation of test results is standardized with an optical reader and computer for rapid and consistent identification.

MICROBIAL IDENTIFICATION AND CHARACTERIZATION: THE NEEDS OF THE PHARMACEUTICAL INDUSTRY

Microbes are often important in the pharmaceutical manufacturing industry either because they are part of the manufacturing process, or because they interfere with or contaminate the process. Therefore, the need to identify or to characterize microbes may be motivated by any of the issues listed below.

1. The need for quality control testing of stock cultures and inoculum cultures used in production. Microorganisms are often subject to subtle changes as they are subcultured. These changes may be temporary phenotypic changes or irreversible genetic changes. In either case these subtle changes may have a large deleterious impact on fermentation reproducibility.

2. The need to use microbes in R&D programs. Microorganisms frequently are used to produce antibiotics, enzymes, or recombinant proteins. It is often important to know the catabolic properties of different strains. This information is useful in developing fermentation growth media, in comparing and distinguishing closely related strains in a culture collection, and in learning about the presence of catabolic enzymes that specific strains may possess.

3. The need to describe and distinguish strains for patent applications.

4. The need for routine environmental monitoring of manufacturing rooms and clean rooms to prevent contamination.

5. The need to track down and eliminate sources of bacterial contamination.

6. The need to meet the requirements of regulatory agencies, including issues related to GMP manufacturing, worker safety, or environmental release.

The Biolog MicroStation provides a simple and practical technology for identifying and characterizing microorganisms.

MICROBIAL IDENTIFICATION: THE MAGNITUDE OF THE PROBLEM

No one knows how many microbial species currently exist on our planet; in fact, we do not even have a sound basis for confidently estimating the number within an order of magnitude (Bull et al. 1992; May 1988). For example, there are only about 2,600 bacterial species named thus far, whereas the number of named protozoan species is approximately 260,000. One would expect there to be many more species of bacteria than protozoa since the number of species generally is inversely correlated with the size of the organism. However, the issue is complicated because there is no precise definition of a bacterial species; thus, bacteria may not be speciated as finely as other organisms that have distinct morphological features or mating behaviors.

For economic reasons the microbes that have received the most consideration are those affecting human beings. Examples include bacteria and yeast causing disease in humans, animals, or plants important to humans, and those causing spoilage of food or contamination of drinking water.

There are about 500 species of bacteria and yeast that have been found to cause human disease out of approximately 3,500 named species. And yet, prior to the introduction of the Biolog System, test kits for the identification of microbes could identify only about 60 percent of even this restricted subset. These test kits typically are composed of a panel of 20 to 30 tests and can identify 50 to 100 species.

THE BIOLOG SYSTEM

The Biolog System (Bochner 1989a,b) is designed around the use of 96-well MicroPlates containing 95 (for bacteria) or 94 (for yeast) tests per panel. Because it has a much larger number of tests per panel, the Biolog System is capable of identifying a much greater range of microbial species.

In the spring of 1989, Biolog introduced the GN MicroPlate panel for the identification of a very wide range of gram-negative species. Two years later, the complementary GP MicroPlate panel for gram-positive bacteria was introduced. Then in 1993 the YT MicroPlate was introduced for the identification of yeast and the capabilities of the GP MicroPlate were expanded to include the important lactic acid bacteria. With these three test kits combined, a microbiologist can identify over 1,100 species of aerobic bacteria and yeast using simple, standardized protocols. This includes virtually all human and animal pathogens, most of the important plant pathogens, and most of the environmental species that are commonly encountered in human environments.

Very recent innovations at Biolog have led to the introduction of SF MicroPlates for sporulating and filament-forming microorganisms such Actinomycetes and Fungi. These MicroPlates currently can be used for strain characterization, but species identification databases are also planned.

A Major Simplification of Technology

To use the Biolog System, all one needs to identify most of the 1,118 species is a Gram stain capability, two basic culture media, and Biolog's three MicroPlate panels. Biolog MicroPlates are prefilled and dried panels that are easily inoculated with a cell suspension. No color-developing reagents, oil overlays, or follow-on tests are needed. All tests in the panels are carbon source utilization tests. In the bacteria identification panels oxidation of a carbon source is detected as an increase in the respiration of cells in the well, leading to irreversible reduction of a tetrazolium dye. A positive utilization reaction is indicated when a purple color forms in a well. This test chemistry has many advantages over the pH-change chemistries used in most other test kits. The YT MicroPlate for yeasts is slightly different from the GN and GP MicroPlates for bacteria in that it contains both oxidation and assimilation tests as part of the carbon source utilization panel.

The utilization pattern that results (Figure 2.1) can be read either by eye or with an automated microplate reader (Figure 2.2). With 95 tests that can be either "positive" or "negative," there are 2^{95} (or 4×10^{28}) utilization patterns that theoretically could result. A simple codebook is not sufficient to interpret the patterns; instead, a proprietary software program is available from Biolog that will run on virtually any IBM-compatible PC. This software matches any pattern to the 1,100 plus species library in about one second. Details of the method of pattern matching will be given later.

Identification of an Unknown Bacterium with the Biolog System

The general identification procedures are diagrammed in Figures 2.3 and 2.4. To identify an unknown bacterium, the microbiologist first performs a Gram stain. If the bacterium is gram negative, it is cultured on either BUGM (Biolog Universal Growth Medium, Biolog, Inc.) with 5 percent sheep blood added or on TSA (Tryptic Soy Agar, commonly available) with 5 percent sheep blood added. A cell suspension is prepared at a specified density and is inoculated into the GN

Figure 2.1. Metabolic fingerprint on the Biolog GN MicroPlate.

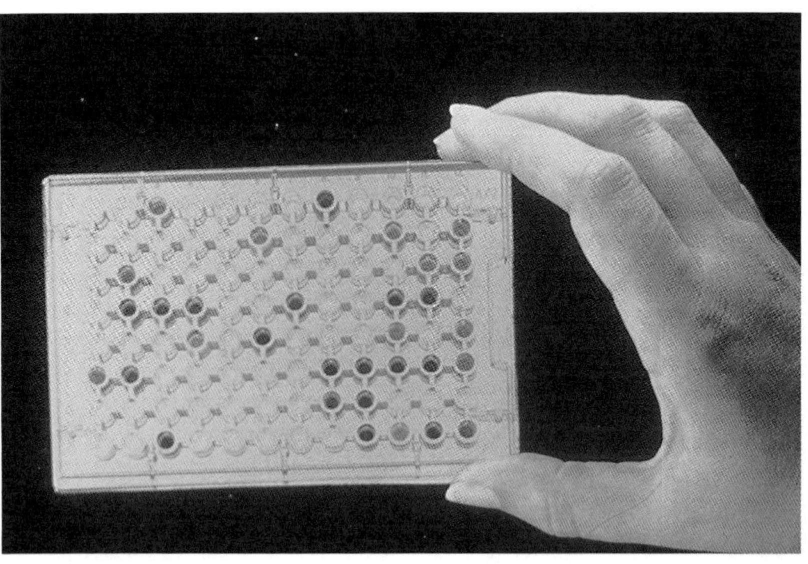

Figure 2.2. The Biolog MicroStation.

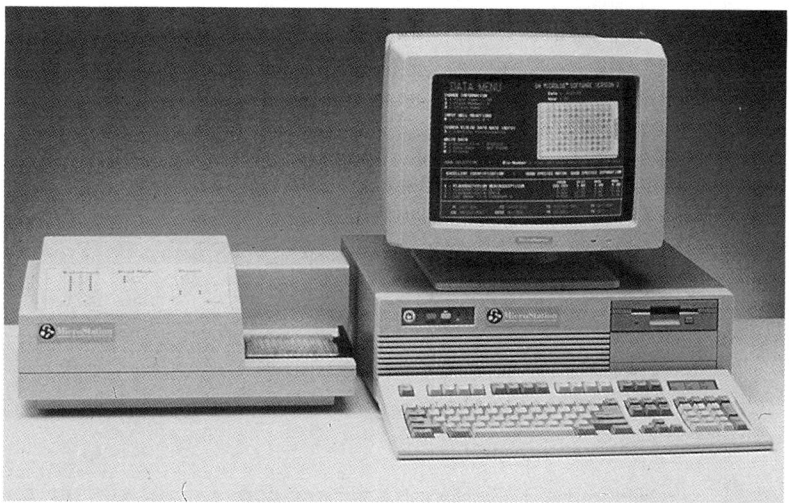

Figure 2.3. Flowchart of culturing and inoculation procedures: (1) Perform a Gram stain and determine whether the bacterium is gram negative or gram positive. If it is gram positive, look also for spores, since spore-forming bacilli require special handling. (2) If the bacterium is gram negative, culture it on BUGM or TSA supplemented with 5 percent sheep blood and inoculate into a GN MicroPlate. (3) If the bacterium is a nonspore-forming gram positive, culture it on BUGM supplemented with 5 percent sheep blood and inoculate into a GP MicroPlate. (4) If the bacterium is a spore-forming gram positive, culture it on BUGM supplemented with 1 percent glucose and inoculate into a GP MicroPlate.

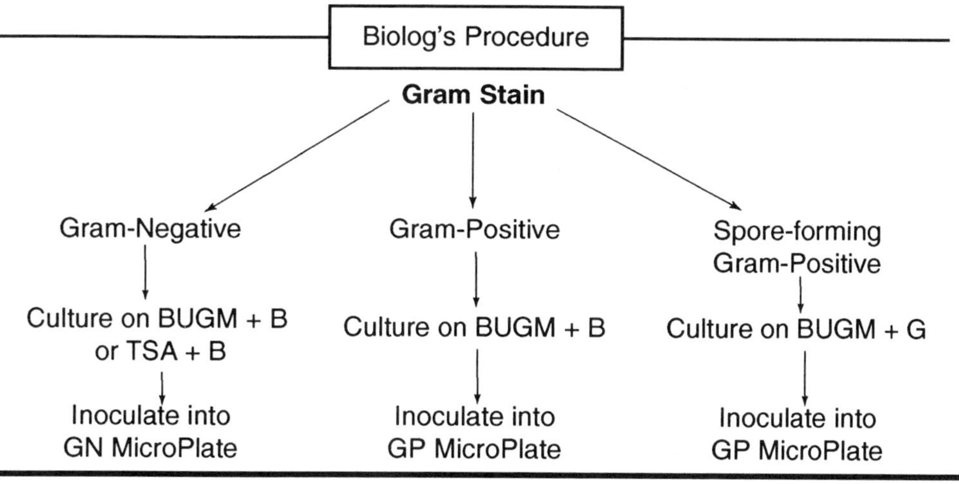

Figure 2.4. Four simple steps for identifying a bacterium.

1 Swab colonies from agar surface and suspend in normal saline.

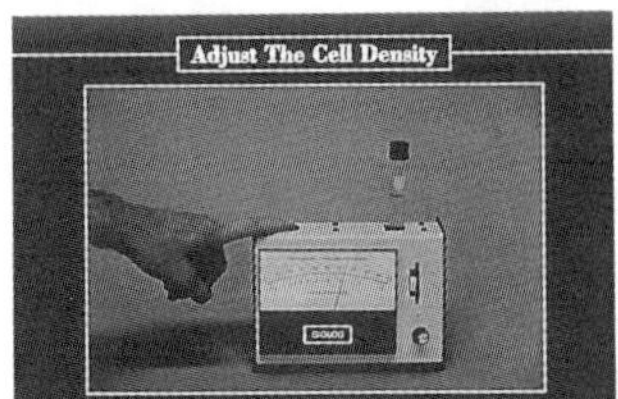

2 Adjust suspension density.

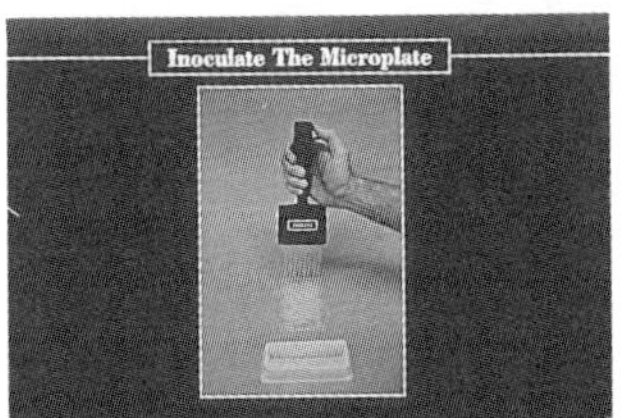

3 Pipette suspension into MicroPlate and incubate for 4 hours or overnight.

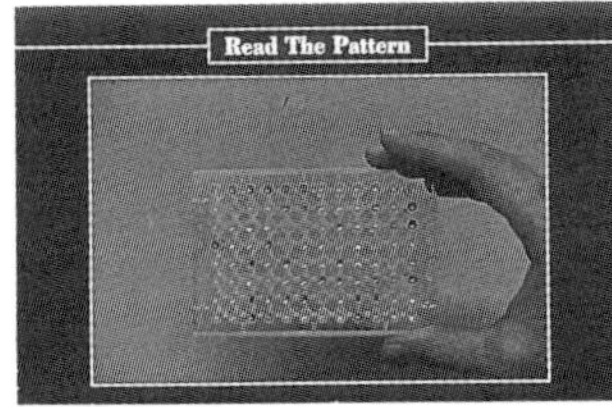

4 Read results (visually or automatically) and identify the bacterium using Biolog's extensive database software.

MicroPlate. If the bacterium is gram positive, the microbiologist must distinguish between spore-formers (i.e., *Bacillus* species), which are cultured on BUGM with 1 percent glucose, or any other gram positives (rods or cocci), which are cultured on BUGM with 5 percent sheep blood. All gram positives are inoculated into the GP MicroPlate. To identify an unknown yeast, the isolate is cultured on BUY Agar (Biolog, Inc.) and inoculated into the YT MicroPlate.

There are a few exceptions to the general protocol for special groups of microbes, such as fastidious gram negatives *(Neisseria, Haemophilus)*, oligotrophic gram negatives *(Rhizobium, Methylobacterium)*, gram-positive plant pathogens *(Clavibacter, Curtobacterium)*, and some lactic acid bacteria *(Lactobacillus)*. However, with the possible exception of *Methylobacterium*, none of these genera are likely to be encountered in a pharmaceutical manufacturing environment.

Methylobacterium isolates can be recognized by their unusual and characteristic pink pigment.

The Importance of Culture Media

The selection of culture media is a very important, underappreciated issue. It is important in two ways. One issue is recovery of microorganisms from the environment. Ideally, one would like to have a single culture medium that would recover and grow all microorganisms at their maximum growth rate. Although an ideal culture medium does not currently exist, there are some very good options available that are superior to traditionally utilized but poorly optimized media.

A second issue, alluded to in the previous section, is the importance of culture media in the accurate identification of microorganisms using carbon utilization testing methods. When microbes are grown on an agar medium, they induce the synthesis of some enzymes and they repress the synthesis of other enzymes. When the microbes are subsequently inoculated into a panel of carbon source utilization tests, they may or may not have appropriate enzymes to perform the utilization, and they may or may not have induced the synthesis of the needed enzymes. Some species adjust quickly, and within a few minutes or hours will make the necessary enzymes to give "positive" reactions for all of the carbon sources that they can utilize. However, other species do not adjust quickly or even within 24 hours. If they were grown on a suboptimal medium, they will give fewer "positive" reactions than they are capable of and they may therefore be misidentified as another species of bacteria with fewer catabolic capabilities. An example is shown in Figure 2.5. In this example *Staphylococcus auricularis* and *Staphylococcus cohnii* were cultured on either TSA with blood or BUGM with blood prior to inoculation into the GP MicroPlate. It is clear that many more positive reactions are seen after growth on BUGM with blood.

The ability of a medium to support rapid growth and the ability of a medium to give a maximum number of "positive" carbon source utilization tests must each be considered. Tryptic Soy Agar is a very good medium that supports the growth of a very wide range of gram-negative and gram-positive species, including the *Staphylococcus* species shown in Figure 2.5. Although both TSA and BUGM work well for carbon source testing of gram negatives, BUGM is clearly superior for gram positives and, therefore, it is superior overall.

Many culture media are formulated with glucose since it is a commonly used carbon source for microorganisms. However, it is generally ill-advised to use media containing glucose prior to carbon source testing. In many microorganisms glucose represses synthesis of other catabolic enzymes (catabolite repression). Furthermore, many bacteria form acids as by-products of glucose utilization. Typical culture media are not adequately buffered, and the drop in pH can weaken or even kill the bacteria after they grow on the agar medium.

One clear exception to this generalization are *Bacillus* species. *Bacillus* are unique among common aerobic bacteria in that they form spores when they are not given an excess of nutrients, including glucose. Prior to testing *Bacillus* strains in the GP MicroPlate, they must be subcultured, preferably two or more times, on BUGM with 1 percent glucose added. This gets a predominance of the cell population out of the sporulation state and into the vegetative state. Fortunately,

Figure 2.5. Example illustrating better carbon source utilization reactions when *Staphylococcus* species are grown on BUGM- rather than TSA-based culture media.

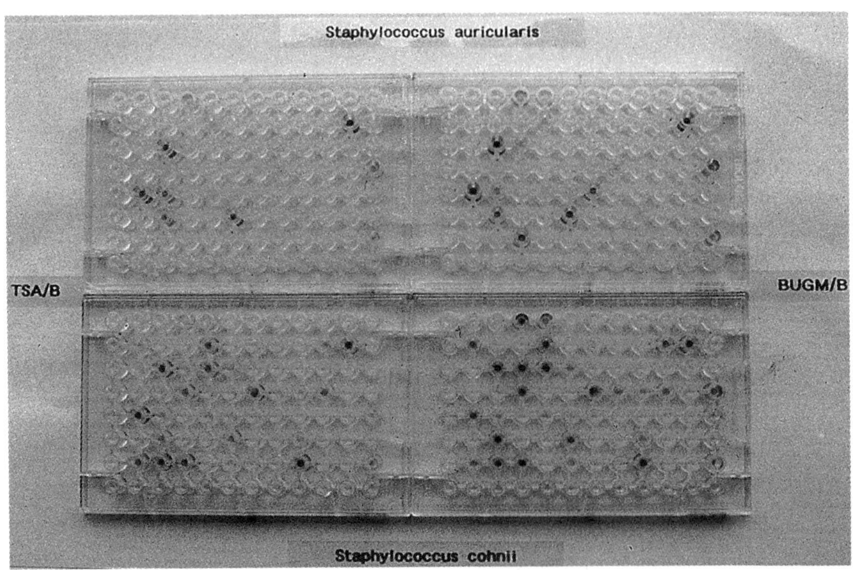

glucose is not converted to acid by *Bacillus*, and it does not cause catabolite repression in *Bacillus*.

Recommended Culture Media

For convenience it is useful to divide the microorganisms that one would expect to encounter in typical human environments into three groups: copiotrophic bacteria, oligotrophic bacteria, and yeasts and other fungi.

Copiotrophic bacteria include the typical fast-growing bacteria forming a clearly visible colony within 24 hours at a moderate incubation temperature (e.g., 30°C). They are called copiotrophs because they generally grow fastest in a nutrient-rich culture medium.

Oligotrophic bacteria include slower-growing bacteria that often require 48 hours or longer to form small colonies. Unlike the copiotrophs, they grow best on media that have low levels of nutrients and obligate oligotrophs will not grow on nutrient-rich media. Oligotrophic bacteria are commonly found in low nutrient environments, such as air, water, and soil. To complicate matters further, copiotrophic bacteria often can survive in oligotrophic environments by changing to an oligotrophic state. Sometimes, they cannot be recovered directly from the environment onto a nutrient-rich medium, but first must be recovered onto an oligotrophic medium, and then gradually subcultured onto a copiotrophic medium.

Yeasts will typically grow best on nutrient-rich media containing high levels of carbohydrates. Although they can grow on a wide range of media, they tend to outgrow bacteria in environments that are acidic and environments containing high levels of sugars. Other fungi will grow on a wide variety of media.

With these general considerations in mind, four basic culture media are most practical to use.

1. **Tryptic Soy Agar** is perhaps the most widely used and readily available culture medium. It is an excellent choice for recovering and growing a wide variety of copiotrophic bacteria. Better still is TSA with 5 percent sheep blood added. The sheep blood adds some additional nutrients, but, more importantly, the blood incorporated into the agar permits the microbiologist to observe the various haemolytic properties of the colonies, which is a very useful aid in recognizing differences between seemingly similar colonies. Haemolytic reactions can also provide useful information toward the identification of the bacterium. The principal drawback to TSA with blood is that it is not a good choice for carbon source utilization testing of many gram-positive species as discussed above.

2. **BUGM** was developed by Biolog to provide the closest thing to a universal culture medium for copiotrophic bacteria. It has the advantage that it works very well for carbon source utilization testing of both gram-negative and gram-positive bacteria. For general use it should be made with 5 percent sheep blood added, and for special use with gram-positive spore-forming bacteria *(Bacillus)*, it should be made with 1 percent glucose added. The only disadvantage of BUGM with blood compared to TSA with blood is that BUGM partially haemolyses the blood, so it is not an ideal medium for observing haemolytic reactions.

3. **R2A** was developed as an optimized medium for recovering bacteria from water (Reasoner and Geldreich 1985). Our laboratory compared R2A with several other popular media used in water, soil, and environmental survey microbiology (e.g., nutrient agar, plate count agar) and found it to be superior. It provides for excellent recovery and growth of both obligate and transiently oligotrophic bacteria.

4. **BUY** was developed by Biolog as an optimal culture medium for recovery, growth, and carbon source utilization testing of yeasts. We have demonstrated that it works very well with 267 diverse species of yeast, including species of importance in clinical, industrial, environmental, and food microbiology.

Our overall recommendations for microbiologists attempting to recover and identify a wide range of environmental microorganisms would be to use a combination of these media. To recover initial isolates, use BUGM with 5 percent sheep blood (or TSA with 5 percent sheep blood) for copiotrophs, R2A for oligotrophs, and BUY for yeasts. Bacteria should then be subcultured onto BUGM with 5 percent sheep blood *(Bacillus* are subcultured onto BUGM with 1 percent glucose) prior to carbon source utilization testing. With the exception of pink

Methylobacterium strains, bacteria that will not grow when transferred from the oligotrophic R2A medium onto BUGM with 5 percent sheep blood probably are not in the Biolog database. BUY medium can be used for both recovery and carbon source utilization testing of yeast isolates.

Problems Most Frequently Encountered

Some of most common problems that are encountered in attempting to identify environmental microbes are discussed below.

Problems in Obtaining Pure Cultures

Identification of bacteria or yeast by carbon source utilization testing always requires that one start with a pure culture. The importance of this should not be underestimated. Often it is not a trivial matter to tell how many types of microbes are present on an agar plate, because colony morphologies can be very similar. In examining agar plates, look very carefully at the colonies from different angles in a work area that has good lighting. Using a low-power magnifying lens is very helpful and is strongly recommended. Even with great care and attention to detail, "mixed" cultures can be missed.

Under standard microbiological practice, it is generally assumed that when one obtains a well-isolated, symmetric colony, that such a colony arises from a single cell and is, therefore, "pure." However, there are instances, particularly with staphylococci that have been freshly isolated, when simply streaking for isolated colonies may not produce a pure culture. The apparent reason for this is that staphylococci tend to grow in bunches sometimes referred to as "grape clusters." Isolates from the environment may contain two or more species of staphylococci that grow next to each other and become physically attached. Thus, streaking with a needle or stick may not suffice to physically separate these clusters, and the result is that isolated but "mixed" colonies are formed on the agar plate.

This is not an easy problem to solve and it is compounded by the fact that many of the *Staphylococcus* species have very similar colony morphologies. There is no foolproof way to tell that your isolate is "mixed," but if you have a culture that identifies with a low similarity to a *Staphylococcus* species (SIM index less than 0.5 after 24-hour incubation in the GP MicroPlate), there is a good possibility that it is not a pure culture. Our best suggestion is to leave streaked plates out on the benchtop for about a week and then carefully examine the isolated colonies with a magnifying lens. If the colony is "mixed," you typically see that one side of the colony is slightly different in color (e.g., cream vs. gray) or in texture (e.g., smooth vs. wrinkled). Agar media with blood (e.g., BUGM with 5 percent sheep blood) tend to enhance morphological differences. If you see evidence of a mixed culture, use a fine sterile needle and pick cells from the various edges of the colony for restreaking.

Problems with Gram Staining

Performing Gram stains is not as simple and routine as many people assume. The most common problem is that gram-positive bacteria may decolorize and be misclassified as gram-negative bacteria. This can happen because some species decolorize easily (notably *Bacillus* species), the cells are too old, or the Gram stain reagents are not good. An improved Gram stain reagent was recently introduced by Carr Scarborough Microbiologicals (Decatur, GA). The converse problem of gram-negative strains appearing to be gram positive should not occur.

In performing Gram stains, if the cells appear to be gram positive or indeterminate, it is important to look for spores, since spore-forming bacilli require special handling. With fresh cultures of some *Bacillus* strains, spores may be very difficult to observe. If the colonies can be stored for several days, they will usually take on a crusty appearance and the spores should be easy to observe.

After following the recommendations above, if the Gram stain results are still ambiguous, some alternatives are to (1) test for sensitivity to vancomycin using impregnated disks (von Graevenitz and Bucher 1983) (most gram negatives are resistant to disks containing 5 µg of vancomycin; most gram positives are sensitive to vancomycin), (2) test for growth on MacConkey plates and Columbia CNA plates (most gram negatives will grow on MacConkey but not on CNA; most gram positives are inhibited on MacConkey but grow on CNA), or (3) test for L-alanine aminopeptidase activity (Manafi and Kneifel 1990) using commercially available test strips impregnated with a colorimetric substrate for this enzyme (most gram negatives have this enzyme; most gram positives do not).

Problems with Mucoid Strains or Strains That Form Pellicles

Fast-growing rod-shaped bacteria that form mucoid colonies are likely to be *Klebsiella* or *Bacillus* species. *Bacillus* species can also form crusty surfaces or pellicles (skinlike clumps). These species can be difficult to test for two reasons.

First, it may be difficult to prepare a uniform cell suspension. If an isolate that forms mucoid or pellicular clumps is encountered, use a swab to break up the clumps on the surface of the agar plate. Then pick the small clumps onto the swab and crush the swab against the inside surface of the test tube above the surface of the saline. This should allow you to break up most of the clumps. Once the swab is immersed, the clumps fall off and become impossible to crush. If some clumps remain, let the saline suspension stand for a few minutes and allow the clumps to settle to the bottom of the tube. Then pour off the uniform cell suspension at the top of the tube into a filling reservoir.

Second, when the strain is inoculated into a MicroPlate, you may get a so-called "false positive" reaction in which all wells including the A-1 reference well turn purple. The reasons for the "false positive" reactions in *Klebsiella* and *Bacillus* are very different and are dealt with in different ways.

Most *Klebsiella* strains, and some *Enterobacter, Serratia,* and *Salmonella* strains, secrete a capsular polysaccharide that they can later reutilize as a carbon source. Because the polysaccharide sticks to the outer surface of the cells, it is transferred into the saline suspension along with the cells, and then into every well of the MicroPlate. When it is reutilized, the cells reduce the tetrazolium dye and all the wells (including the A-1 reference well) turn "positive."

The "false positive" reaction of *Klebsiella* is delayed and is not observed until well after 4 hours of incubation. Presumably this is because it takes many hours for the cells to begin to reutilize the polysaccharide. Therefore, if you are running a strain that might be a *Klebsiella*, we recommend that the pattern be read after 4 hours of incubation. If the pattern does not match sufficiently to give an identification, wait 2 more hours and read it again (i.e., after 6 hours), using the 4-hour database. If it still does not match sufficiently, continue the incubation overnight and try to read it once more.

Typically, after overnight incubation, the A-1 well (i.e., the reference well) will turn "false positive" due to utilization of the polysaccharide. To solve this problem, simply repeat the test, diluting the inoculum so that the amount of polysaccharide carried over is insufficient to cause a "false positive" reaction. Our recommended procedure is to first prepare a cell suspension at the usual density, and then dilute it 1:20 in sterile saline. Inoculate the GN MicroPlate with this dilute cell suspension, incubate overnight, and read the pattern using the 24-hour database. This procedure should diminish or completely eliminate false positive reactions and enable you to obtain a good identification.

Bacillus strains turn the wells "false positive" because they are in the process of sporulating. Unlike the delayed reaction observed with *Klebsiella*, the "false positive" reaction of *Bacillus* species occurs rapidly and is very clear before 4 hours have elapsed.

The solution here is to subculture the *Bacillus* on a very nutrient-rich medium, specifically BUGM with 1 percent glucose. This tends to shift the majority of the cell population from the sporulation state into the vegetative state, and should help diminish the "false positive" color. When subculturing the isolate, use the standard 4-quadrant streaking method. Incubate the plates for 16 hours rather than 24 hours and pick cells from the isolated colonies in the 4th zone. Because these bacteria are less crowded, they have more food diffusing to them and they are less likely to have begun to sporulate. If more cells are needed to obtain the proper inoculum density, pick from the 3rd zone; but again, try to pick cells from the edge of the growth since these cells have the most food available. By picking cells from the 4th and 3rd zones, you should find that the "false positive" background color is reduced substantially, and that the patterns are easier to read. If the "false positive" color persists, try subculturing the strain one or two more times on BUGM with 1 percent glucose.

What If the Strain Cannot Be Identified?

As discussed at the start of this chapter, many species of bacteria and yeast remain to be described and named. Microbiologists obtaining environmental isolates are likely to encounter strains that form a distinctive fingerprint that does not match any species in the Biolog database. In anticipation of this likelihood, we have designed advanced software to allow the microbiologist to store and retrieve fingerprints, thereby creating a customized user database.

If a fingerprint that is saved, or one that is similar is encountered again, the software will retrieve it as a match to the previous isolate. In many cases where a microbiologist is trying to track the occurrence of an isolate, this is sufficient, and a species name (which may not exist anyway) is unnecessary. This can provide a very useful "strain-tracking" feature.

Mathematical Analysis and Clustering

Biolog's software uses straightforward mathematical approaches to accomplish species identifications based on pattern matching. Details of these methods are given in the *Biolog Reference Manual*. The same mathematics are the basis for powerful, yet easy-to-use, one-, two-, and three-dimensional clustering algorithms that allow microbiologists to recognize groups of similar isolates and quantitate their relatedness. The latter feature is particularly useful in research applications involving characterization of culture collections.

All analyses use "distance" ("DIST") as a unit of measure, where the "distance" between two strains is essentially the number of tests that give a different result between the two strains that are compared. An illustrative example is shown in Figure 2.6. To simplify the example, I have used only values of 0 percent, 50 percent, and 100 percent, corresponding to "negative," "borderline," and "positive" reactions. Of course, real species matrices, obtained by averaging the patterns of numerous strains, could have any values ranging between 0 percent and 100 percent. The carbon source utilization pattern of these two hypothetical species are quite similar and differ only with tests A5 and G1, which are 100 percent positive for species A and 0 percent positive for species B; tests C8 and D3, which are 0 percent positive for species A and 100 percent positive for species B; and test E12, which is 0 percent positive for species A and 50 percent positive for species B. The DIST value that we would assign to these two patterns would be

$$\frac{100+100+100+100+50}{100} = 4.50$$

Because there are nearly 100 tests included in the analysis, the DIST value approximates the percentage of carbon source utilization tests by which the two patterns differ. In this example these two patterns differ in 4.50/95 = 4.74 percent of their carbon source utilization tests.

Pattern matching is based primarily on a Bayesian analysis, which estimates the probability that a specific pattern is a reasonable match with the various patterns stored in the species database. If one assumes that the pattern must match one of the species in the database, then the probability for any given species will be a "similarity" number ranging from 0 to 1 (i.e., 0 percent to 100 percent probability); the summation of the similarities for all the species in the database will equal 1.

The final judgment on whether any given pattern is an adequate species match is determined by a weighted similarity number that we term the "similarity index" or "SIM." In general, a SIM value greater than 0.5 is required for a species match. The reason that SIM is weighted is that there is a very real possibility that the pattern may not match any species in the database. From empirical studies we have determined that a species match must have a DIST value of 6.5 or less. SIM values are weighted downward as the DIST values increase. The summation of the similarities for all the species in the database will always be a number less than 1, and it can approach 0 when the DIST values are large (i.e., greater than 15). An example of an identification result showing how the DIST and SIM parameters are used is shown in Figure 2.7.

Figure 2.6. Example of species matrices and calculation of DIST.

Species A	1	2	3	4	5	6	7	8	9	10	11	12
A	0	0	0	50	[100]	0	100	0	100	50	0	0
B	100	100	0	100	0	0	100	100	0	0	50	50
C	0	100	0	0	0	0	0	[0]	0	50	0	0
D	50	0	[0]	0	50	100	0	0	0	0	0	0
E	0	0	0	50	0	0	0	0	0	0	0	[0]
F	0	0	0	0	0	0	100	100	50	0	0	0
G	[100]	50	0	0	50	0	0	0	0	0	0	0
H	0	0	0	50	50	0	0	0	0	0	100	100

Species B	1	2	3	4	5	6	7	8	9	10	11	12
A	0	0	0	50	[0]	0	100	0	100	50	0	0
B	100	100	0	100	0	0	100	100	0	0	50	50
C	0	100	0	0	0	0	0	[100]	0	50	0	0
D	50	0	[100]	0	50	100	0	0	0	0	0	0
E	0	0	0	50	0	0	0	0	0	0	0	[50]
F	0	0	0	0	0	0	100	100	50	0	0	0
G	[0]	50	0	0	50	0	0	0	0	0	0	0
H	0	0	0	50	50	0	0	0	0	0	100	100

Figure 2.7. Example of Biolog software screen display.

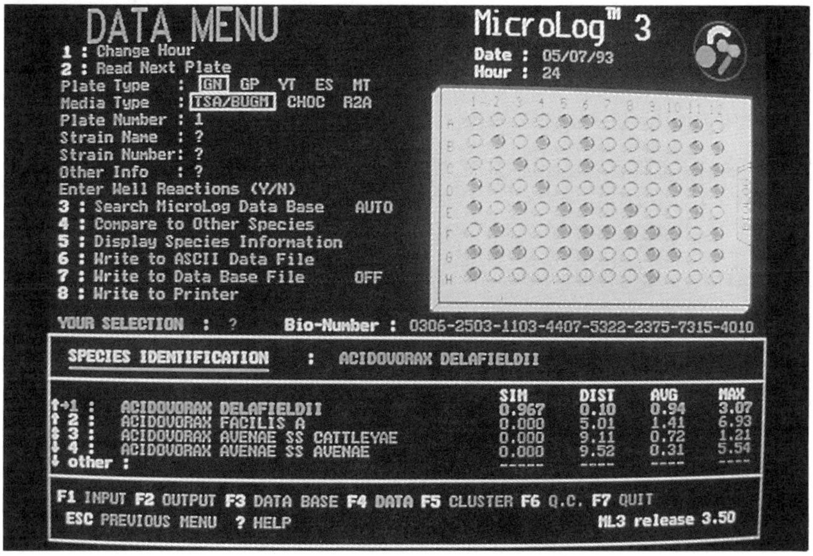

Cluster diagrams can also be generated based on the DIST values. Dendrograms (one-dimensional cluster diagrams) are generated using a modified "UPGMA" analysis developed by Biolog. The algorithm uses DIST values (instead of the similarity coefficients commonly used in a standard UPGMA analysis) to generate the branching structure of the dendrogram. A DIST value of 14 is usually used as a "cutting level" to separate genera. This value has been determined empirically to be an appropriate one for most genera of gram-negative bacteria. After a preliminary dendrogram is generated, a second pass is made through the dendrogram, and all branch points are rotated to their most preferred rotation to best maintain proximity of related strains or species. If the distance between strains or species adjacent to cut points is less than 14, they are joined with a dashed line that indicates their actual distance.

By mathematically removing the one-dimensional constraints of the dendrogram, two-dimensional and three-dimensional representations of the DIST relationships can also be generated. All of these cluster diagrams are only approximations of the true relationships. With 95-discriminatory tests it would take 95-dimensional space to provide clustering that is accurate in all detail. However, as a practical matter, the one-, two- and three-dimensional cluster diagrams give a very satisfactory representation in nearly all cases.

Advanced Applications of Strain Characterization

The Biolog System provides a very simple and efficient means for characterizing individual strains, large collections of isolates, and even microbial communities.

This can be done by employing one or more MicroPlates for testing and then utilizing the software features to construct a custom library and to perform mathematical clustering analyses. Examples of studies where this is potentially useful include the following:

1. Searching for unique or defining strain characteristics

2. Searching for characteristics that differentiate two or more strains

3. Performing epidemiological studies to look at outbreaks of infectious microbes in animal populations or outbreaks of contamination in sterile manufacturing environments

4. Subgrouping collections of strains into groups with the most similar metabolic properties

5. Analyzing strains to determine their range of growth substrates and/or their content of catabolic enzymes

6. Analyzing mixed microbial populations or communities taken directly from the environment

In addition to the GN, GP, and YT MicroPlates, Biolog also manufactures ES, MT, and SF MicroPlates. The ES MicroPlate has a set of 95 carbon source utilization tests specifically selected for fine discrimination or epidemiological analysis of two important pathogenic bacteria: *Escherichia coli* and *Salmonella typhimurium*. The MT MicroPlate is identical to the GN MicroPlate, except that the carbon sources have been omitted from the plate. This gives users the flexibility to add any set of carbon sources, thereby creating a custom panel tailored to special applications. The SF-N and SF-P MicroPlates are for microorganisms that do not reduce or are inhibited by the tetrazolium redox dye. They are primarily useful for sporulating and filament-forming microorganisms, such as Actinomycetes and Fungi, but they may also prove useful for oligotrophic microorganisms. Other procedural modifications can be made to accommodate the use of the various MicroPlates with most halophilic marine bacteria as well as anaerobic bacteria.

By combining the use of all of these tools, the microbiologist can work with a very high level of efficiency. It takes only about one minute of labor to set up each panel of 95 tests, and the results can be read directly into computer files, stored, and analyzed in seconds. Even collections containing several hundred strains can be analyzed in great detail by a single microbiologist within a week or two.

VALIDATING THE PERFORMANCE OF THE BIOLOG SYSTEM

Microbiologists at clinical laboratories and at pharmaceutical manufacturing facilities are often required to validate the performance of their bacteria identification system and to define protocols to insure that procedures are working properly and that the staff maintains an adequate level of training and competence. For simple, routine checking, Biolog provides sets of specific strains of bacteria and yeast that can be used in quality control (QC) testing of its microbiology test panels.

Currently, there are no set rules that define validation protocols for microbial identification systems. Typically, the regulatory agencies are looking for a commonsense approach to validation that demonstrates a good knowledge of the issues and problems posed by environmental monitoring, good technical competence in microbiology, good record keeping to permit retrospective traceability of problems, and constant vigilance to insure that the quality of the technical work remains high.

One approach to validation in a pharmaceutical manufacturing setting is to take a representative set of bacteria and periodically test all microbiologists on staff to demonstrate that they can correctly identify them. The set of bacteria should be chosen to reflect the breadth of bacteria that are typically found in the environment being monitored. For example, in pharmaceutical manufacturing facilities this is likely to include both gram-negative (e.g., *Alcaligenes xylosoxydans, Pseudomonas aeruginosa, Pseudomonas pickettii, Serratia marcescens,* and *Xanthomonas maltophilia*) and gram-positive (e.g., *Bacillus circulans, Micrococcus luteus,* and *Staphylococcus aureus*) species. A good way to select the set is either to go by past experience, or to consult with another microbiologist that has successfully addressed the issue.

Biolog is now working closely with the FDA to better define protocols and SOPs to help validate the use of the system for species identification. Microbiologists interested in this issue should contact Biolog Technical Service to obtain the most up-to-date information.

HOW ACCURATE IS THE BIOLOG SYSTEM?

Determining the accuracy of the Biolog System is very complex; there is no simple answer. Some of the issues involved in establishing system accuracy have been addressed in a recent editorial (Miller 1991). The ultimate answer, however, is that it is simply not possible to assign an overall accuracy percentage to any microbial identification system.

Evaluations that are published in the literature inevitably rely on strain collections that are inherently biased both by their composition and by the method used to obtain identifications. Usually the strains have been analyzed previously by "classical identification methods," many of which are poorly defined and subject to error. Two laboratories can use "classical methods" and arrive at different answers. The most scientifically acceptable means of speciation involves performing many reciprocal DNA hybridizations, but this technique is still much too cumbersome to be used by most laboratories. Polyphasic approaches are also excellent, but they are not practical in situations where strains must be identified rapidly and at low cost.

Because there is currently no practical "gold standard" on which one can base an identification, it is often impossible to distinguish whether a particular strain is an "atypical" isolate or whether it belongs in a new, undescribed species. Also, some species are currently assigned to the wrong genus, so it can even be fallacious to assume that a correct genus level identification means that the identification method was reasonably accurate. For example, if a strain of *Pseudomonas cocovenenans* (which is not really a *Pseudomonas*) were misidentified as *Burkolderia gladioli*, this would actually be a less serious error than misidentifying it as *Pseudomonas putida* (see Figure 2.8, GN groups 7 and 9). Furthermore, some genera are problematic for all identification systems because taxonomists have not yet agreed on how the species within them should be delineated. Examples include *Acinetobacter, Aeromonas, Bacillus, Corynebacterium, Enterobacter,* and *Pseudomonas.*

KEEPING UP WITH CHANGES IN TAXONOMY AND NOMENCLATURE

The pace of change in our knowledge of microbial taxonomy has begun to accelerate within the last decade, creating a potentially burdensome problem for microbiologists wishing to keep up with current knowledge. The principal impetus behind the revolution is based on the pioneering work of Woese (1987) and Stackebrandt and coworkers (1988) to reconstruct taxonomic relationships based upon 16S and 23S rRNA sequences. This approach has picked up momentum with the advent of relatively rapid methods of amplifying and sequencing these genes from microorganisms.

For some groups of microorganisms, there has been little change in our thinking, but for other groups the changes have been substantial. First of all, the simple division of all bacteria into two classes, gram-negative bacteria (Gracilicutes) and gram-positive bacteria (Firmicutes), is viewed as inaccurate. It is now accepted that there are other classes. For example, the genera *Flavobacterium, Cytophaga,* and *Bacteroides* actually belong in a distinct branch of evolution from Gracilicutes, even though they stain as gram negative. Similarly the radiation-resistant genus *Deinococcus* belongs in a distinct branch of evolution from Firmicutes, even though it stains as gram positive.

Many genera have been subdivided and reorganized. Of particular interest to environmental microbiologists are changes that have or will soon occur in the nomenclature of *Pseudomonas* and *Bacillus.* Many former species of *Pseudomonas* have been reassigned to new genera such as *Comamonas* (De Vos et al. 1985), *Burkholderia* (Li et al. 1993; Yabuuchi et al. 1992), *Acidovorax* (Willems et al. 1990), *Hydrogenophaga* (Willems et al. 1989), *Shewanella* (MacDonnell and Colwell 1985), *Xanthomonas* (Swings et al. 1983), *Sphingomonas* (Yabuuchi et al. 1990), and *Methylobacterium* (Patt et al. 1976). The true *Pseudomonas* and *Xanthomonas* species are in the II-gamma evolutionary branch, whereas the other new genera are in the III-beta and IV-alpha branches (Stackebrandt et al. 1988), being no more than remotely related. The most recent taxonomic data (Priest 1993) also indicates that the genus *Bacillus* needs to be reorganized into at least five genera because the species are so diverse.

To some microbiologists performing species identifications, correct nomenclature may seem an esoteric point. However, it can be important to have an accurate perspective of the relatedness among species. In the example given in the previous section, it is terribly misleading to think of *cocovenenans* as a *Pseudomonas* species. If one isolates a *Pseudomonas cocovenenans* that has interesting or useful properties, one would do better to look for other potentially useful strains by examining a truly related species such as *Burkholderia gladioli,* rather than an unrelated species such as *Pseudomonas putida.*

The Biolog System is constantly updated so that microbiologists using the system will have the latest nomenclatural and taxonomic information readily available. Figure 2.8 shows a compilation of estimated taxonomic relationships for over 800 species of bacteria. To obtain this compilation, genera and species were first assigned to groups based on information obtained from taxonomic publications. Biolog's carbon source utilization data was then used in conjunction with the clustering software described above to generate a comprehensive series of dendrograms. Although our understanding of taxonomic relationships continues to evolve, the information contained within this analysis should serve as a useful interim guide.

CONCLUSION

In the past, microbiological studies of the environment have been severely limited by tedious and cumbersome technologies. It has not been possible for microbiologists to identify a diverse range of microorganisms in an efficient and cost-effective manner. We also have been hampered by an incomplete enumeration of microbes in the environment. For these reasons substantial environmental studies have been largely deferred. Biolog's new technology and future developments that evolve from it will allow diverse microbiological studies to progress at a much faster pace.

Figure 2.8. Dendrograms including over 800 species of bacteria.

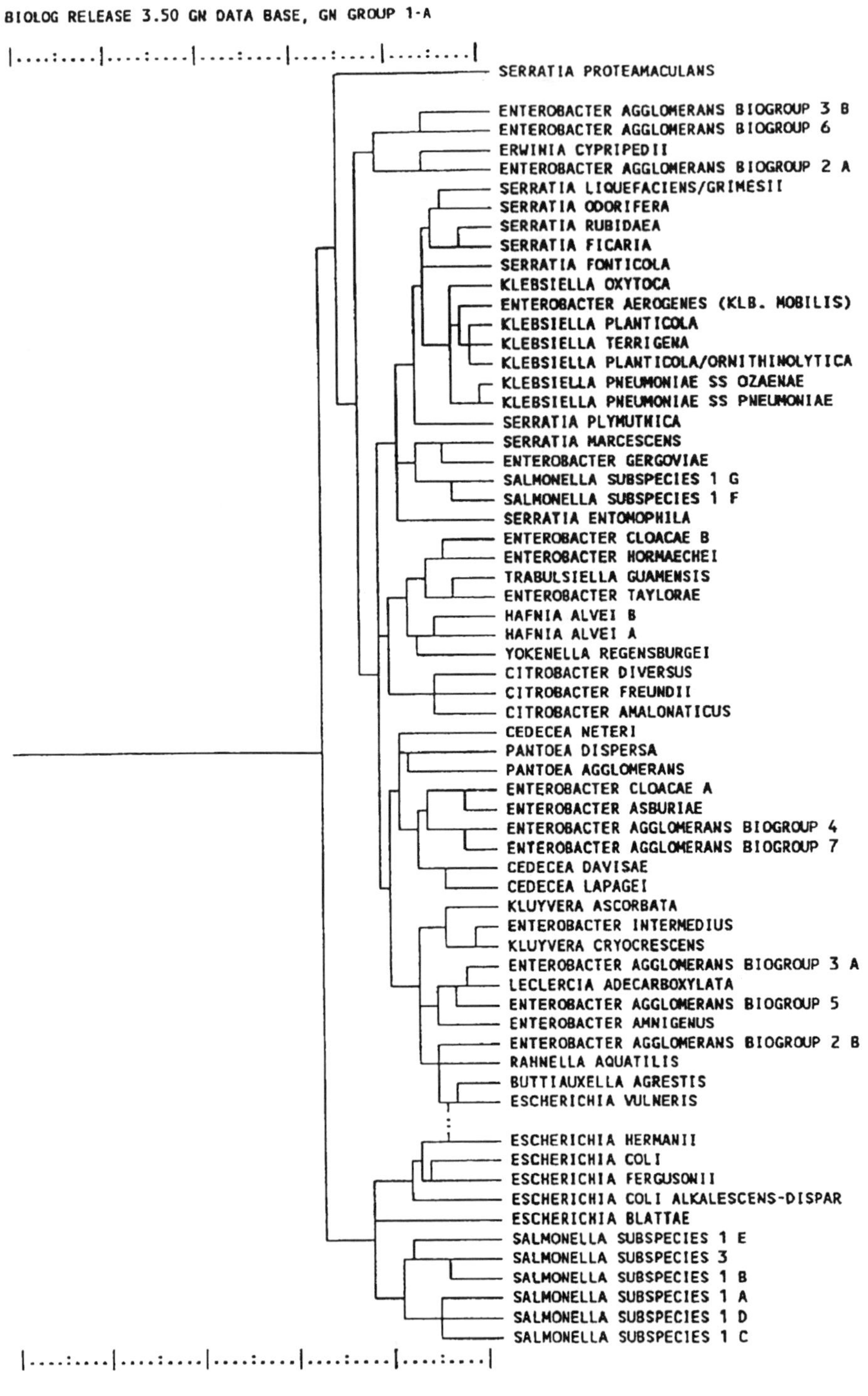

BIOLOG RELEASE 3.50 GN DATA BASE, GN GROUP 1-B

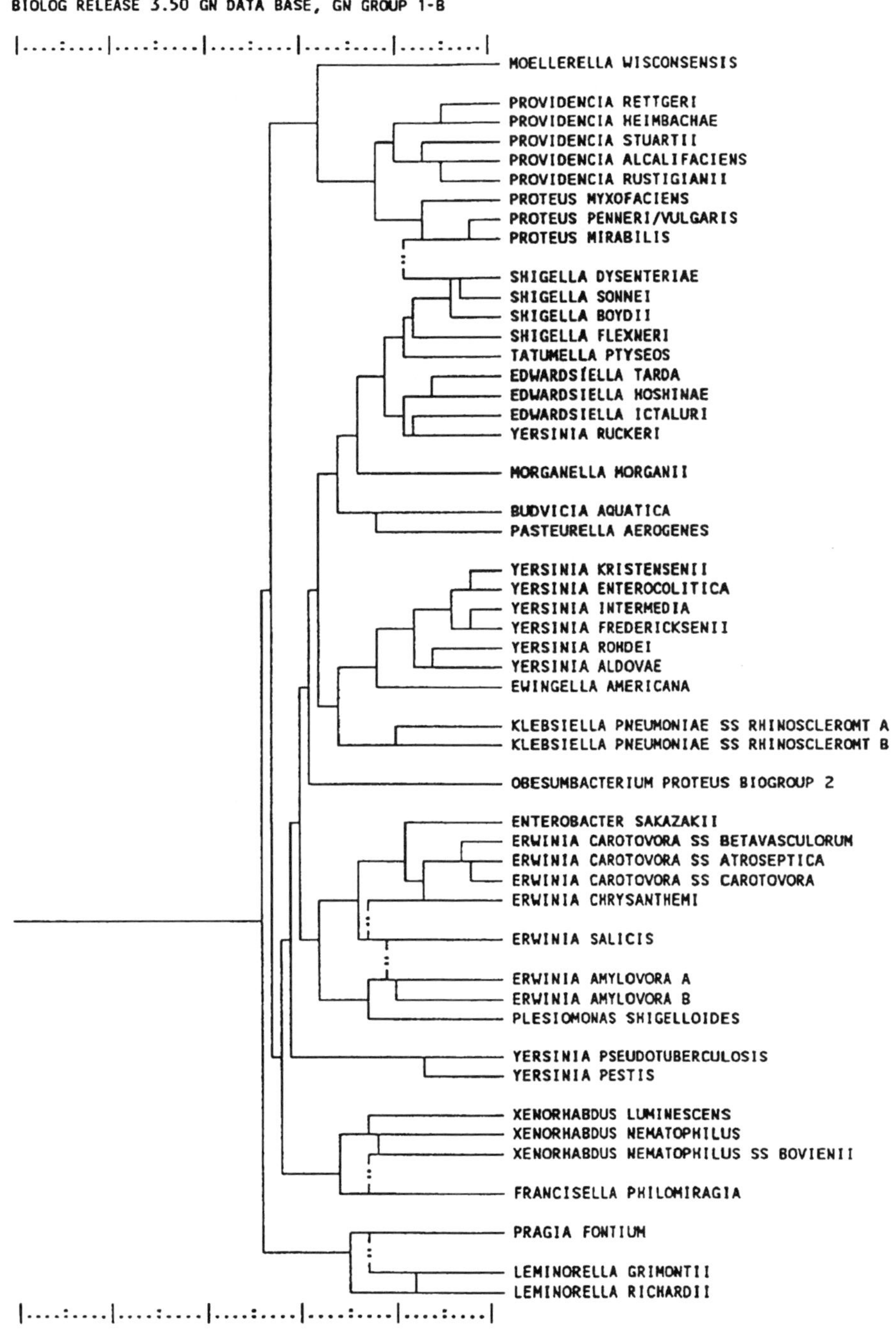

BIOLOG RELEASE 3.50 GN DATA BASE, GN GROUP 2

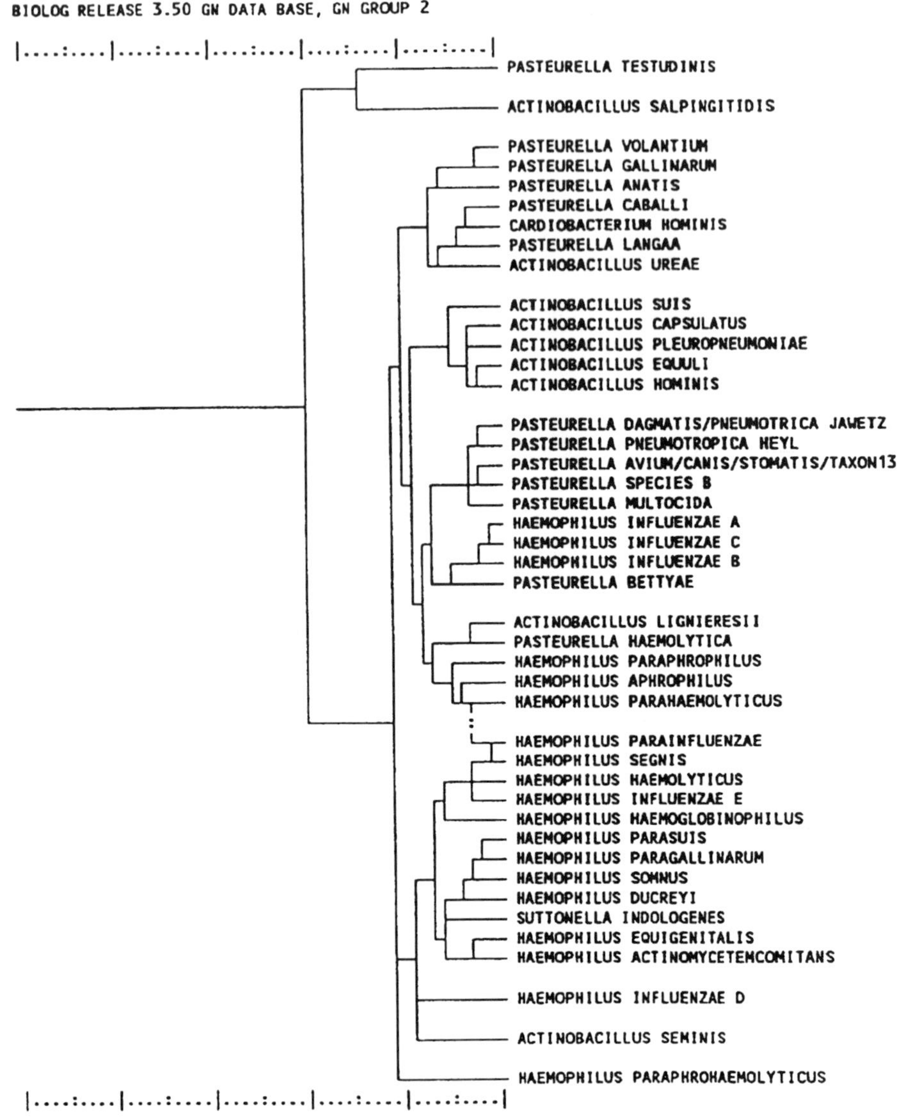

BIOLOG RELEASE 3.50 GN DATA BASE, GN GROUP 3

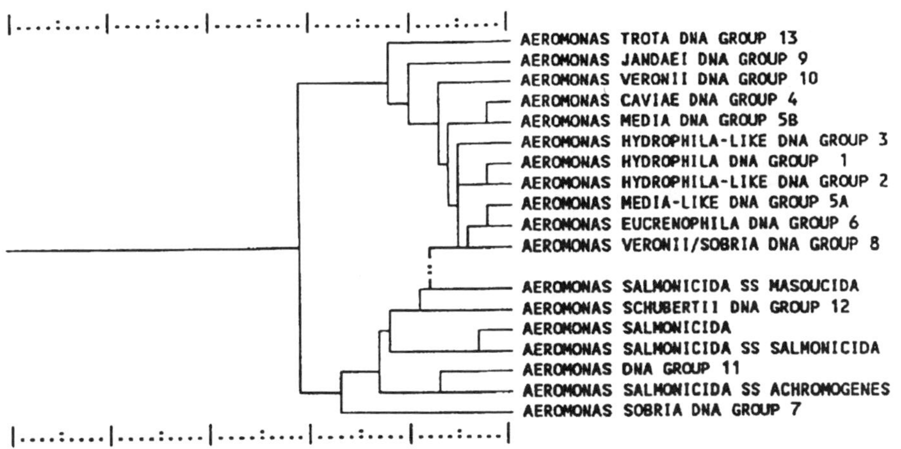

BIOLOG RELEASE 3.50 GN DATA BASE, GN GROUP 4

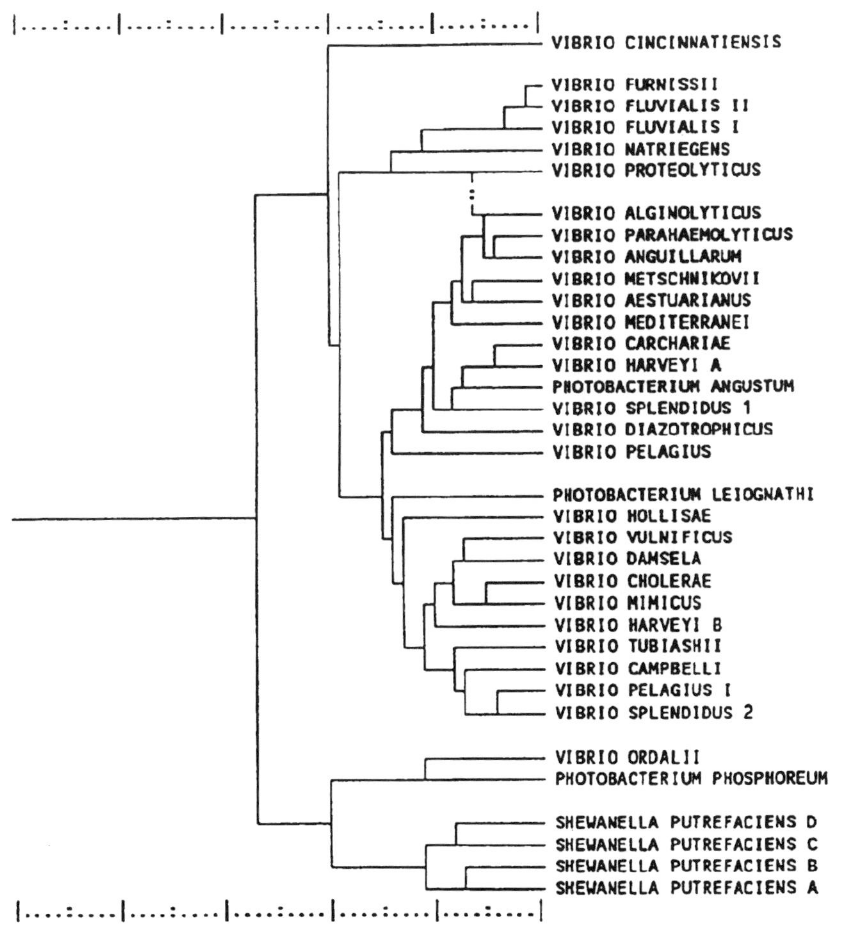

BIOLOG RELEASE 3.50 GN DATA BASE, GN GROUP 5

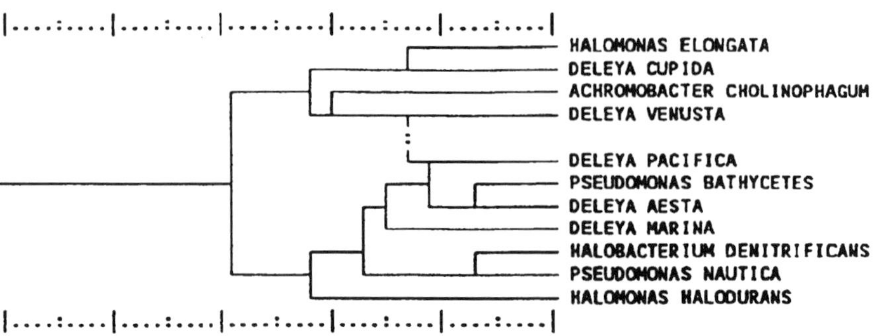

BIOLOG RELEASE 3.50 GN DATA BASE, GN GROUP 6

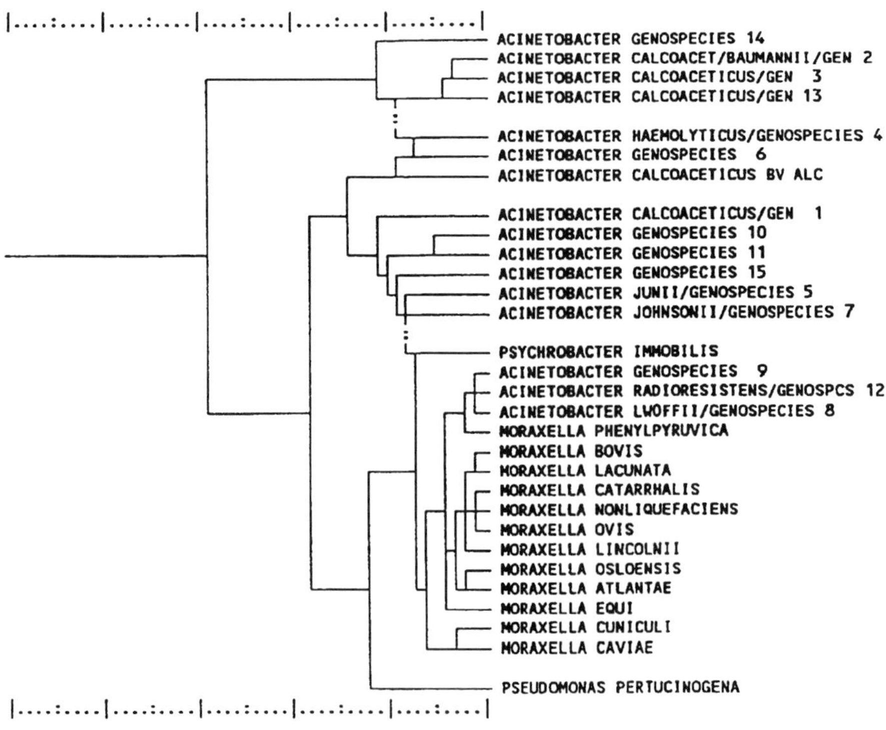

BIOLOG RELEASE 3.50 GN DATA BASE, GN GROUP 7

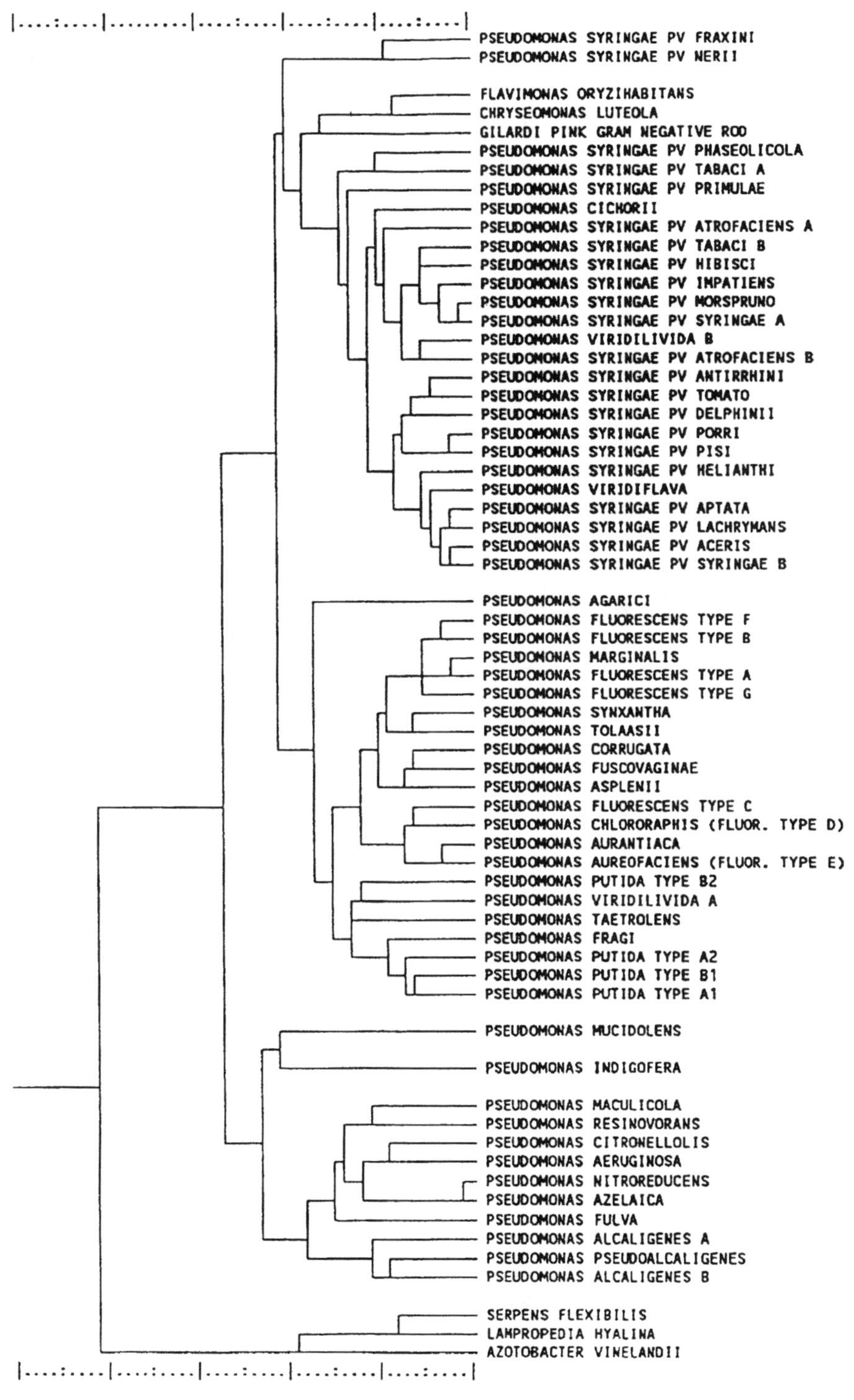

BIOLOG RELEASE 3.50 GN DATA BASE, GN GROUP 8

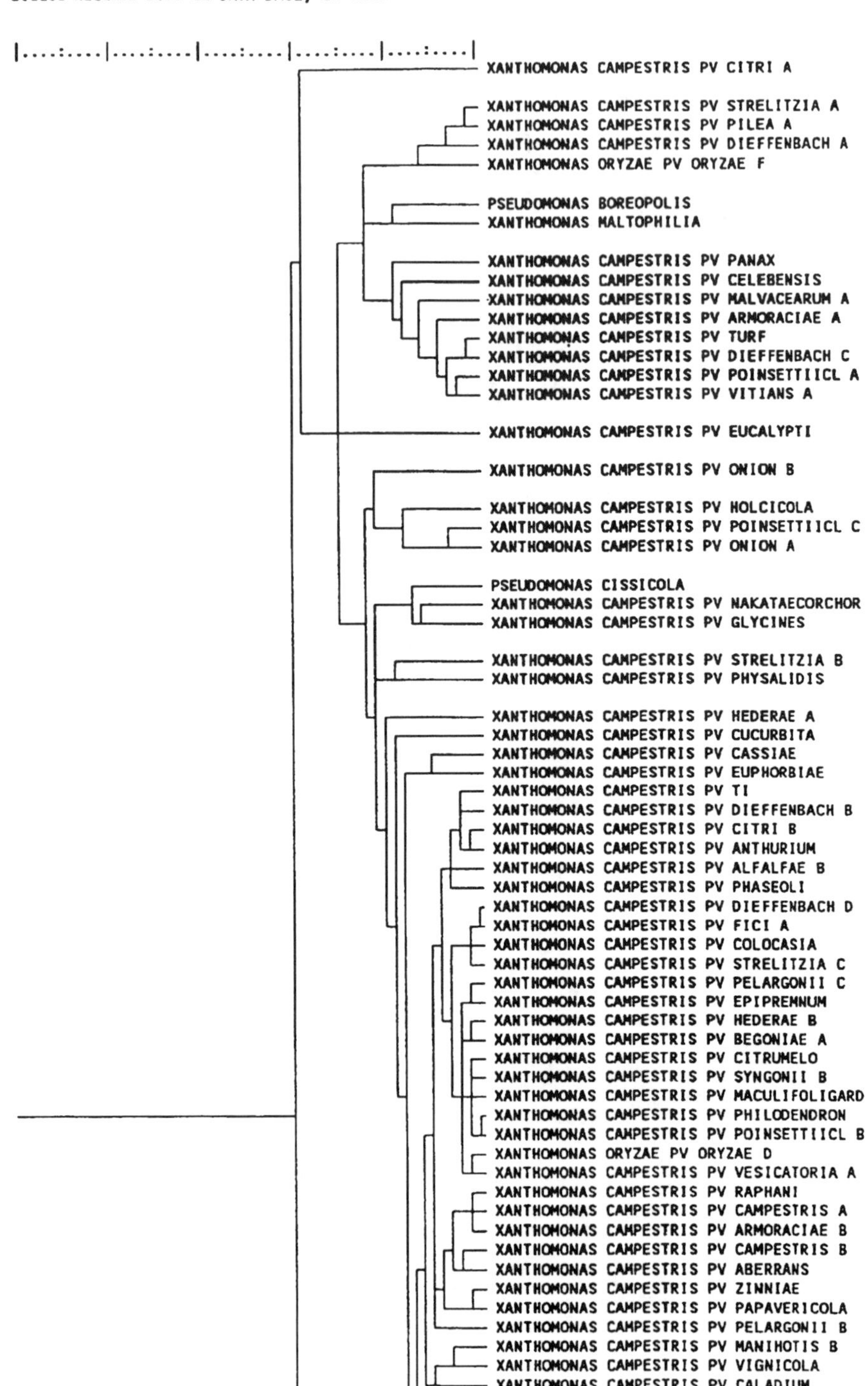

Group 8 continued on next page.

Group 8 continued.

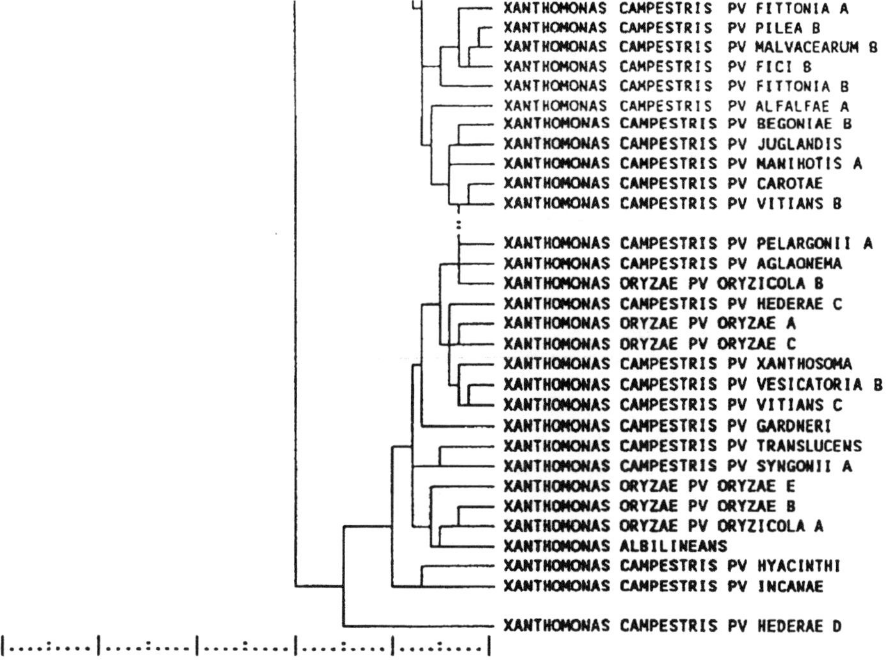

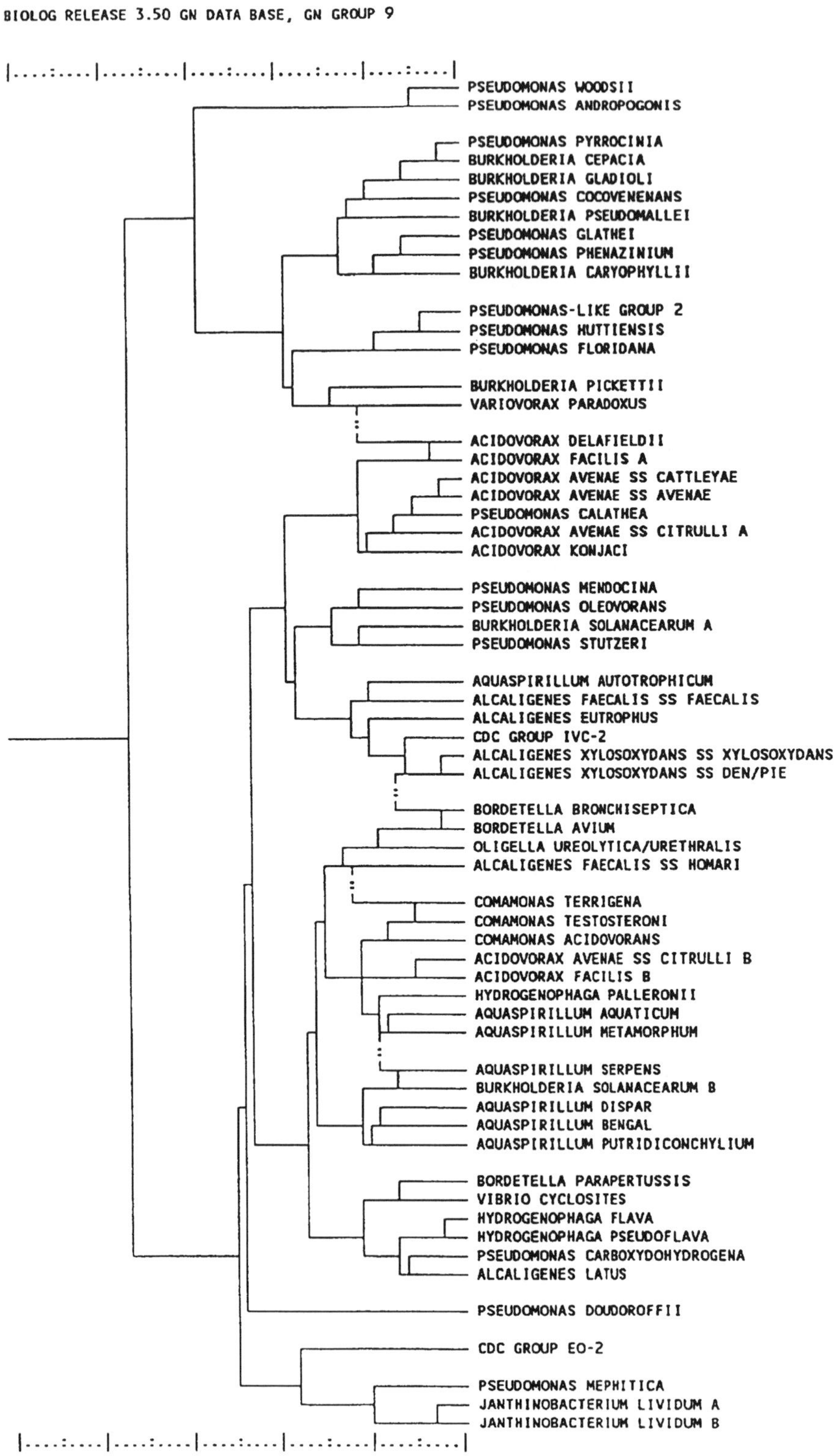

BIOLOG RELEASE 3.50 GN DATA BASE, GN GROUP 9
PSEUDOMONAS WOODSII
PSEUDOMONAS ANDROPOGONIS
PSEUDOMONAS PYRROCINIA
BURKHOLDERIA CEPACIA
BURKHOLDERIA GLADIOLI
PSEUDOMONAS COCOVENENANS
BURKHOLDERIA PSEUDOMALLEI
PSEUDOMONAS GLATHEI
PSEUDOMONAS PHENAZINIUM
BURKHOLDERIA CARYOPHYLLII
PSEUDOMONAS-LIKE GROUP 2
PSEUDOMONAS HUTTIENSIS
PSEUDOMONAS FLORIDANA
BURKHOLDERIA PICKETTII
VARIOVORAX PARADOXUS
ACIDOVORAX DELAFIELDII
ACIDOVORAX FACILIS A
ACIDOVORAX AVENAE SS CATTLEYAE
ACIDOVORAX AVENAE SS AVENAE
PSEUDOMONAS CALATHEA
ACIDOVORAX AVENAE SS CITRULLI A
ACIDOVORAX KONJACI
PSEUDOMONAS MENDOCINA
PSEUDOMONAS OLEOVORANS
BURKHOLDERIA SOLANACEARUM A
PSEUDOMONAS STUTZERI
AQUASPIRILLUM AUTOTROPHICUM
ALCALIGENES FAECALIS SS FAECALIS
ALCALIGENES EUTROPHUS
CDC GROUP IVC-2
ALCALIGENES XYLOSOXYDANS SS XYLOSOXYDANS
ALCALIGENES XYLOSOXYDANS SS DEN/PIE
BORDETELLA BRONCHISEPTICA
BORDETELLA AVIUM
OLIGELLA UREOLYTICA/URETHRALIS
ALCALIGENES FAECALIS SS HOMARI
COMAMONAS TERRIGENA
COMAMONAS TESTOSTERONI
COMAMONAS ACIDOVORANS
ACIDOVORAX AVENAE SS CITRULLI B
ACIDOVORAX FACILIS B
HYDROGENOPHAGA PALLERONII
AQUASPIRILLUM AQUATICUM
AQUASPIRILLUM METAMORPHUM
AQUASPIRILLUM SERPENS
BURKHOLDERIA SOLANACEARUM B
AQUASPIRILLUM DISPAR
AQUASPIRILLUM BENGAL
AQUASPIRILLUM PUTRIDICONCHYLIUM
BORDETELLA PARAPERTUSSIS
VIBRIO CYCLOSITES
HYDROGENOPHAGA FLAVA
HYDROGENOPHAGA PSEUDOFLAVA
PSEUDOMONAS CARBOXYDOHYDROGENA
ALCALIGENES LATUS
PSEUDOMONAS DOUDOROFFII
CDC GROUP EO-2
PSEUDOMONAS MEPHITICA
JANTHINOBACTERIUM LIVIDUM A
JANTHINOBACTERIUM LIVIDUM B

BIOLOG RELEASE 3.50 GN DATA BASE, GN GROUP 10

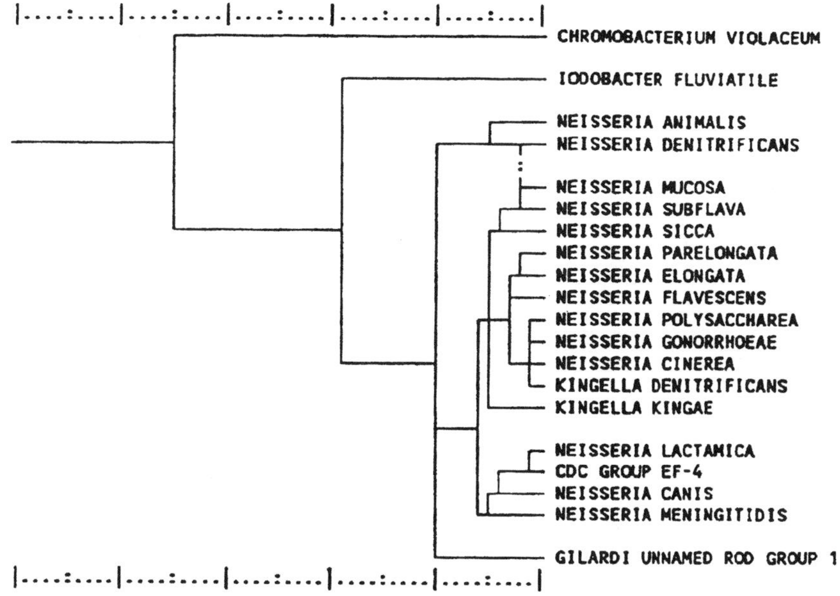

BIOLOG RELEASE 3.50 GN DATA BASE, GN GROUP 11

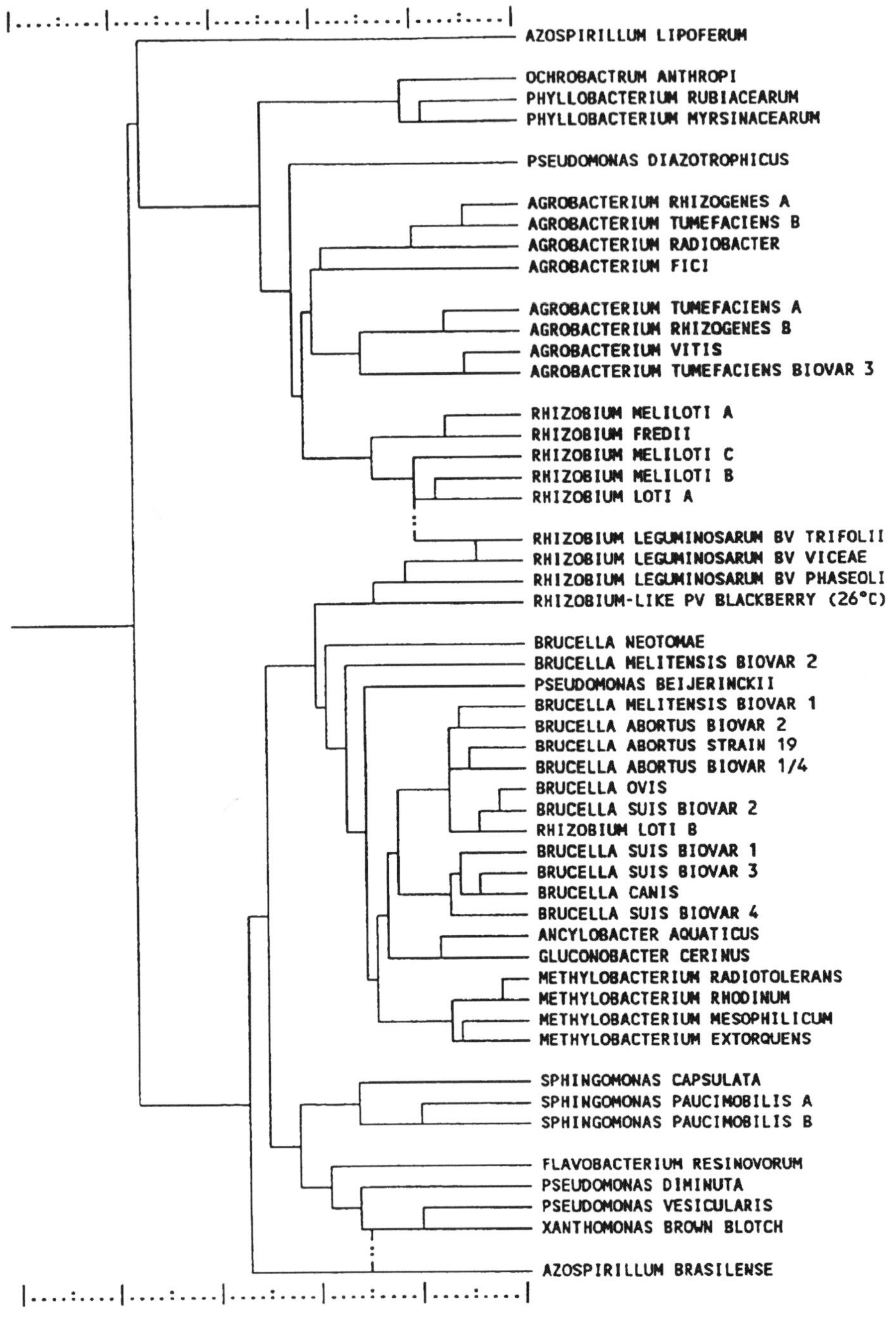

BIOLOG RELEASE 3.50 GN DATA BASE, GN GROUP 12

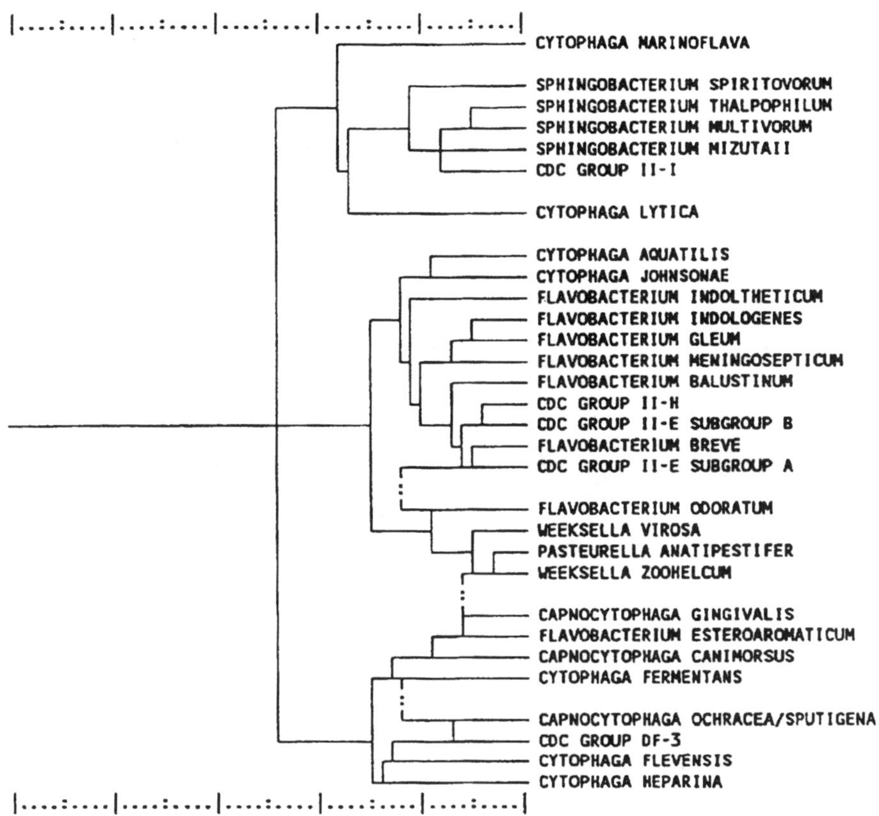

BIOLOG RELEASE 3.50 GP DATA BASE,   COCCI

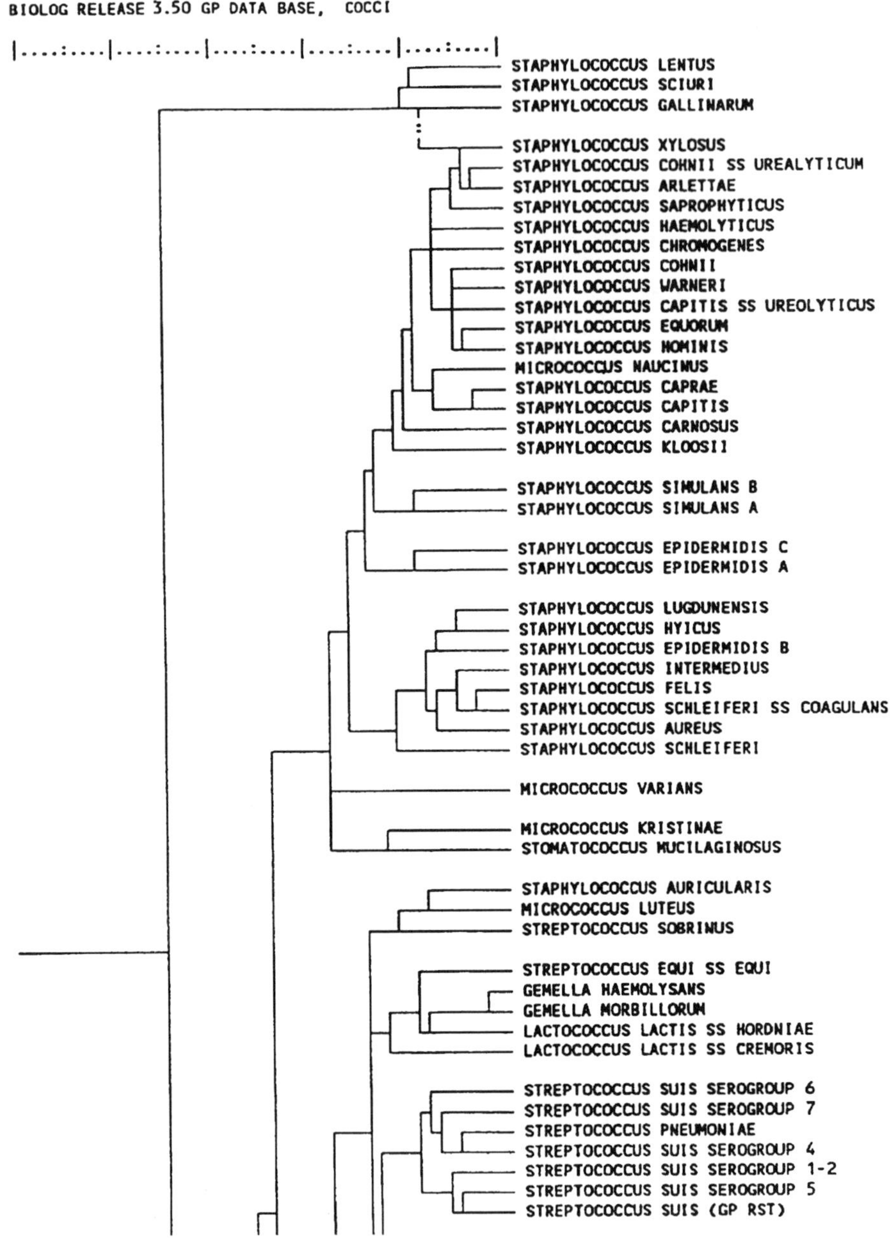

Group Cocci continued on next page.

Group Cocci continued.

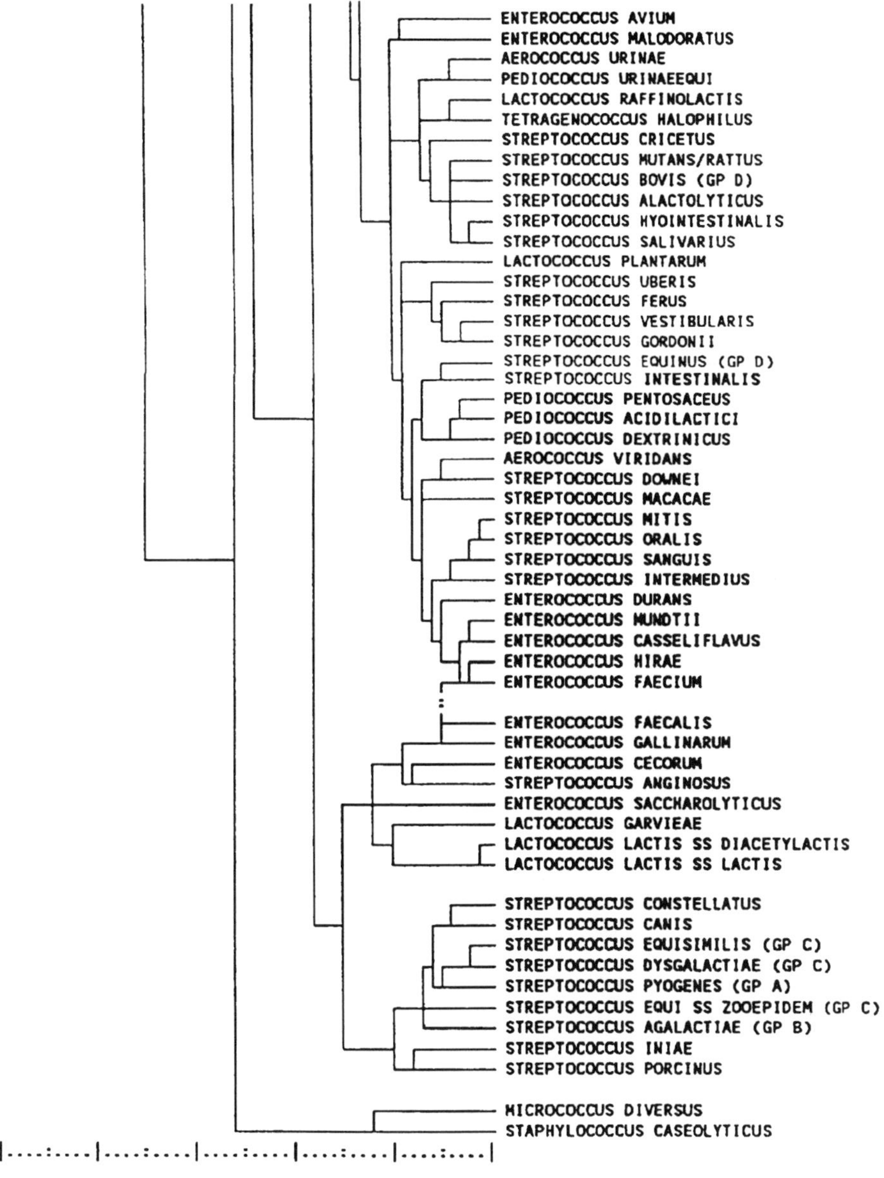

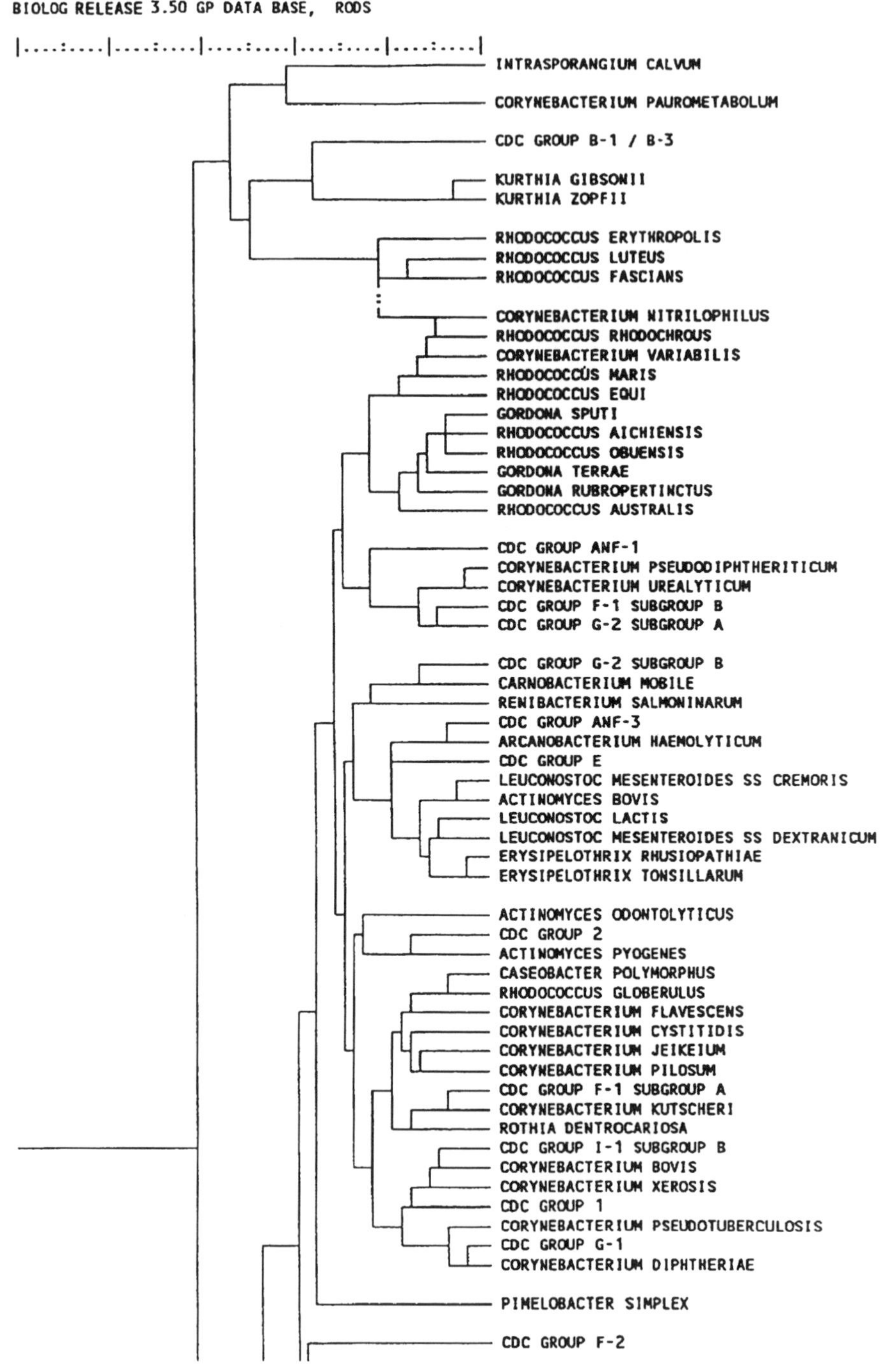

Group Rods continued on next page.

Group Rods continued.

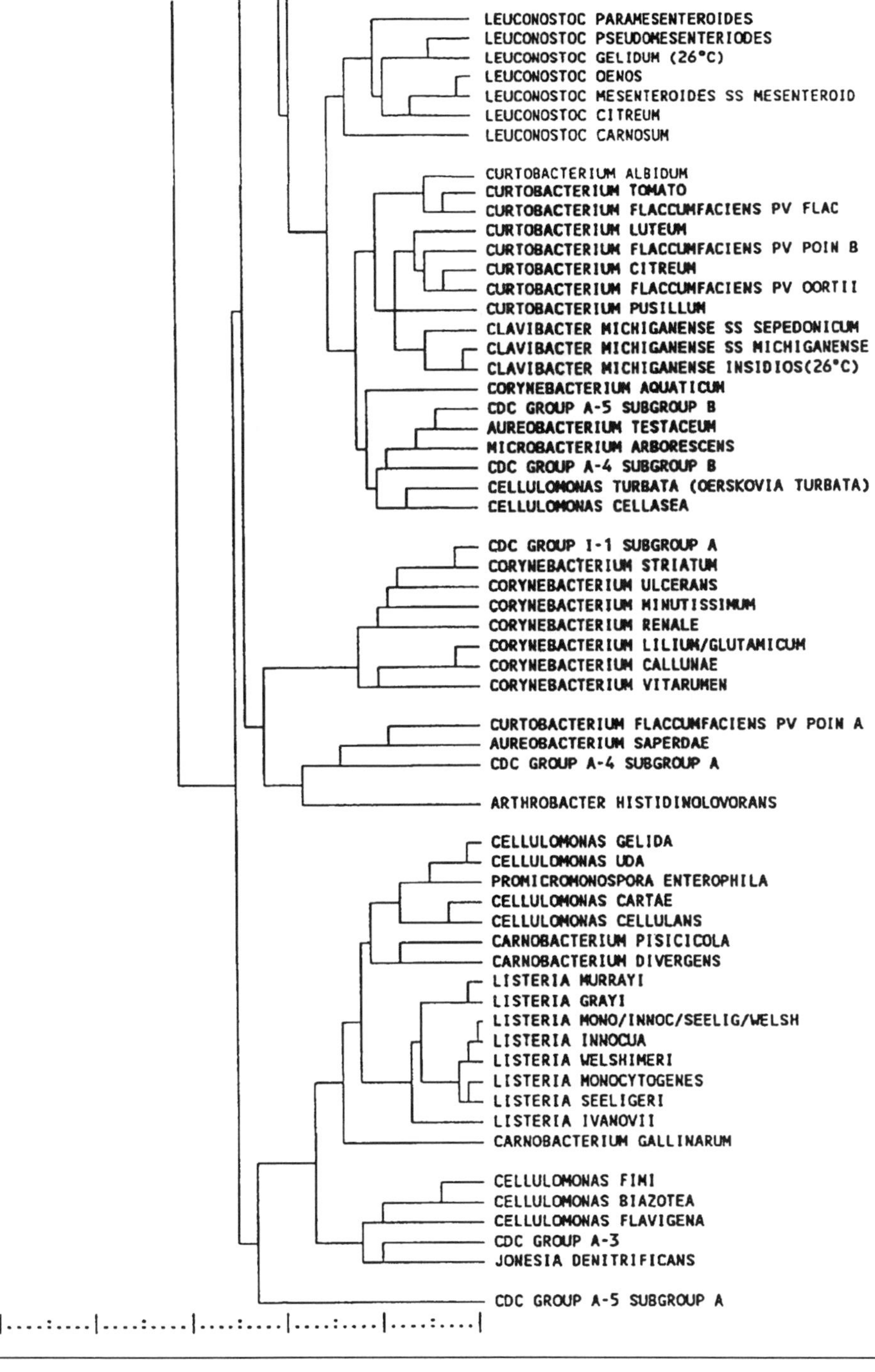

BIOLOG RELEASE 3.50 GP DATA BASE,   SPORE FORMING RODS

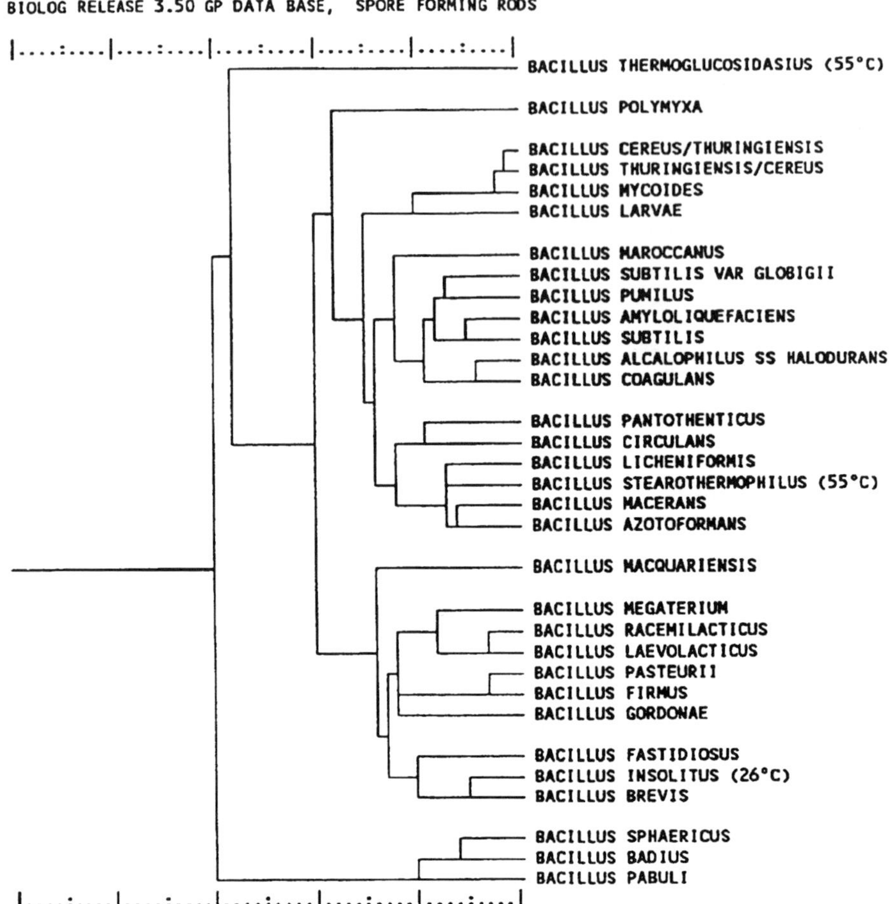

BIOLOG RELEASE 3.50 LA DATA BASE, LACTOBACILLUS

REFERENCES

Bochner, B. R. 1989a. Breathprints at the microbial level. *ASM News* 55:536–539.

Bochner, B. R. 1989b. Sleuthing out bacterial identities. *Nature* 339:157–158.

Bull, A. T., M. Goodfellow, and J. H. Slater. 1992. Biodiversity as a source of innovation in biotechnology. *Ann. Rev. Microbiol.* 46:219–252.

De Vos, P., K. Kerster, E. Falsen, B. Pot, M. Gillis, P. Segers, and J. De Ley. 1985. *Comamonas* Davis and Park 1962 gen. nov., nom. rev. emend., and *Comamonas terrigena* Hugh 1962 sp. nov., nom. rev. *Int. J. Syst. Bacteriol.* 35:443–453.

Li, X., M. Dorsch, T. Del Dot, L. I. Sly, E. Stackebrandt, and A. C. Hayward. 1993. Phylogenetic studies of the rRNA group II pseudomonads based on 16S rRNA gene sequences. *J. Appl. Bacteriol.* 74:324–329.

MacDonell, M. T.. and R. R. Colwell. 1985. Phylogeny of the Vibrionaceae, and recommendation for two new genera, *Listonella* and *Shewanella*. *Syst. Appl. Microbiol.* 6:171–182.

Manafi, M., and W. Kneifel. 1990. Rapid methods for differentiating gram-positive from gram-negative aerobic and facultative anaerobic bacteria. *J. Appl. Bacteriol.* 69:822–827.

May, R. M. 1988. How many species are there on earth? *Science* 241:1441–1449.

Miller, J. M. 1991. Evaluating biochemical identification systems. *J. Clin. Microbiol.* 29:1559–1561.

Patt, T. E., G. C. Cole, and R. S. Hanson. 1976. *Methylobacterium*, a new genus of facultatively methylotrophic bacteria. *Int. J. Syst. Bacteriol.* 26:226–229.

Priest, F. 1993. Systematics and ecology of *Bacillus*. In Bacillus subtilis *and other gram-positive bacteria*, edited by A. L. Sonenshein, J. A. Hoch, and R. Losick. Washington, DC: ASM Press.

Reasoner, D. J., and E. E. Geldreich. 1985. A new medium for the enumeration and subculture of bacteria from potable water. *Appl. Environ. Microbiol.* 49:1–7.

Stackebrandt, E., R. G. E. Murray, and H. G. Truper. 1988. Proteobacteria classis nov., a name for the phylogenetic taxon that includes the "purple bacteria and their relatives." *Int. J. Syst. Bacteriol.* 38:321–325.

Swings, J., P. De Vos, M. Van den Mooter, and J. De Ley. 1983. Transfer of *Pseudomonas maltophilia* Hugh 1981 to the genus *Xanthomonas* as *Xanthomonas maltophilia* (Hugh 1981) comb. nov. *Int. J. Syst. Bacteriol.* 33:409–413.

von Graevenitz, A., and C. Bucher. 1983. Accuracy of the KOH and vancomycin tests in determining the Gram reaction of non-Enterobacterial rods. *J. Clin Microbiol.* 18:983–985.

Willems, A., J. Busse, M. Goor, B. Pot, E. Falsen, E. Jantzen, B. Hoste, M. Gillis, K. Kersters, G. Auling, and J. De Ley. 1989. *Hydrogenophaga*, a new genus of hydrogen-oxidizing bacteria that includes *Hydrogenophaga flava* comb. nov.

(formerly *Pseudomonas flava*), *Hydrogenophaga palleronii* (formerly *Pseudomonas palleronii*), *Hydrogenophaga pseudoflava* (formerly *Pseudomonas pseudoflava* and *Pseudomonas carboxydoflava*), and *Hydrogenophaga taeniospiralis* (formerly *Pseudomonas taeniospiralis*). *Int. J. Syst. Bacteriol.* 39:319–333.

Willems, A., E. Falsen, B. Pot, E. Jantzen, B. Hoste, P. Vandamme, M. Gillis, K. Kersters, and J. De Ley. 1990. *Acidovorax*, a new genus for *Pseudomonas facilis, Pseudomonas delafieldii*, E. Falsen (EF) Group 13, EF Group 16, and several clinical isolates, with the species *Acidovorax facilis* comb. nov., *Acidovorax delafieldii* comb. nov., and *Acidovorax temporans* sp. nov. *Int. J. Syst.Bacteriol.* 40:384–398.

Woese, C. R. 1987. Bacterial evolution. *Microbiol. Rev.* 51:221–271.

Yabuuchi, E., Y. Kosako, H. Oyaizu, I. Yano, H. Hotta, Y. Hashimoto, T. Ezaki, and M. Arakawa. 1992. Proposal of *Burkholderia* gen. nov. and transfer of seven species of the genus *Pseudomonas* homology group II to the new genus, with the type species *Burkholderia cepacia* (Palleroni and Holmes 1981) comb. nov. *Microbiol. Immunol.* 36:1251–1275.

Yabuuchi, E., I. Yano, H. Oyaizu, Y. Hashimoto, T. Ezaki, and H. Yamamoto. 1990. Proposals of *Sphingomonas paucimobilis* gen. nov. and comb. nov., *Sphingomonas parapaucimobilis* sp. nov., *Sphingomonas yanoikuyae* sp. nov., *Sphingomonas adhaesiva* sp. nov., *Sphingomonas capsulata* comb. nov., and two genospecies of the genus *Sphingomonas*. *Microbiol. Immunol.* 34:99–119.

3

Substrate Utilization and the Automated Identification of Microbes

Lucia Clontz

AAI, Inc.

HISTORY OF VITEK®

The VITEK® system (VITEK) is an established rapid and fully automated microbial identification/susceptibility system (MIS) used for a wide variety of industries. The company that developed this system is recognized for introducing the world's first fully automated MIS.

VITEK was developed in the late 1960s by a joint venture between McDonnell Douglas and NASA to detect and identify pathogens directly from urine samples from astronauts in space. The original system was called MLM (microbial load monitor). In the late 1970s the first automated system was marketed by McDonnell Douglas for clinical microbiology applications and was originally called "Auto Microbic System." This system could identify nine common urinary tract pathogens directly from urine specimens. In 1977, a new company was started, Vitek Systems, from a division of McDonnell Douglas. The product soon became known as the VITEK, meaning "life technology." Additional test kits were later introduced that required use of pure microbial culture for inocula. These added technologies surfaced in the early to mid 1980s, a time when Vitek Systems continued developments in the area of industrial microbiology that led to the creation of a separate industrial division.

In the late 1980s, McDonnell Douglas decided to exit the market, and in 1989 Vitek Systems became part of the bioMérieux Group which had previously acquired API, best known as the developer of rapid microbiology diagnostic test kits. In the early 1990s, Vitek Systems changed its name to bioMérieux Vitek, Inc. to reflect its membership as one of over twenty subsidiaries worldwide. Today there are over 3000 VITEK systems in place worldwide and aboard space station *Freedom*.

SYSTEM COMPONENTS

The VITEK system is an integrated modular system that consists of five basic components.

The system is based on microbial growth in microwells of thin plastic identification cards (ID Cards), approximately the size of a standard playing card (Figure 3.1). There are 30 wells per ID Card; each well contains a different dehydrated biochemical substrate configured for the specified use. Once a test kit has been inoculated with a suspension of the sample organism, biochemical reactions are monitored hourly. Optical density readings of the samples being tested are automatically compared to a standard database to produce microbial identification results.

The *filler/sealer* unit has a vacuum chamber that forces the inoculum to flow into the test ID Card and uses a hot sharp blade to cut the inoculum transfer tubes

Figure 3.1. Gram-negative identification card diagram.

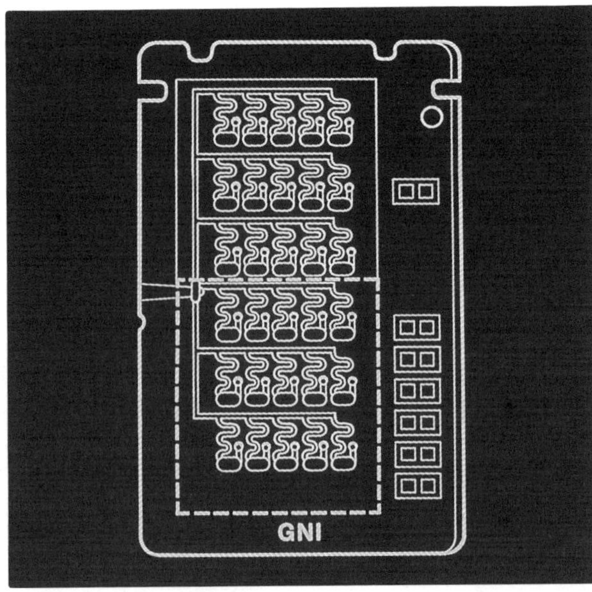

and seal the port of the test ID Card. This procedure creates a totally enclosed system.

The *reader/incubator* component is capable of testing up to 30, 60, 120, 240, 360, or 480 test kits, which are loaded onto the trays that have a capacity of 30 ID Cards each. The ID Cards are held in a temperature-controlled environment (35°C–37°C) and are automatically scanned once every hour by the photometric sensors for changes in color or turbidity. This information is transferred to the computer for subsequent analysis.

The *computer* processes test results data, collects results from the reader/incubator, and stores them during incubation time. It also manages extra data management options, including storing a product's database and retrieving information for trend analysis. The main database is the primary react file (PRF), composed mostly of clinical, environmental, food, and other isolates; it offers a high confidence value for the normalized probability reported. Once an ID Card has been incubated a sufficient length of time for an identification to be attempted, the biochemical test results are matched to the react file (database), which is a matrix of percent probabilities of positive tests for each biochemical test and each organism identified by the ID Card. The probabilities in the react file are derived from the test results of representative strains of each organism identified by the test ID Card.

If a match cannot be obtained through the PRF, a window is opened for additional test evaluation at a lower confidence value for the probability reported. This database is called the supplemental react file (SRF) and is only available to customers with industrial microbiology applications. This program also allows for the creation of a customized database for the identification of atypical organisms encountered in the laboratory and environment, since newly described or rare organisms may not be included in the system's database.

Improvements are continuously developed to expand the capabilities of the system, and selected organisms are added to the database as strains become available at no additional charge to the customer.

The *data terminal/keyboard* component allows the operator and the machine to communicate through typed commands and choices of menu.

The *printer* facilitates written data transmission from the system. It also provides hard copies of final test results automatically as well as interim results at the request of the operator. The report generated presents the individual biochemical test results and lists the one or two most likely organism identifications, along with their respective normalized percent probabilities. The bionumber printed on the report is a 9-, 10- or, 11-digit number derived from the test ID Card biochemical profile.

Preparations for Identification

Although the VITEK system is a highly sophisticated MIS, the test setup is very simple. The main steps are subculturing for the isolation of a pure culture; Gram stain; oxidase, coagulase, or catalase tests; and preparation of the organism suspension in saline. A Gram stain and colonial morphology are performed so that the appropriate test ID Card is used. The interpretation of test results requires the judgment and skill of personnel knowledgeable in the area of microbial identification, since additional testing occasionally is required. For some applications the source of the specimen, colonial and microscopic morphology, serology, antimicrobial susceptibility, and patient drug therapy can be important considerations. Once an isolated colony has been achieved from the culture plate or tube, the operator's manual involvement with the VITEK is limited to the preparation of the organism suspension, inoculation, and loading of the ID Card into the VITEK.

INOCULATION OF TEST KITS AND INTERPRETATION OF RESULTS

The organism to be identified must be a pure culture no more than 24 hours old. Isolates are ideally subcultured for two passages onto tryptic soy agar or blood agar plates. Protocols are outlined with each test kit for the isolation of a pure culture and detailed information on test methodology.

Once the choice of appropriate test kit has been made, the ID Card is marked with an instrument-read sample number and any appropriate external test marks. The ID Card is then placed in a holder with a transfer tube inserted inside a test tube containing a diluted suspension of the isolated organism (Figure 3.2) adjusted to the appropriate McFarland standard, using the VITEK colorimeter (Figure 3.3). Test kits and their holders are placed in the filler/sealer section of the instrument where the diluted sample is vacuum-drawn into the test kit wells (Figure 3.4). Tape on both side surfaces of the test kit prevents the escape of microbial contaminants into the laboratory's environment and provides for optimal

---

Figure 3.2. After diluent is dispensed into test tube containing sample, card is placed in holder with the transfer tube inside test tube.

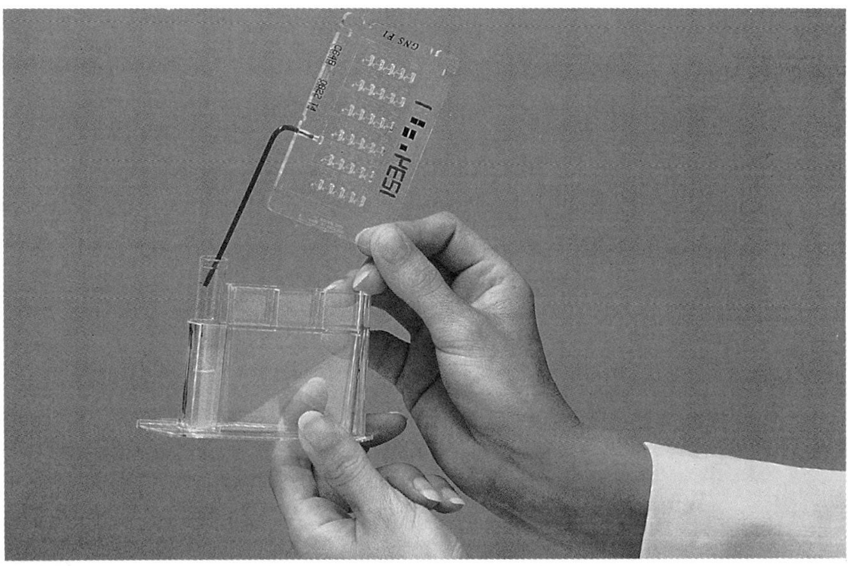

Figure 3.3. VITEK colorimeter.

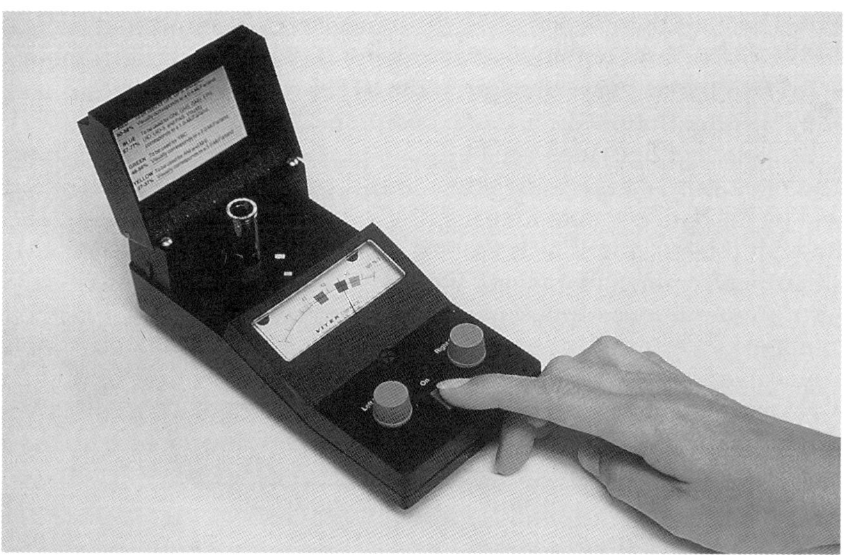

Figure 3.4. Cards/holders are placed in the filler/sealer where the diluted sample is vacuum-drawn into the card wells, and the card is sealed.

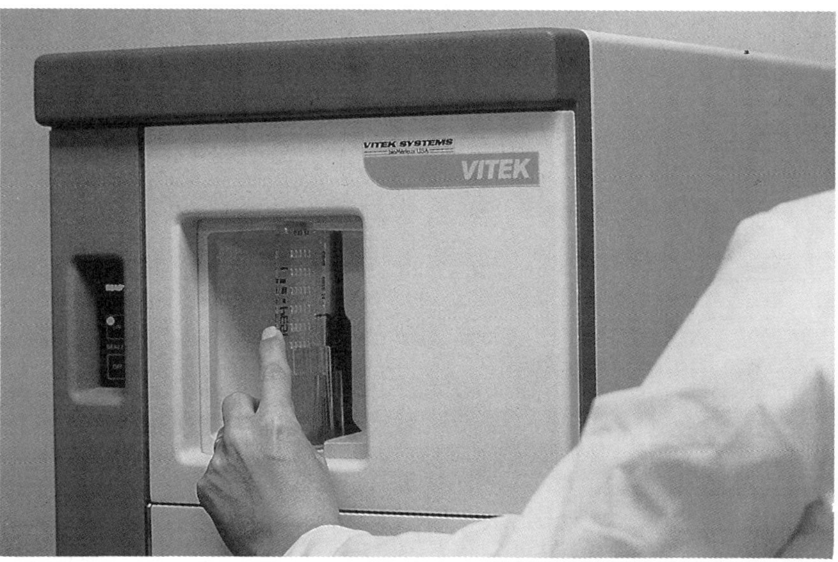

organism oxygen requirements. Use of the ID Card removes subjective errors in reading the standard scheme, reduces the chance for contamination, and provides for a uniform inoculation of all wells. Once inoculation of the ID Card is completed, the test kit is then sealed off at the transfer tube's connection.

The test kits, in any combination or order of ID Card, are inserted onto a tray and loaded into the reader/incubator of the VITEK system (Figure 3.5).

The VITEK-programmed computer determines when a well is positive based on the attenuation of light measured by an optical scanner. These patterns are analyzed automatically. At the completion of the incubation cycle, an identification is printed by the data terminal for each ID Card in the reader/incubator.

Although these test ID Cards employ many conventional biochemical tests, their unique environment combined with short incubation may produce results that may differ from those appearing in published material of other methods.

Prompted by keyboard entry of the sample numbers, a preliminary status report is available as early as one hour after loading or at any time during the test period. No additional biochemicals need to be added after the test kits have been loaded in the system. A final report is printed automatically for each test kit at the end of its cycle.

Special Messages

Several types of special messages may appear on the final report. These special messages will either explain why an identification cannot be made or act as a qualifying statement for the identification. Examples include the following:

Figure 3.5. The cards, in any combination of types, are inserted in a tray for loading into the reader/incubator.

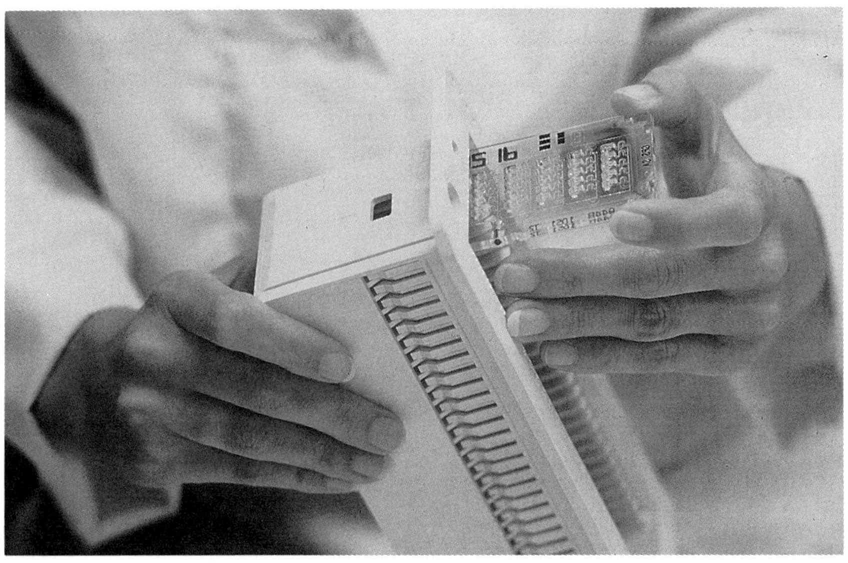

- Nonviable organism

- Unidentified organism

- Good confidence, marginal separation

- Questionable biopattern

These messages also refer to notes in the VITEK *Technical Bulletins Manual* by bioMérieux Vitek, Inc., which assists the operator in the resolution of the problem or refers to additional testing needed for a complete identification to the species level.

TYPES OF SYSTEMS AVAILABLE

An extensive range of tests for industrial applications is available, as well as custom-defined ones. Identification ID Cards that are automatically compared to a standardized database are the gram-negative identification (GNI), gram-positive identification (GPI), *Bacillus* identification (BAC), anaerobe identification (ANI), yeast identification (YBC—yeast biochemical card), *Neisseria/Haemophilus* identification (NHI), and nonfermenter identification (NFC). In addition, two growth-based test Cards, assay (ASC) and bioburden (BIO), are available.

The database has been derived from a large collection of clinical and industrial isolates as well as a number of stock strains provided by reference laboratories and isolates solicited from customers. All identifications were confirmed through conventional biochemical and serological testing. Biopatterns were established for each species in the database, by evaluating the biochemical results from the different ID Cards for all strains of each species contained in the database. The probability number, printed along with the identification on the status report, is an expression of how close the unknown organism's biopattern resembles those stored in the database for the species. The accuracy of the probability number for each species is a function of the frequency of occurrence for each species in the database.

Other tests include over 40 antimicrobial agents available on the susceptibility Cards for relative strength or efficacy testing. Results from these types of tests include an interpolated, calculated minimum inhibitory concentration (MIC) as well as classification ratings as established by the National Committee for Clinical Laboratory Standards.

Gram-Negative Identification

The GNI Card is intended for the automated identification of microorganisms of the family Enterobacteriaceae and it contains 99 species in its database. In addition, a selected group of glucose nonfermenting gram-negative bacteria, *Actinobacillus urea* (*Pasteurella urea*), *Pasteurella haemolytica*, *Pasteurella multocida*, and members of the family Vibrionaceae can be identified. The VITEK GNI Card is a Food and Drug Administration (FDA) cleared product and has also received Final Action Status by the Association of Analytical Chemists (AOAC) International for the identification of *Salmonella*, *Escherichia coli*, and other Enterobacteriaceae.

The GNI Card performs a series of conventional and nonconventional biochemical tests. It is based on established biochemical methods of Edwards and Ewing (1972 and 1973) and Gilardi (1978 and 1983).

Most of the tests are conventional, such as hydrogen sulfide, the carboxylases, urea, citrate, and tryptophan (TDA), which have been adapted for use in the VITEK system. The miniaturized concept of the GNI Card has uniquely reproduced aerobic and microaerophilic conditions suited for each test provided. The ability of an organism to utilize glucose fermentatively determines into what group the computer places the unknown organism for identification.

The setup involves a Gram stain, a cytochrome oxidase test, as well as matching the inoculum to a McFarland standard with the VITEK colorimeter prior to inoculating the ID Card. Identification requires 2–13 hours of incubation in the reader/incubator for the Enterobacteriaceae and Vibrionaceae, and 4–18 hours for *Pseudomonas aeruginosa*. The other nonfermenting gram-negative bacteria require an incubation of 4–18 hours for identification.

Limitations

The following are some of the limitations of the GNI Card as viewed by the vendor.

- Occasionally, isolates from MacConkey agar have been found to express slower biochemical reactions.

- Identification of *Acinetobacter calcoaceticus* bio. *lwoffii* is considered presumptive since it is based primarily on negative results on the GNI Card.

- Only specific biotypes of *Ochrobactrum anthropi* (*Achromobacter* species Vd), *Pseudomonas mendocina*, *Pseudomonas pickettii* (*thomasii*), *Pseudomonas stutzeri*, and *Pseudomonas vesicularis* are identified by the GNI Card. Other strains of these taxa are reported as an unidentified organism.

- Identification of *Pseudomonas pseudomallei*, *Yersinia pestis*, and *Ochrobactrum anthropi* (*Achromobacter* species Vd) are based on a small number of organisms and their identification should be considered presumptive.

Gram-Positive Identification

The GPI Card is intended for the automated identification of clinically significant *Streptococci*, *Staphylococci*, and a selected group of gram-positive cocci and gram-positive bacilli. Among the 48 species of bacteria that may be identified by this test kit are: *Streptococcus pyogenes*, *Group G streptococcus*, *Enterococcus faecalis*, *Streptococcus bovis*, *Gemella morbillorum*, *Streptococcus pneumoniae*, *Staphylococcus aureus*, *Staphylococcus epidermidis*, *Listeria* species, *Actinomyces pyogenes*, *Arcanobacterium haemolyticum*, *Corynebacterium xerosis*, and *Corynebacterium* species.

The VITEK GPI Card is an FDA-cleared product and also has received First Action Status by AOAC International for the identification of *Listeria* species when used in conjunction with the GNI Card.

Media used on the GPI Card are based on conventional biochemical tests that have been adapted for use in the microaerophilic environment of the test ID Card.

The setup involves performing catalase, coagulase, and Gram stain tests. Positive catalase, coagulase, and beta hemolytic results from the streaking of the isolate onto blood agar plates are coded on the ID Card. Incubation time is 2–15 hours and is based on the organism's growth rate.

Limitations

The following are some of the limitations of the GPI Card as viewed by the vendor.

- Isolates with colonies that are extremely adherent and cannot be uniformly suspended in 0.45 percent saline solution to produce a #0.5 McFarland standard with the VITEK colorimeter are not suitable for use on the GPI Card.

- If the organism is a strict aerobe not in the database and it does not grow in the microaerophilic environment of the GPI Card, no result will be produced and the "insufficient growth" message will be printed on report.

- *Micrococcus* species cannot be identified due to inaccurate test results. The operator must perform a Gram stain and if the result is gram-positive cocci in tetrads, an alternate method of identification, such as the API system, must be chosen.

- *Lactobacillus* is not in the GPI database at the present time, but may be identified using the ANI Card.

Yeast Identification

The YBC is intended for the automated identification of the most frequently isolated clinical yeasts. The organisms identified are as follows:

Candida species: *albicans, famata, glabrata, guilliermondii, humicola, krusei, lambica, lusitaniae, parapsilosis, paratropicalis, pseudotropicalis, rugosa, stellatoidea, tropicalis,* and *zeylanoides*

Cryptococcus species: *candidum, laurentii, luteolus, neoformans, terreus,* and *uniguttulatus*

Rhodotorula species: *glutinis, pilimanae,* and *rubra*

Geotrichum species: *candidum* and *penicillatum*

Saccharomyces species: *cerevisiae*

Trichosporon species: *beigelii* and *pullulans*

Blastoschizomyces species: *capitatus*

Hansenula species: *anomala*

Phichia species: *ohmeri*

Prototheca species: *wickerhamii* and *zopfii*

Sporobolomyces species: *salmonicolor*

Yarrowia species: *lipolytica*

The YBC Card, an FDA-cleared product, contains 26 biochemical broths and 4 negative control broths; it reproduces an environment favorable for the growth of fermentative and nonfermentative yeasts. All of the tests are based on established biochemical methods of Wickerham and Burton (1948). They are conventional tests, such as carbohydrate assimilation, urea hydrolysis, resistance to cycloheximide, and nitrate reduction.

The setup for this ID Card involves preincubation of test kits in an off-line incubator at 30°C for 24 hours (in some cases 48 hours), and a single reading at 24 hours (and in some cases a second reading at 48 hours), in the reader/incubator.

Limitations

The following are some of the limitations of the YBC Card as viewed by the vendor.

- The database is limited. Only those taxa listed previously are acceptable for identification by the YBC Card.

- Test for *Trichosporon pullulans* can yield inaccurate results. The nitrate test for this organism is usually slow and false identification of *Trichosporon beigelii* may occur.

- Rare yeast strains have been found that do not grow well in the special YBC glucose media, but grow well on other types of media, such as Sabouraud dextrose agar or potato dextrose agar. For these organisms standard methods or another type of automated or manual method of identification should be used.

- Microscopic examination of morphology, such as germ tube production and capsule formation, should be considered for further assurance of speciation results.

Bacillus Identification

The BAC Card and database are intended for the automated identification of microorganisms of the family Bacillaceae, including *alvei, cereus, circulans, coagulans, firmus, laterosporus, lentus, licheniformis, macerans, megaterium, polymyxa, pumilus, sphaericus, stearothermophilus, subtilis, subtilis (globigii)*, and *thuringiensis*.

The BAC Card contains biochemical broths for conventional and nonconventional tests and it reproduces aerobic and microaerophilic conditions suited for each test provided. This ID Card is based on established methods of Ruth Gordon (1975).

The ability of an organism to grow at 55°C and the utilization of glucose are determining factors for the speciation of unknowns. The normal time for incubation in the reader/incubator is 6–15 hours. Organisms that grow at 55°C require 18–24 hours external incubation in an airtight humidity chamber prior to loading into the system.

The setup is simple. The usual Gram stain, a catalase test, and preparation of a suspension matching a McFarland #0.5 standard with the VITEK colorimeter are the only added steps needed.

Limitations

The following are some of the limitations of the BAC Card as viewed by the vendor.

- Cultures that can grow at 55°C must be incubated at this temperature in an external incubator prior to loading into the system.

- Only cultures suspected of belonging to the genus *Bacillus* should be inoculated into the BAC Card.

Anaerobe Identification

The ANI Card was developed for the identification of medically important anaerobic and microaerophilic bacteria of human origin. This test ID Card is an FDA–cleared product. The ANI react file is divided into four sections: gram-negative bacilli (30 species), gram-positive bacilli (44 species), gram-negative cocci (24 species), and gram-positive cocci (22 species). Members of the anaerobic and microaerophilic bacteria taxa identified include

Actinomyces species: *bovis* and *pyogenes*

Arcanobacterium species: *haemolyticum*

Bacteroides species: *fragilis*, *oralis*, and *vulgatus*

Bifidobacterium species

Capnocytophaga species

Clostridium species: *baratii*, *botulinum*, *cadaveris*, *dificilis*, *perfringens*, *sporogenes*, and *tetani*

Corynebacterium species: *diphtheriae*

Eubacterium species: *aerofaciens*

Fusobacterium species: *mortiferum*

Gemella species: *morbillorum*

Lactobacillus species: *acidophilus* and *casei*

Porphyromonas species: *gingivalis*

Propionibacterium species: *acnes*

Staphylococcus species: *saccharolyticus*

Streptococcus species: *intermedius*

Veillonella species: *parvula*

The tests used on the ANI Card are based on microbial degradation of specific chromogenic substrates that are detected by a variety of indicator systems. The main difference in the use of this ID Card is that this test kit is never actually loaded into the system. The operator performs a Gram stain and a spot indole test, besides the usual preparation of a suspension equal to a certain McFarland standard with the VITEK colorimeter.

Cards are incubated for four hours at 35°C–37°C in an off-line non-CO_2 incubator. After incubation is complete, positive and negative results are visually determined by comparison of the color reactions in the wells to a positive color guide in the benchtop off-line viewer or handheld viewer. The results are manually entered into the system. Patterns are analyzed automatically and an identification report is printed by the system's printer.

Limitations

The following are some limitations of the ANI Card as viewed by the vendor.

- Subjective errors associated with the visual interpretation of test results may occur. Therefore, the interpretation of results and use of the ANI Card require a competent technician who must have experience in microbial identification, as well as specimen information and proper handling.

- Gram stain reaction, colonial and cellular morphology, as well as growth aerobically or in CO_2 should be considered when using the ANI Card, since aerotolerance may be necessary for the separation and confirmation of certain organisms.

- The use of organisms not listed in the test protocols will lead to aberrant results.

- The accuracy of the ANI Card is based on statistical use of multiple tests contained on the ID Card. Errors in the interpretation of a test will impact the overall delineation of an identification.

- The time between preparation of inoculum, ID Card inoculation, and placement in the incubator should not exceed 20 minutes.

Bioburden Determination

The bioburden test Card is intended for the automated determination of microbial populations in a liquid sample. A wide range of aerobic and facultative anaerobic microorganisms may be enumerated with this test Card. The bioburden test Card estimates bacterial contamination by the dilution method. This test Card consists of 30 wells, each containing dried growth media, sealed with oxygen-permeable tape.

Once the test Card is loaded in the machine, the instrument detects the presence or absence of growth in each well according to light attenuation measured by an optical scanner. Rapid-growing organisms can be enumerated in less than 24 hours. Incubation will automatically be extended to 48 hours for slow growers. A final report is issued when 5 hours have elapsed following the most probable number (MPN) determination, or if all 30 wells are positive, or when 48 hours have elapsed from test initiation.

The relationship between the number of positive wells and the MPN is illustrated in Figure 3.6. An inoculum containing between 5–80 organisms/ml provides the most accurate microbial population determination.

Figure 3.6. MPN graph for bioburden determination.

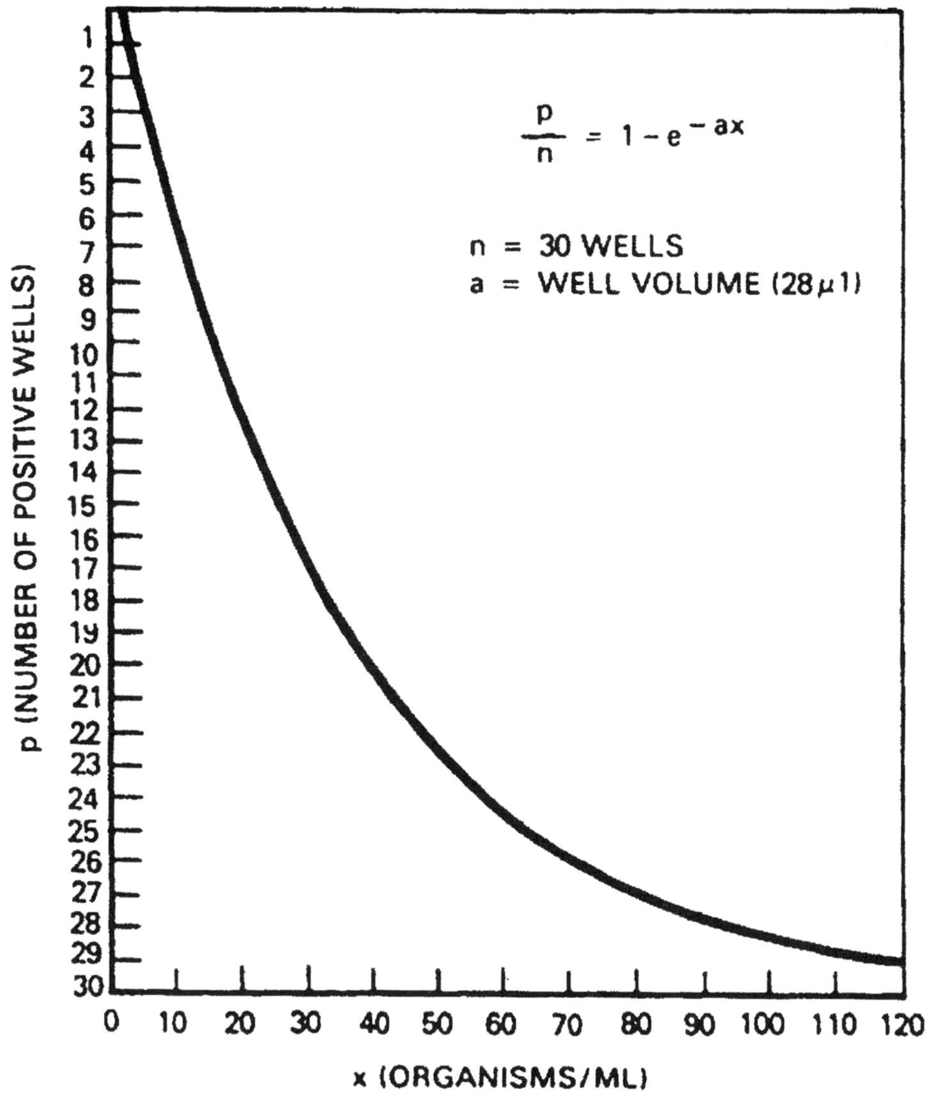

$$\frac{p}{n} = 1 - e^{-ax}$$

Limitations

The following are limitations of the bioburden test Card as viewed by the vendor.

- Visually opaque liquids must be diluted before use with this test Card.

- The presence of inhibitory substances in samples may result in low count estimates.

- Visually turbid or pigmented (especially blue-green) liquid samples may seriously reduce the sensitivity of the card.

- Test sensitivity is based on statistical calculations. Therefore, the absence of growth on the bioburden test Card should be interpreted as <1.2 organisms/ml, and not an indication of sample sterility.

Assay Card

The assay test Card is intended to be used as a bacterial growth test kit to measure the relative strength or efficacy of samples such as antibiotics, vitamins, disinfectants, or preservatives. The assay test Card has three sections of 10 wells that do not contain media. This allows for the simultaneous testing of three samples. Test samples are inoculated into the test Card along with the challenge organism suspension and growth medium. The readings for light attenuation are converted into percent change values based on the initial reading or baseline. The hourly percent change values are calculated and stored by the computer. Twenty-four hours of data may be collected on each test Card or a predetermined incubation time; data for each section of the test Card is then analyzed.

Limitations

While the setup and principle of test are simple, interpretation of test results requires the judgment and skill of personnel with knowledge in the area of bioassays.

Neisseria/Haemophilus Identification

The *Neisseria/Haemophilus* Identification (NHI) test Card is intended for the automated identification of significant *Neisseria* and *Haemophilus* species. The organisms identified are: *Neisseria gonorrhoeae*, *Neisseria meningitis*, *Neisseria lactamica*, *Moraxella* species, *Branhamella catarrahalis*, CDC M-groups, *Kingella* species, *Haemophilus influenzae*, *Haemophilus parainfluenzae*, and other *Haemophilus* species. The NHI test Card is an FDA–cleared product.

NHI utilizes 16 conventional and single-substrate chromogenic tests. The tests used in the NHI test Card are based upon microbial degradation of specific substrates detected by a variety of indicator systems.

The setup involves performing Gram stain, cytochrome oxidase, and catalase tests. A spot indole test for the *Haemophilus* species group is required and the test of the ability to grow on modified Thayer-Martin agar may be helpful with the identification of *Neisseria* species. There is no addition of reagents to the NHI test Card after incubation.

After 4 hours of incubation at 35°C, positive and negative results are determined by visually comparing the wells to a positive color guide in the benchtop off-line viewer or in the handheld viewer. Results are manually entered into the system. Patterns are analyzed automatically and an identification report is printed by the system's printer.

Limitations

The following are some of the limitations of the NHI test Card as viewed by the vendor.

- Gram stain reaction, oxidase production, growth on selective agars, and morphological characteristics should be considered when using this test Card.

- The accuracy of the NHI test Card is based upon the statistical use of the multiple tests contained in the test Card. The delineation of an identification by using a single test is subject to the error inherent in the test alone. The NHI test Card must be used in conjunction with the VITEK.

- Subjective errors associated with the visual interpretation of test results may occur.

Nonfermenter Identification

The nonfermenter Card (NFC) is intended for the automated identification of oxidase positive and some oxidase negative gram-negative nonfermenting bacilli. In addition, some oxidase positive glucose fermenting non-Enterobacteriaceae can be identified by the NFC Card. Members of the family Vibrionaceae are also identified using the NFC Card. Oxidase positive *Pasteurella aerogenes* is included as well. Among the 42 species of bacteria that may be identified by this test kit are *Acenitobacter haemolyticus, Aeromonas caviae, Bordetella bronchiseptica,* CDC group IVC–2, *Chromobacterium violaceum, Chryseomonas luteola, Flavimonas oryzihabitans, Flavobacterium meningosepticum, Pseudomonas aeruginosa, Pseudomonas fluorescens, Vibrio alginolyticus,* and *Vibrio cholerae.*

The NFC Card is automated assimilation testing based on the work of Stanier (1966) in the use of traditional and nontraditional substrates as a sole carbon source for the growth and characterization of the aerobic pseudomonads.

There are 30 wells in the NFC Card; 29 of them contain a chemically defined minimal medium with 28 individual organic compounds (substrate wells) and one contains a medium control. The growth medium contains chemicals that do not support growth without an available carbon source. Carbon assimilation results in a nutritional pattern that helps to separate and identify organisms to the species level.

Identification using the NFC Card requires 6–18 hours of incubation in the system. The setup includes isolation of organism on blood agar, performing a cytochrome oxidase test, and a Gram stain.

Limitations

The following are some of the limitations of the NFC Card as viewed by the vendor.

- The database is derived solely from organisms grown on tryptic soy agar with 5 percent sheep blood. The use of any other medium may result in biopatterns not consistent with the NFC Card.

- The McFarland No. 1 turbidity standard must be met to ensure the accuracy of the test Card.

- The database is almost exclusively for the oxidase positive gram-negative nonfermenting bacilli and some oxidase positive non-Enterobacteriaceae. *Stenotrophomonas maltophilia* and *Acinetobacter* species are the only oxidase negative organisms in the database.

- Some oxidase negative gram-negative fermenting bacilli are biochemically similar to *Agrobacterium tumefaciens*; therefore, one must confirm all calls of *Agrobacterium tumefaciens* with an oxidase positive reaction.

- Identifications for *Ochrobactrum anthropi* and CDC EF-4 are based on a limited number of isolates.

- Although the NFC Card employs many conventional assimilation tests, the unique environment of this test Card combined with the short incubation may produce results that may differ from those appearing in published material of other methods.

DATA MANAGEMENT AND INTERFACE OPTIONS

Data Management System Options

The information management system (IMS) is available for the generation of cumulative, quality control and customized reports when utilized with the VITEK computer. VITEK test results and off-line results can be combined for such reports.

The Data Trac™ program is available for generations of various statistical reports when used with the VITEK CC1 computer with bioLiaison™, a graphical user interface with state-of-the art Windows™-style features. Organism occurrence, biochemical patterns, product report, cumulative data, data logbook, data trending, and quality control are some of the reports that can be generated.

Supplemental React File

The SRF allows for the setup of a customized database by industrial customers for the identification of rare organisms not included the system's database. If an unidentified organism with a specific bionumber was given a code at an earlier test, this same organism when again isolated will be keyed out with the SRF and that specific code will be printed on report.

Computer Interface Options

The bidirectional computer interface (BCI) is an option that enables the automatic flow of information between the computer and a laboratory information system computer. The unidirectional computer interface (UCI) is an option available that streamlines the flow of information from the computer to the information management system (LIMS) computer by electronically transferring the data.

APPLICATIONS

The VITEK system is being used in the pharmaceutical, food, dairy, beverage, cosmetic, meat/poultry, and seafood industries as well as government and reference laboratories. Its short testing times make it very attractive for pathogen detection and confirmation applications.

Pharmaceutical Industry

One of the most popular uses of the VITEK system in the pharmaceutical industry is for the identification and data tracking of environmental isolates. If the system is equipped with the optional IMS or Data Trac™ software, trend analysis can be performed, as well as easy access for filing and retrieving of data based on demographics chosen, such as area/room, sample type, operator, and so on.

One of the applications in trend analysis is the tracking by occurrence of certain types of organisms in a particular environment. Reports can be generated reflecting types of organisms and their percentages of occurrence in relation to the total number of isolates. The IMS or Data Trac™ main menu offers reports by organism occurrence, workload, biochemical patterns, product report, cumulative by category, cumulative by MIC, percent susceptible, trending and logbook. Therefore, several of these types of reports can be combined for use in an environmental monitoring program evaluation.

Other applications of the VITEK in pharmaceutical plants and laboratories include monitoring data for water systems, bioburden samples from finished products and raw materials, and useful applications during equipment cleaning validations and routine cleaning procedures. In cases where evaluation of the source of contamination is needed, organism identity is a crucial key for the determination of potential causes and prevention procedures.

The SRF is particularly useful in the areas of environmental monitoring. Isolates that yield bionumbers of low percentage probability or that are unidentified can undergo identification by classical manual methods or alternate rapid identification systems; the result is then manually entered in the SRF for that particular bionumber, allowing for expansion of the database. The isolates that consistently and repeatedly demonstrate the same bionumber are then given that particular identification.

The rapid identification of human-borne pathogens such as *Salmonella* species, *Escherichia coli*, and *Staphylococcus aureus* for product release testing has made the instrument a key in the pharmaceutical quality control (QC) laboratory. These organisms undergo a two-hour identification test, expediting product release.

In sterile manufacturing facilities characterization of isolates establishes a valid database, demonstrates that cleaning and sanitizing are effective, and establishes a potential relationship between organisms found during sterile media fills and found during product sterility testing. Such correlations are extremely valuable during an investigation of sterility test failures.

The VITEK system has definitely contributed to the time savings and organization of reports in this industry. Reduction of paperwork, which can free staff to perform other duties, is also an important factor to consider.

Food and Other Related Industries

The food processing industry has increased the need for the rapid detection of pathogens in order to meet the demand for rapid turnaround of QC samples. The conventional manual biochemical identification methods are no longer acceptable, especially for the identification of organisms such as *Salmonella* species, *Listeria* species, and other pathogens. The AOAC conventional biochemical method, including serological testing for the identification of *Salmonella* species can take up to seven days. The VITEK system is able to meet these demands by processing a large number of samples quickly.

Most of the organisms of concern in this industry are of the Enterobactericeae, *Klebsiella*, and *Serratia* groups, found in ground beef, dairy products, processed chickens, frozen desserts, commercial poultry feeds and shellfish. Also, *Listeria* species are of great concern, especially with fresh produce items, and *Bacillus* species are indigenous to confectionary flavorants.

Listeria spp. have been found in fresh produce, including broccoli, cabbage, carrots, cauliflower, potatoes, and tomatoes. Epidemics of food-borne listeriosis with high fatality rates have demanded a faster and more accurate method of detection for this type of organism. In 1981 an epidemic was linked to raw cabbage, lettuce, celery, and tomatoes as suspected vehicles of transmission. A study was conducted to determine *Listeria* contamination of 10 types of fresh produce obtained from two Minneapolis area supermarkets from October 1987 through August 1988. The VITEK system was chosen as the automated system in this study (Heisick 1989). Standard manual methods may require up to seven days for identification. The GPI ID Card enables identification of *Listeria* species within 24 hours, with added time needed for identification to species level. Recently, a method that uses a GPI ID Card and GNI ID Card has enabled rapid identification of *Listeria* to the species level and has received first action approval by the AOAC.

The system has other applications in this industry. The VITEK system for microbial identification may be combined with the Bactometer®, a fully automated impedance technology system for quality indicator tests such as total counts, coliforms, yeast, and mold as well as shelf life determinations. The VITEK may also be combined with the VIDAS®, a fully automated enzyme-linked fluorescent assay (ELFA) technology system for pathogen screening of *Salmonella*, *Listeria*, *Escherichia coli*, *Listeria monocytogenes*, or Staphylococcal enterotoxins. The Bioassay test Card has been used for qualitative determination of antimicrobial activity of food additives and to evaluate the effectiveness of sanitizers in food plants.

Other Applications

There are many other areas of applications for the VITEK system, such as hospitals, research and routine diagnostic laboratories, and any other type of industry that deals with microorganisms in production (fermentation process manufacturing) or as contaminants. VITEK can expedite identification work and quantitation of microbial population with a very high level of accuracy.

In animal facility laboratories the VITEK is used to identify routine isolates such as *Pasteurella multocida* and *Bordetella bronchiseptica*, and also to monitor the cages that house the animals for an acceptable living environment. According to one source, use of the Bioburden Card reduced water sampling costs by 50 percent.

SYSTEM EVALUATION—A USER'S PERSPECTIVE

Several collaborative studies have been carried out for the evaluation of the performance of the VITEK as a rapid automated MIS.

Gram-Negative Identification

One study specifically evaluated the performance of the GNI Card to identify members of the Enterobacteriaceae family. Eighty-four isolates previously isolated from foods were used in this study (Knight et al. 1990). The conclusion was that the VITEK system correctly identified 96.7 percent of *Salmonella* spp., 97 percent of *Escherichia coli*, and an average of 93.8 percent of other enteric genera. Some of the isolates examined were considered atypical in their conventional biochemical patterns and one *Salmonella* isolate was unidentified due to an error in the off-line oxidase reaction interpretation. The overall conclusion was that the probable causes for the unidentified and misidentified organisms were either due to mixed inocula or due to cultures being older than the optimum 24-hour organism preparation. The recommendation was favorable for the GNI Card being a suitable alternative to conventional biochemical methods for the identification of *Escherichia coli* and of presumptive generic identification of other Enterobacticeae from foods. It was also noted that the lowest percentage of correct identification was for *Hafnia*. This has been improved since the time of the original study.

Other studies have shown that the GNI Card has correctly identified 79.6–95 percent of nonenterobacter gram-negative organisms tested, and that *Pseudomonas aeruginosa* and *Acinetobacter* spp. are the most commonly misidentified types of organisms. One suggestion (also a recommendation by bioMérieux Vitek, Inc.) was that unidentified organisms should be further evaluated by a retest with a 18–24 hour culture and inoculum checked for purity.

In another study conducted for identification of Enterobacteriaceae in foods (Bailey et al. 1985), 98.2 percent of the isolates were correctly identified. It was also noted that food isolates frequently have been exposed to one or more types of chemical stress during processing and often exhibit different biochemical profiles from clinical isolates. Therefore, it is important that an enrichment-recovery step is carried out prior to isolation for purity and testing, so that the correct phenotype is expressed.

Environmental Isolates

A study conducted by a pharmaceutical company evaluated multiple genera of gram-negative bacilli, gram-positive cocci, gram-positive bacilli, and yeasts. Sources of these organisms included ATCC and environmental isolates. The conclusion of the study was that the VITEK System correctly identified ATCC isolates 100 percent of the time and environmental isolates 80 percent of the time.

Listeria Identification

A study conducted for identification of *Listeria* (Knight et al. 1988) concluded that of 236 positive isolates, 7 were unidentified according to the GPI Card. These isolates were then confirmed to be *Listeria* by conventional biochemical tests. Also, confirmation of *Listeria* by the GPI Card occurred between 6 and 8 hours in 95 percent of the isolates, the maximum time being 16 hours and the minimum time 4 hours. This study also suggests that *Listeria* cells are injured by the use of chemical sanitizers and that a 7-day incubation time may be necessary to allow the repair of injured cells.

Another study for the evaluation of the GPI Card involving over 400 isolates of food and naturally occurring *Listeria* spp. concluded that the GPI Card compared favorably with the combination GPI/GNI Card method, with a 97 percent correlation. The GPI Card with two off-line biochemical reactions had a 100 percent correlation with conventional methods. One problem with the use of the GPI Card for the identification of *Listeria* is that this ID Card does not contain the rhamnose biochemical that is required for identification of *Listeria monocytogenes*. Therefore, this reaction must be run as an off-line test. As an alternative to an off-line test, the GNI Card can be used in conjunction with the GPI Card and the xylose and rhamnose reactions are then read from the GNI Card.

Gram-Positive Identification

The GPI Card was evaluated for the identification of *Staphylococcus* spp. One study (Bannerman et al. 1993) involved 500 clinical isolates for a comparison of the GPI Card with conventional methods: results were in 89 percent agreement. One problem noted for the lower than expected percentage obtained was that the system requires use of a pure culture. This may pose a problem for staphylococcal isolates, which at initial isolation (18–24 hours), colonies of different strains and even species are indistinguishable. This study recommends improvement in the identification of *Staphylococcus hominis*, *Staphylococcus lugdumensis*, and *Staphylococcus capitis* subsp. *ureolyticus* in a future upgrade and has been completed in a recent upgrade. The bionumber was also noted as having limited epidemiological value because of the frequent occurrence of different strains giving the same bionumber.

Other studies have shown that *Streptococcus pyogenes* and species of group D *Streptococci* and *Enterococci* yield high identification accuracy. However, identification of other *Streptococcus* species has not been so successful.

Yeast Identification

Recent studies show that the YBC Card correctly identifies 93.5–98 percent of the isolates tested. Another study included 205 strains of 8 species of yeasts that had been included in the YBC database: 96.9 percent of the isolates tested were correctly identified. The lowest correlation (88.9 percent) occurred for *Sporobolomyces salmonicolor* and equivocal results occurred for *Prototheca wickerhamii* and *Sporobolomyces salmonicolor*.

A low percent of incorrect identifications occurred for only 2 species, *Hansenula anomala* and *Candida paratropicalis*. Overall, the YBC Card identifies less common but clinically important yeasts with good accuracy.

Bioassay Card

The applicability of using the Bioassay Card to quantitate antimicrobial activity of known concentrations of disinfectants against stock cultures was examined by different groups. The method allows for the evaluation of disinfectants in a minimum of 24 hours, depending on the organism, and the software graphs the results. Problems were experienced with the yeasts as a result of the 37°C incubation temperature of the reader/incubator. The temperature limitation also applies to other ID Cards used for the identification of bacteria and, therefore, identification of an organism that cannot grow well at 37°C, such as *Methylobacterium* sp., cannot be achieved by the VITEK System.

The newer software versions show improved *Bacillus* sp. identification (68 percent for T 05.1 compared to 83 percent for T 06.1). One publication (Stager and Davis 1992) states that the VITEK system needs improvement for the identification of some *Staphylococcus* and *Streptococcus* species, in addition to accuracy improvement for the identification of unusual and uncommon bacteria and yeasts. Recommendations for shortened identification times for the nonglucose fermenting gram-negative bacilli, gram-positive bacteria, and yeasts, as well as on-line incubation for the YBC and ANI ID Cards were also made for an increased utilization of this system.

COMPARISON TO OTHER AUTOMATED SYSTEMS

Many studies have been carried out for the evaluation of individual systems, but not for a complete and comprehensive comparative evaluation of the currently available systems on the market. These individual studies are performed in different parts of the country, with different types of organisms and strains and specific methodologies. Therefore, they cannot really be considered comparative studies; even the accuracy of some of the same systems has been observed to vary among laboratories for the same type of organism being tested. We do know, however, that all of these systems can accurately identify common clinical isolates and that rare biotypes and unusual organisms are often unidentified or misidentified. But what level of accuracy can one consider to be acceptable?

The data show that all of the systems currently being used generally perform at greater than 90 percent accuracy when compared to conventional methods and when used to test common clinical isolates. The real question is, which system is best for my application? A thorough review of each system's database, performance, price range, upgrade capabilities, technical expertise of prospective operators, technical support, and methodologies are some of the points to be considered.

The MIDI Microbial Identification System (Microbial ID Inc., Newark, Del.) is valuable for the differentiation of phenotypically similar organisms. The MIDI system analyzes microbial fatty acids found in the cell wall and cell membranes, whereas other systems use conventional but automated biochemical reactions. It does not have the temperature limitation associated with the VITEK system, but the technique for isolate identification is more complex. The organism must undergo a series of chemical extractions and the system (gas chromatography) involves methods that most microbiologists have not used.

Like the MIDI system, the Biolog system (Biolog, Inc. Hayward, Calif.) has programs that offer dendrograms to demonstrate the relationships of strains and species being evaluated. Biolog has a large database but further evaluation of its gram-negative, gram-positive, and *Bacillus* identification kits is needed.

The consensus is that no single system is ideal for all applications. Some identifications, especially of wild strains, vary depending on the system used. The best solution is to use only one system for trend analysis. It is common for laboratories to use different systems for different applications and types of isolates. The good news is that all of the systems are undergoing upgrades constantly; as their capabilities expand, accuracy and system utilization should increase.

QUALITY CONTROL

bioMérieux Vitek, Inc. and its products are registered with the FDA and is currently operating in compliance with the Good Manufacturing Practices Regulation (21 CFR Part 820). An ISO 9001 registration audit was conducted in November 1994 and bioMérieux Vitek, Inc. was recommended for ISO 9001 certification. The company's quality system was assessed and found to be compliant with the ISO 9001 standard based on the auditors' findings, which encompass the design, development, manufacture, distribution, and servicing for automated and manual in vitro diagnostic devices and reagents for medical and industrial laboratories.

Each shipment of test kits received at the laboratory site, although carrying a certificate of analysis, should be tested to ensure product quality. bioMérieux Vitek has special recommendations that can be incorporated into a user's quality control protocol. The procedure includes a simple visual inspection of the test kits as well as testing one test Card from each shipment for microbial contamination and performance.

bioMérieux Vitek offers a list of recommended QC organisms to be used and their optimum maintenance conditions (see Table 3.1 for list of package insert QC organisms). The 0.45 percent saline used to prepare the organisms' suspensions should be checked for sterility at least once a week, to ensure that the test Cards are inoculated with a pure culture. Once a QC program is established, the system allows the user to test and store the QC data and test Card lot inventories, which can later be retrieved in a number of report formats.

Table 3.1. bioMérieux Vitek Quality Control Organisms

(to be used with the VITEK test ID Cards)

ANI

Bacteroides ureolyticus	ATCC 33387
Bacteroides vulgatus	ATCC 8482
Propionibacterium acnes	ATCC 11827
Porphyromonas gingivalis	ATCC 33277
Bacteroides distasonis	ATCC 8503
Peptostreptococcus micros	ATCC 33270

BAC

Bacillus licheniformis	ATCC 12759
Bacillus sphaericus	ATCC 4525

GNI

Proteus mirabilis	ATCC 7002
Pseudomonas aeruginosa	ATCC 27853
Serratia odorifera	ATCC 33077
Klebsiella pneumoniae	ATCC 13883
Acinetobacter calcoaceticus bio. *anitratus*	ATCC 19606
Yersinia kristensenii	ATCC 33639
Bordetella bronchiseptica	ATCC 10580

GPI

Streptococcus pyogenes	ATCC 19615
Enterococcus faecalis	ATCC 29212
Erysipelothrix rhusiopathiae	ATCC 19414
Enterococcus durans	ATCC 6056
Streptococcus bovis	ATCC 9809
Staphylococcus xylosus	ATCC 29971
Streptococcus equi	ATCC 9528

Table 3.1 continued on next page.

Table 3.1 continued.

NFC

Bordetella bronchiseptica	ATCC 10580
Acinetobacter baumannii	ATCC 19606
Pseudomonas aeruginosa	ATCC 27853
Aeromonas caviae	ATCC 15468
Pasteurella aerogenes	ATCC 27883

YBC

Candida albicans	ATCC 14053
Yarrowia lipolytica	ATCC 9773
Cryptococcus albidus	ATCC 34140
Candida humicola	ATCC 9949

NHI

Neisseria gonorrhoeae	ATCC 49981
Neisseria meningitidis	ATCC 13102
Neisseria lactamica	ATCC 23970
Haemophilus parainfluenzae IV	ATCC 7901
Branhamella catarrahalis	ATCC 25238
Haemophilus influenzae	ATCC 9006

SYSTEM VALIDATION

Validation is the **documentation** to ensure that a specific process, method, or equipment consistently produces a result (product) that conforms to predetermined specifications and quality attributes. Validation can also be viewed as documented data providing **proof** that a system/equipment functions as designed and intended.

Why Validate? ("If it is not in writing, it is a rumor.")

If a company complies with cGMPs and GLPs, equipment and system validation is enforced. Even though a system has been extensively tested prior to marketing and more tests were performed prior to shipment to a certain facility, one must assure that the equipment was properly and safely installed at the site where it will be used.

Validation will ensure the equipment is functioning as expected and will initiate the creation of SOPs for calibration, operation, preventative maintenance, QC and personnel training. It is also an opportunity for the company/laboratory personnel to become familiar with the different functions and features of the equipment.

Validation—Step by Step

The first step toward validation is the creation of a Validation Team, which should include the functional group (user), quality assurance, preventative maintenance, metrology, and computer systems validation representatives.

The next step is the creation of a Validation Protocol, a written plan that describes how to conduct the validation and sets parameters to assess the system's performance. A validation protocol should include the following sections:

- Purpose

- Application

- Abstract

- Procedure

- Personnel identification (initial identification)

- Protocol approval

- Final evaluation

- Validation approval

- Exhibits

The procedure section of a validation protocol usually encompasses the following:

Installation Qualification

A verification and documentation of equipment and components and supporting information, such as equipment identification (model number, location, etc.), utility requirements, and safety features.

Operation Qualification

A verification and documentation that the equipment can operate within preestablished parameters. In this phase of the validation, operational SOPs are created, personnel training documentation takes place, preventative maintenance evaluation occurs, and the installation checklist from bioMérieux Vitek is prepared.

Performance Qualification

The phase of the validation for verification that the equipment can perform reliably and repeatedly within the predefined criteria established in the validation protocol.

Since the VITEK system is automated and computerized, the validation protocol must include assessment of the performance of the system's hardware and software after installation. The extent of this part of the validation will depend on the capabilities of the system under evaluation and how extensively the computer validation specialist wants to verify the system's output.

The performance qualification phase can be divided into two sections: evaluation for **accuracy** and **reproducibility.**

This author recommends that before initiating the actual performance validation phase, the laboratory should conduct a prevalidation assessment. The study can include testing of isolates from the laboratory culture collection, in duplicate or triplicate.

This exercise will enable the operator(s) to become familiar with the system, identify problem areas, and address any specific needs prior to the initiation of the actual performance qualification work. The organism cultures chosen for the accuracy and reproducibility study can be based on the preassessment evaluation and should include QC organisms as recommended by bioMérieux Vitek.

The following is a suggested protocol for the accuracy and reproducibility study.

Accuracy.   Prepare three separate inocula of each chosen organism. Test each preparation using the appropriate test kit. Compare the VITEK results to the ATCC stock culture identification. If discrepancies exist, investigate the results by use of alternate conventional method (acceptance criteria: a VITEK confidence level of at least 85 percent).

Reproducibility.   Using the same test organisms from the accuracy study, prepare one inoculum per organism. Test the preparations in triplicate to evaluate the reproducibility of the system using the appropriate test kit (acceptance criteria: all three identifications from the same preparation must be identical and have a VITEK confidence level of at least 85 percent).

The following are organisms suggested for use in the accuracy/reproducibility study. Each group of organisms includes at least one organism recommended by bioMérieux Vitek to be used for QC purposes. The choice of organisms must reflect the types of isolates that are objectionable in a particular industry and may include environmental isolates. One concern about using environmental isolates in this part of the study is that wild strains or injured organisms may not achieve the level of accuracy or reproducibility established in the protocol, which does not reflect the system's performance.

- **Gram-positive identification**

    Staphylococcus aureus (ATCC 6538)

    Streptococcus faecium (ATCC 10541)

    Erysipelothrix rhusiopathiae (ATCC 19414)

- **Gram-negative identification (fermenters)**

    Escherichia coli (ATCC 8739)

    Salmonella enteriditis (ATCC 13076)

    Klebsiella pneumoniae (ATCC 13883)

    Enterobacter cloacae (ATCC 13047)

- **Gram-negative identification (nonfermenters)**

  Bordetella bronchiseptica (ATCC 4617)

  Pseudomonas aeruginosa (ATCC 9027)

  Pseudomonas cepacia (ATCC 25416)

- **Anaerobe identification**

  Propionibacterium acnes (ATCC 11827)

  Clostridium sporogenes (ATCC 11437)

  Clostridium perfringens (ATCC 13124)

- ***Bacillus* identification**

  Bacillus subtilis (ATCC 6633)

  Bacillus sphaericus (ATCC 4525)

  Bacillus licheniformis (ATCC 14580)

- **Yeast identification**

  Candida albicans (ATCC 10231)

  Cryptococcus laurentii (ATCC 18803)

  Saccharomyces cerevisiae (ATCC 9763)

For the other types of test kits available such as the bioburden Card and the Assay Card, validation is necessary also. A test protocol can be created to address the accuracy and reproducibility of recovery of known-level inocula preparations. Results obtained can be compared to results obtained by standard plate count methods. For an acceptable recovery one should expect the results to be at least within 50 percent of the standard plate count results. As a rule of thumb, triplicate tests are the minimum number of replicates to be performed for validation purposes.

Change Control

A system must be in place to address any modification, revision, or repair to the system once validated. This change must be documented and evaluated by a designated team. Depending on the nature of the change, revalidation, in full or part, may be required.

REFERENCES

Bailey, J. S., N. A. Cox, J. E. Thomson, and D. Y. C. Fung. 1985. Identification of Enterobacteriaceae in foods with the AutoMicrobic System. *J. Food Protection.* 48:147–149.

Bannerman, T. L., K. T. Kleeman, and W. E. Kloos. 1993. Evaluation of the Vitek Systems gram-positive identification card for species identification of coagulase-negative Staphylococci. *J. Clin. Microbiol.* 31:1322–1325.

Edwards, P. R., and W. H. Ewing. 1972. Identification of Enterobacteriaceae, 3rd ed. Minneapolis: Burgess Publishing Co.

Ewing, W. H. 1973. *Differentiation of Enterobacteriaceae by biochemical reactions*, rev. DHEW Publication No. (CDC) 75-8270. Atlanta: DHEW.

Gilardi, G. L. 1978. *Glucose nonfermenting gram-negative bacteria in clinical microbiology*. West Palm Beach, FL: CRC Press, Inc.

Gilardi, G. L. 1983. *Identification of glucose non-fermenting gram-negative bacteria*, rev. New York: Department of Laboratories, Hospital for Joint Diseases and Medical Center.

Gordon, R. E. 1975. *The genus* Bacillus. Agriculture Handbook Stock Number 202-275-2091. Washington, DC: U.S. Dept. of Agriculture.

Heisick, J. E., D. E. Wagner, M. L. Nierman, and J. T. Peeler. 1989. *Listeria* spp. found on fresh market produce. *Appl. Environ. Microbiol.* 1925–1927.

Knight, M. T., D. W. Wood, J. F. Black, G. Gosney, R. O. Rigney, J. R. Agin, C. K. Gravens, and S. M. Farnham. 1990. Gram-negative identification card for identification of *Salmonella, Escherichia coli,* and other Enterobacteriaceae isolated from foods: Collaborative study. *J. Assoc. Off. Anal. Chem.* 73:729–733.

Knight, M. T., J. F. Black, and D. W. Wood. 1988. Industry Perspectives on *Listeria monocytogenes. J. Assoc. Off. Anal. Chem.* 71:682–683.

Stager, C. E., and J. R. Davis. 1992. Automated systems for identification of microorganisms. *Clin. Microbiol. Rev.* 5:302–327.

Stanier, R. Y., N. J. Palleroni, and M. Doudoroff. 1966. The aerobic *Pseudomonads:* A taxonomic study. *J. Gen. Microbiol.* 43:159–271.

Wickerham, L. J., and K. A. Burton. 1948. Carbohydrate assimilation tests for the classification of yeasts. *J. Bacteriol.* 56:363–371.

4

Anaerobes

Roger W. Kelley
Tracey C. Schipp
Kelley & Associates

Anaerobes present a special challenge to developers of commercial identification systems. While most identification kits are designed for a group of organisms with similar morphological and biochemical characteristics, such as Enterobacteriaceae, the category of "anaerobes" is based solely on the ability to obtain energy in the absence of oxygen. Therefore, anaerobe identification kits must deal with a large number of phenotypes, including gram negatives and gram positives, cocci and rods, spore-formers and nonspore-formers, fast growers and slow growers, nutritionally fastidious, and saccharolytic and nonsaccharolytic species. Such a wide range of taxa makes it difficult for a manufacturer to derive a small subset of tests capable of identifying these organisms.

Before reviewing the current commercially available systems, we begin with a discussion of the different principles of how the systems operate and the major variables that can affect their performance.

PRINCIPLES OF SYSTEM PERFORMANCE

There are two basic types of commercially available identification kits: microbiochemical and rapid enzymatic systems. The microbiochemical systems are intended to be miniaturized versions of conventional tube tests. The systems are growth dependent and typically require 24–48 hours incubation. The tests included on the kits are predominantly the ones used in conventional identification schemes; the media formulations in the cupules or microtiter wells are very similar to the formulations used in tube tests. Minor modifications of formulations such as changing the pH indicator or altering the buffering capacity are frequently required when miniaturizing a test. These changes are made in order to achieve easy-to-read color reactions, to better replicate conventional methods that may have been based on longer incubation times, and to provide manufacturing stability during the drying process.

The ability of these miniaturized systems to mimic conventional results is largely dependent on the type of microorganisms that are being tested. Results from kits designed for rapidly growing fermentative aerobic or facultative bacteria usually correlate very highly to conventional tube tests. Kits devised for yeast based only on growth, and not acid production, also closely agree. For other organisms, such as anaerobic bacteria, which may require a variety of media supplements as well as nonoxidized media components, the correlation is not nearly as good.

In contrast, rapid enzymatic systems are not growth dependent, include many tests that have no conventional counterparts, and require four hours (or less) incubation time to obtain a final reaction. The rapid enzymatic systems typically employ a variety of sugar or amino acid substrates bound to a chromophore (or fluorophore). If the unknown organism contains the appropriate enzyme, the chromophore-substrate complex is cleaved and a positive reaction is detected. Mono and disaccharides are usually covalently linked to either ortho- or para-nitrophenol. When linked together, these substrates are colorless. Once cleaved, the free nitrophenol is yellow. Single amino acids or dipeptides are usually linked to β-naphthylamine. To detect a completed reaction, amino acid β-naphthylamine substrates require the addition of a reagent, usually a cinnamaldehyde compound, to cause a color reaction to occur (pink/red/purple) because β-naphthylamine is colorless in both the bound and free states. Because the design of the VITEK® AMS cards for anaerobes does not allow the addition of a developing reagent, amino acid substrates are linked to para-nitroaniline, which is yellow in its free state.

In addition to the nitrophenol or β-naphthylamine substrates, rapid systems usually employ one or more modified, miniaturized tests to detect reactions, such as urease, indole, nitrate, or catalase production, and acid production from sugars. The formulations used to detect these reactions, especially acid production from sugars, are usually very different from conventional tests in order to be able to detect a positive reaction in four hours. Thus, the correlation to conventional tests is frequently marginal. The detection of phosphatase (acid and/or alkaline) is also used in many systems. Phosphate is usually bound either to p-nitrophenol (yellow when cleaved) or to indoxyl. Once cleaved, indoxyl oxidizes to indigo blue, which forms an insoluble blue or blue-green precipitate.

As mentioned above, the test results from a rapid system may not correlate well to an overnight conventional test. There also may be considerable variation in results with the same substrate between one vendor's system versus that of another vendor. Because formulations vary, a strain that gives a positive reaction for a particular substrate with one system will not necessarily be positive in a different system. Thus, it is important to always keep in mind that a negative reaction does not mean that an organism does not have a particular enzyme, just that it cannot be detected in a specific commercial system. Confirmation of the presence of a specific enzyme is further confused by the fact that some amino acid arylamidases are capable of hydrolyzing several different amino acid β-naphthy-lamide substrates with differing degrees of activity.

How an organism reacts with a specific substrate is dependent on many variables. The major variables are listed below. Bear in mind that the goal of the manufacturer is to create a set of tests that provide the maximum possible separation of species. The goal is *not* to create formulations that are optimal for detecting a particular enzyme. In fact, in some cases it may be better to formulate a reaction to deliberately suppress the activity of a specific enzyme from a particular species.

1. *Substrate concentration.* Selection of the substrate concentration to be used for a particular test is influenced by several factors. The higher the concentration of substrate, the more dramatic the color change when the chromophore is released. Unfortunately, many of the substrates are expensive and so the manufacturer's goal is to use as low a concentration as possible. High concentrations of some substrates can also cause a solubility problem that can make manufacturing difficult. At low substrate concentrations, however, if an enzyme is present only at low levels or if the medium conditions are suboptimal, a positive reaction may be undetectable visually.

2. *Organism concentration.* The higher the inoculum concentration, the more enzyme is present to create positive reactions. High inoculum concentrations are also very important for tests where metabolic activity is required, such as acid production from sugars. A concentrated cell suspension may be essential to get a clear pH shift within four hours. This is counterbalanced by the fact that it can be very difficult to obtain sufficient growth of some slow-growing species to create highly turbid suspensions. From a user's point of view, the less inoculum required, the better. In some instances, the use of lower concentration inocula may actually improve the performance of a system. When one species contains a highly active enzyme and another contains one that is only weakly active against a particular substrate, if the cell concentration is high enough, both species will be positive for the reaction. At lower cell concentrations, though, the species that is only weakly active may appear negative in a four-hour time frame versus a positive reaction for the other species.

3. *pH of reaction.* Enzymes from different species can have very different pH optimums, or in some instances, such as acid and alkaline phosphatases, there may different enzymes for different pH ranges. By altering the pH of the formulation, it is possible for the manufacturer to change the pattern of observed reactions.

4. *Buffering capacity.* This is one of the most critical variables for miniaturized versions of conventional tests based on acid production from sugars, since a four-hour incubation period is only minimally adequate for some species to produce sufficient acid to cause a visibly distinct color change. If a test is formulated with a very low buffer strength, even weak acid

producers may be positive within four hours; at a higher buffer strength, the same species may appear negative.

5. *Additives.* For some organism/enzyme combinations the addition of detergents or surfactants can alter observed enzyme activity. At higher concentrations these additives can significantly inhibit various enzymes. At lower concentrations detection of enzyme activity can be enhanced for some species, possibly due to the increased release of intracellular or membrane-bound enzymes.

6. *Incubation time.* Most rapid commercial systems are designed to be incubated for four hours. The amount of time actually required to obtain a strong positive reaction, however, varies significantly. Many reactions are complete in 60 minutes or less. Others, such as acid production from sugars, may take much longer, and four hours may not be long enough to obtain a positive reaction for some species. The choice of four hours as a standard incubation time was primarily driven by marketing considerations. For a clinical laboratory, the goal is to set up the unknown isolate and obtain a result early enough in the day so that the physicians are still on duty and can make any necessary changes in therapy based on the results. Thus, an incubation time of longer than 4–5 hours does not provide any real advantage versus an overnight system.

7. *Growth medium.* The medium that the organism is grown on has minimal effect on the detection of most enzymes. Sugar fermentation activity, however, can be affected in some species, depending on what enzymes are induced or repressed.

VARIABLES AFFECTING SYSTEM PERFORMANCE

System performance is affected by a wide range of variables, some of which are controlled by the manufacturer and some by the user. Many of the variables related to media formulation are discussed above. Variation in media formulations can make large differences in how easy the systems are to read. The more clear-cut the differentiation between a weak positive and a negative reaction, the easier the system is to use, thus increasing reproducibility and performance. Other significant variables determined by the manufacturer are test selection and database development. The number of tests is frequently limited by the actual design of the test system. In other systems, such as MicroScan's that uses a 96-well microtiter plate, the actual number of tests is a trade-off between performance, minimizing the amount of time and culture required for inoculation, and the cost of the substrates. Unlike conventional overnight commercial systems, where the cost of the media is a very minimal part of the cost of goods, chromogenic substrates can be very expensive. Therefore, there are always compromises made between selecting the substrates with the best differential abilities versus keeping the cost of the system competitive.

Although most commercial systems contain similar numbers of species in their databases, and all systems contain all of the most common species, the quality of the databases can vary widely. Within a system's database, some species may be represented by hundreds of strains, while others that are rare may be represented by only a few strains, many of which may be reference cultures. This lack of available strains for rare species is one of the major reasons why commercial systems usually do not perform well at identifying uncommon species, especially in the earliest versions of the database. Deliberate alteration of database percentages by the manufacturer can also result in misidentification of uncommon species. Most vendors use Bayesian probability calculations to determine which species an unknown isolate most resembles. In some instances, due to the limited number of tests in a kit, a common species and an uncommon one may not be separated by any good differential tests. Thus, misidentifications could easily occur. Because the manufacturers have decided that it is better to misidentify an uncommon species that will seldom be isolated rather than a common one, they sometimes alter the calculations to achieve this result. For example, if a common species is 100 percent positive for a particular test but an uncommon one is positive only 5 percent of the time, the manufacturer might enter 0 percent in the database instead of 5 percent. This assures that if the unknown strain is positive for that test that it will be identified as the common species only, even though 5 percent or more of the uncommon strains will be misidentified. It is also possible, when using Bayesian equations, to factor the actual frequency of isolation of a particular species into the determination of the identification probability. Thus, if two species have very similar biochemical patterns, the program might ordinarily assign equivalent identification probabilities; for example, 60 percent for species 1 and 40 percent for species 2. However, if species 2 is normally isolated 20 times more frequently, then the manufacturer can use this factor—instead of species 2 being listed as a second choice, it will be listed as the first choice with a higher percent match.

When a manufacturer first develops an identification system, they do so using a specific inoculum concentration and specific media to grow the organisms. Once in the field, though, these variables are controlled by the user. One should always follow the manufacturer's instructions with regard to media since the use of different media may change the pattern of induced or repressed enzymes and cause misidentifications. It is also critical to train analysts carefully in inoculum preparation, since over or under inoculation can drastically alter the detected reactions and change identification accuracies. Reading chromogenic reactions has also been problematic for many users. The difference between a weak positive and a negative reaction can be very subtle for some color reactions. Whenever using a system for the first time, it is strongly recommended that the analysts who will be reading the results be trained directly by a manufacturer's representative rather than relying on the package insert alone. Once analysts are appropriately trained, however, chromogenic systems can give very reproducible results.

COMPARISON OF JOURNAL ARTICLES

Comparing the performance of various systems by reviewing published articles is very difficult, since the results can be easily skewed depending on how the study is conducted. As a potential user, some of the major variables to be evaluated before accepting someone else's conclusions, or to be determined prior to conducting one's own experiments, are as follows:

1. *Source of strains.* The majority of species and strains tested by the manufacturers are of human clinical origin. Thus, if the strains used to challenge the system are also of clinical origin, one would expect much better results than if one uses isolates from other environments. Also important is whether the strains are fresh isolates or from a national culture collection such as the American Type Culture Collection (ATCC). Manufacturers also have access to culture collections and almost always include the type strains in their databases. Typically, type strains should not be used to challenge the performance of a system, since they are not necessarily representative of fresh isolates of the species.

2. *Database version.* The first database version of a new identification system usually does not perform as well as subsequent versions that contain a greater number of species along with more accurate percentages for the reactions. As a result, the earliest articles evaluating a system may show a poorer performance than later articles because in the interim the manufacturer has updated the database.

3. *The "gold standard."* Proper evaluation of a system cannot be done unless one knows the "correct" identification of the isolate used to challenge a system. The appropriate method is to identify the unknown conventionally and then compare that identification to the one obtained from a rapid identification system. The conventional identification scheme used should be at least as complete with regard to number of tests as a rapid system. Articles that test an unknown on two or more rapid systems and define a correct identification as one where at least two systems agree on the same species name are of marginal value.

4. *Number of species/strains.* The number and distribution of species and strains used to challenge a system has a very large impact on the perceived performance of the system. Identification systems can be challenged in two basic ways, both of which may give substantially different results. The first way is to collect and test strains as they are isolated in the laboratory. This gives a data set of many strains of the most common species and a few strains each of the less common species; it is representative of how the product should actually perform on a day-to-day basis. Because the data set emphasizes the most common species (as does the manufacturer's database), the percentage of accurate identifications should be fairly high. The second way is to test approximately the same number of strains for each species but to test a larger, more diverse set of species. While this does not mimic the actual use of a system, it does give a better picture of how well the system will perform on uncommon species. Since most systems do fairly well on the common species, this can be a better indicator of differences between systems. It will also, however, give lower overall correct matches than if the majority of the challenge organisms are the most common species.

5. *Supplementary tests.* When evaluating the results of a particular system, the article should report the percent correct identifications, both with and without the use of any recommended supplementary tests. To report only one result can be misleading. Manufacturers usually report performance based on the use of supplementary tests. However, the goal of the user is to perform additional tests as infrequently as possible. Ideally, commonly occurring species should not require any supplementary tests at all. If two systems give equivalent results using all the recommended supplementary tests, but one system is 10–15 percent more accurate without any extra tests, then clearly that system is preferable.

6. *Accuracy.* One of the most critical variables is the definition of a correct identification. Surprisingly enough, almost every author seems to adopt a different definition that can make comparison between articles very difficult. As mentioned above, some authors do not conventionally identify their unknowns and define a correct result as one obtained by two or more systems. Even if a conventional scheme is used as a baseline, one must still define what an acceptable identification is for the rapid system. Is >99 percent a correct identification; >95 percent; >90 percent? While some authors set a particular threshold for accepting a result as valid, others simply accept the first result listed as correct regardless of the percentage. One article defined an identification as correct if the correct species name was any of the ones listed, even if it was a second or third choice. A related variable is the number of misidentifications. It is far better for a system to report an isolate as unidentifiable than to give an incorrect identification.

ANAEROBE IDENTIFICATION SYSTEMS

The role of anaerobes in serious clinical infections, such as botulism, tetanus, and gas gangrene, and the importance of lactobacilli in a variety of foods, has been known and studied for over a century. Until relatively recently, however, interest in other species of anaerobes was restricted to a relatively small number of researchers. This changed rapidly beginning in the 1950s, partially due to the advent of better methods for culturing strict anaerobes, including Hungate's roll tube method, glove boxes, and anaerobe jars. It was soon established that there are many different species of anaerobes that are involved in a wide range of clinical infections (Finegold and George 1989) and that anaerobic bacteria are the predominant normal flora in humans as well as most other environments.

The primary schemes for identification of anaerobes were developed at the Centers for Disease Control, Wadsworth Veterans Administration Hospital, and the Virginia Polytechnic Institute (VPI) Anaerobe Laboratory (Dowell and Hawkins 1974; Holdeman et al. 1977, 1978; Summanen et al. 1993). The first two schemes focused primarily on species of human clinical significance, whereas the VPI Anaerobe Laboratory Manual also included many normal flora and environmental species. All three were built around the fermentation of various carbohydrates combined with other tests, such as indole, catalase, gelatin lique-faction, and lecithinase or lipase production. In addition, anaerobe identification was frequently dependent on gas chromatographic analysis of short chain fatty acids produced as metabolic end products of fermentation. With the increased recognition of the importance of anaerobes in clinical samples and normal flora, and the availability of well-defined techniques for identification, many laboratories began attempting to identify anaerobes to the species level. The major impediment to this goal was that accurate identification was very labor and time intensive due to the necessity of preparing special media formulations, prereduction of media, anaerobic cultivation, and the fact that many species grow fairly slowly. The cost of a gas chromatograph was also a factor for some laboratories.

In order to address the cost and labor issues associated with conventional identification methods, commercially available kits for the identification of anaerobes were developed and launched in the early 1970s. The first kits were the API 20A (bioMérieux Vitek Inc., Hazelwood, MO) and the Minitek Anaerobe System (BBL Microbiology Systems, Cockeysville, MD). These kits are basically miniaturized versions of conventional tests that are growth dependent and rely on the metabolic breakdown of carbohydrates and other traditional substrates. These systems require anaerobic incubation for 1–2 days and frequently require various supplementary tests, such as gas chromatography, in order to obtain accurate identifications (Citron 1984). The Flow Laboratories Anaerobe-Tek system and the MicroScan Anaerobe Combo Panel were two additional growth dependent kits developed that were subsequently withdrawn from the market. More recently, a variety of rapid identification systems requiring only four hours of aerobic incubation were introduced. These include the Rapid ID 32A, the API AN-Ident, and the VITEK ANI (bioMérieux Vitek Inc., Hazelwood, MO), the MicroScan Rapid Anaerobe Panel (Baxter Diagnostics Inc., West Sacramento, CA), and the RapID ANA II (Innovative Diagnostic Systems Inc., Norcross, GA).

API 20A

The API 20A, introduced in 1974, was the first commercial kit designed to identify anaerobes without the time and expense of conventional biochemical testing (Appelbaum et al. 1983). Like other API kits, the 20A consists of microtubes of dehydrated substrates on a plastic strip. A 24–48 hour culture grown on a nonselective, suitably enriched medium is used to prepare an inoculum equal to at least a #3 McFarland standard in the provided basal medium. The bacterial suspension is then pipetted into the 20 microtubes, and the strip is incubated for 24 hours anaerobically at 35–37°C. After incubation any necessary reagents are added, and the test results are read manually. Several authors have noted that it may be difficult to interpret shades of brown or brown-purple that occur on the strip after the addition of bromcresol purple to the fermentation reactions (Appelbaum et al. 1983; Mangels 1992a). A 7-digit profile number is created from the results of the 21 tests, which is looked up in a codebook. The database contains 16 genera and 77 taxa. Early studies demonstrated fairly good correlation with conventional biochemical tests (Hansen and Stewart 1976; Hanson et al. 1979; Moore et al. 1975). Percent correlation ranged from 70.8 percent to 99.4 percent. Improvements were made and an expanded database was released. Studies after this suggest that most *Bacteroides fragilis* and *Clostridium perfringens* isolates can be identified without additional tests, but identification rates of other anaerobes range from 8 percent to 100 percent. Several supplemental tests, including Gram stain, morphology, reactions on egg-yolk agar, and gas-liquid chromatography (GLC), are often necessary for the accurate identification of many organisms, especially the nonreactive organisms such as the fusobacteria, nonspore-forming gram-positive rods, and the anaerobic cocci (Appelbaum et al. 1983; Karachewski et al. 1985; Murray et al. 1985; Rosenblatt 1985; Summanen and Jousimies-Somer 1988). Karachewski et al. (1985) found GLC necessary for 36 percent of the study isolates.

Minitek Anaerobe II

The Minitek system is a miniaturized biochemical system that utilizes paper disks saturated with substrates. The anaerobe system of 35 available disks was introduced shortly after the API 20A, and the performance of the miniaturized biochemicals compared favorably to conventional methods (Hansen and Stewart 1976; Stargel et al. 1976). In the Anaerobe II system a panel of 18 disks is recommended, but the panels can be easily customized with additional substrates. Disks are dispensed into a specialized disposable plastic tray. The tray is then inoculated with a turbid suspension of the organism equal to a #5 McFarland, by means of the Minitek pipetting device. The tray is incubated for 48 hours anaerobically at 35–37°C. After incubation reagents are added and a 7-digit code is manually generated from the recommended 18 test results. Reactions are relatively straightforward to read. Any difficulties interpreting some shades of yellow-orange in the fermentation reactions may be rectified by adding another drop of phenol red (Appelbaum et al. 1983). The numerical code is then looked up in the codebook. The database includes 20 genera and 114 taxa. The codebook has not been updated since 1986, and therefore, does not reflect current taxonomy. As with the API 20A, several supplemental tests may be necessary for satisfactory identifications. Rosenblatt (1985) demonstrated a significant improvement in performance for all types of anaerobic bacteria when suggested supplemental tests were performed. Without additional tests the Minitek correctly identified only 28–46 percent of isolates to the species level. The rate increased to 63–82 percent with supplemental tests. Other studies have shown higher rates with common isolates such as *Bacteroides fragilis, Clostridium difficile, Clostridium perfringens*, and *Propionibacterium acnes* (Appelbaum et al. 1983; Bate 1986; Head and Ratnam 1988). But again, poor results were seen with nonreactive isolates such as *Fusobacterium* spp. and the anaerobic cocci. Supplemental tests including Gram stain, morphology, and GLC are important characteristics for proper species identification with this kit.

Rapid ID 32A

The Rapid ID 32A is a micro method recently distributed in the United States by bioMérieux. The system is made up of a small strip with 32 cupules providing, at the present, 29 different tests. The remaining three wells are empty, providing for future expansion. The database contains 18 genera and 81 taxa. A bacterial suspension equal to a #4 McFarland is prepared from isolates grown on Columbia agar supplemented with 5 percent sheep or horse blood and vitamin K_3. The wells are inoculated with the enclosed micropipette or by the ATB instrument, and the strip is incubated aerobically for 4 hours at 35–37°C. Required reagents are added and reactions are read according to manufacturer's instructions. Alternatively, the strip may be read automatically by the ATB reader. Both methods have been found to be fairly reproducible, with some doubtful reactions reported when reading manually (Kitch and Appelbaum 1989). The first 24 tests are used to construct a profile number, and the last 5 are additional tests for cases of low discrimination. The 32A is a reliable means of identifying the common anaerobes such as *Bacteroides fragilis*, *Clostridium perfringens*, *Propionibacterium acnes*, and many *Prevotella* spp. The addition of supplementary tests (19 are listed) greatly improves the performance with organisms such as *Clostridium difficile* and *Fusobacterium* spp. Rates of correct identification to the species level range from 44 percent to 96 percent without the additional tests, and they improve to 71–100 percent with the extra tests. The importance of Gram stain results in the interpretation of the results has been greatly stressed (Arzese et al. 1994; Kitch and Appelbaum 1989; Looney et al. 1990; Pattyn et al. 1993). Kitch and Appelbaum (1989) suggested the routine addition of Gram stain, catalase, aerotolerance, and pigmentation data to supplement the primary database. When these four tests were added, the identification of some *Bacteroides* spp., *Fusobacterium* spp., *Clostridium* spp. and *Eubacterium* spp. was improved. The automated reading of the 32A by the ATB system has also been evaluated (Arzese et al. 1994). By using the ATB reader, the problems interpreting doubtful reactions was eliminated, and color interpretation was standardized. This study also suggested the addition of Gram stain, spore or pigment production, and catalase.

AN-Ident

Another API product from bioMérieux, the AN-Ident, uses 21 tests to identify anaerobes. Organisms must be grown on a nonselective enriched blood agar medium, such as enriched Columbia agar; 5 percent sheep blood tryptic soy agar or Schaedler's blood agar is not acceptable. A #5 McFarland suspension is prepared and used to inoculate the test strip. The strip is incubated at 35–37°C aerobically for four hours. As with other test strips, a numerical code is developed after the addition of reagent. The code is looked up in the database containing 17 genera and 84 taxa. The AN-Ident successfully identifies *Bacteroides fragilis*, *Clostridium perfringens*, *Propionibacterium acnes*, and *Veillonella parvula*. Lower rates of identification are seen for other species (Burlage and Ellner 1985; Murray et al. 1985; Quentin et al. 1991). Quentin et al. (1991) found supplementary tests of some sort necessary in 5 percent of cases. A lipase test was found to be almost required for the separation of *Clostridium difficile* and *Clostridium sporogenes*. Most supplemental tests suggested are simple biochemical tests or morphological characteristics, but GLC is listed. Murray et al. (1985) stated that most gram-positive rods did not require GLC for identification since the AN-Ident successfully identified them to genus level. Stenson et al. (1986) used GLC for 4 percent of isolates, most of them *Clostridium* spp. infrequently isolated in the clinical setting.

RapID ANA II

The RapID ANA II is a compact unit with 10 wells molded out of hard plastic. A #3 McFarland inoculum suspension is prepared from a selective or nonselective medium (bile-esculin–containing or sugar-supplemented agar such as Schaedler's cannot be used). All the wells are inoculated in one motion, and then the tray is incubated aerobically for 4–6 hours at 35–37°C. Reagents are added and 18 tests are used to obtain a microcode that is referenced in the code compendium according to Gram reaction. About a dozen supplemental tests are indicated; however, GLC is not among them. The ANA II database contains 20 genera and 95 taxa. It has been noted in several studies that the β-naphthylamine color reactions may be difficult to interpret at first (Adney and Jones 1985; Appelbaum et al. 1985; Celig and Schreckenberger 1991; Downes and Andrew 1988; Karachewski et al. 1985; Murray et al. 1985). Several studies were published on the first generation RapID ANA, which found the kit to be a promising identification method. Some problems were seen with *Clostridium* spp., other than *Clostridium perfringens*, and nonspore-forming gram-positive rods (Downes and Andrew 1988; Marler et al. 1991), but the kits performed well with common clinical isolates and generally performed better than the API 20A and Minitek (Bate 1986; Hussain et al. 1987; Karachewski et al. 1985; Murray et al. 1985; Summanen and Jousimies-Somer 1988). Adney and Jones (1985) challenged the RapID ANA with veterinary isolates. Results were similar to those obtained with human isolates, with the exception of *Fusobacterium* strains. The majority of the misidentifications were in this genus.

The panel was subsequently improved by replacing three substrates and issuing an updated, expanded code compendium. The new panel was designated as RapID ANA II. Two studies are available on the updated panel. Both studies showed good results with *Bacteroides fragilis*, *Clostridium perfringens*, and *Propionibacterium acnes*. The rate of identification of other groups of organisms ranged from 68 percent to 97 percent in the first study (Celig and Schreckenberger 1991) and 55 percent to 72 percent in the second (Marler et al. 1991). The lowest rates were seen with the nonspore-forming gram-positive rods. An additional 10 percent of isolates were identified after supplemental tests were performed in the second study. Similarly, while Celig and Schreckenberger (1991) did not perform supplemental tests, 9.6 percent of the codes obtained suggested additional tests that could have resulted in proper species identification.

VITEK ANI Card

The ANI is a small plastic card with 30 wells containing 28 substrates and one control. It is designed for use with the VITEK AMS unit. The card is inoculated with a bacterial suspension with a turbidity of at least a #3 McFarland by means of the VITEK filling module. After sealing the card is incubated aerobically for four hours at 35–37°C. The card cannot be incubated in the VITEK unit itself since the reader cannot detect reactions in the red-yellow range. After incubation the card is read manually and the results are entered into the ANI program on the VITEK computer. No reagents are necessary, and the computer automatically calculates the numerical code and provides the identification. The test results and identification can be printed out immediately. The VITEK database contains 18 genera and 84 species of anaerobic and microaerophilic organisms. Supplemental tests include fermentations, cell morphology, pigment production, and other simple biochemical tests. *Bacteroides fragilis*, *Clostridium perfringens*, and *Veillonella parvula* are adequately identified with the ANI card, but as is expected, there are problems with other groups of anaerobes (Schreckenberger et al. 1988). Overall, the ANI correctly speciated about 70 percent of the study isolates. When *Bacteroides fragilis*, *Clostridium perfringens*, and *Propionibacterium acnes* are excluded, the rates of identification ranged from 56 percent to 69 percent. Supplemental tests could have improved this to only 60–75 percent. It should be noted that at the time of this book's preparation, no study was yet available that utilized the current software version.

MicroScan Rapid Anaerobe Panel

The MicroScan panel contains 24 dehydrated substrates plus 2 controls, an amide substrate and a β-naphthylamine, as guides for color interpretation. It is inoculated with an organism suspension adjusted to a #3 McFarland using a manual pipetting device and incubated aerobically for four hours (Stager and Davis 1992). Necessary reagents are added and the panels are read either completely manually or with the aid of a Touchscan light box. Software is also provided with the AutoSCAN and WalkAway instruments to automatically read the Rapid Anaerobe panel. There are 13 genera and 58 taxa in the database. GLC is among the tests in the special characteristics described in the codebook. In the single study available the MicroScan panel did not perform well. It identified only 63 percent of the *Bacteroides fragilis* isolates and did very poorly with nonspore-forming gram-positive rods, identifying only 41 percent of the isolates. Other bacterial types were identified at rates ranging from 71 percent to 82 percent. When supplementary tests were performed, the rate of species-level identification improved to 47–86 percent (Stoakes et al. 1990). Database changes have been made since this reference was submitted for publication, and no new data are available.

SUMMARY AND RECOMMENDATIONS

Although it is frequently difficult to compare the performance of different systems, certain common elements are evident. The main similarity is that all of the systems were designed for identification of anaerobes isolated from clinical samples. Because many clinical isolates are also commonly found in the environment (e.g., *Clostridium* species) or as normal human flora (e.g., *Propionibacterium acnes*), the systems should also perform acceptably for the pharmaceutical, food, and cosmetic industries. For other applications such as veterinary microbiology (Adney and Jones 1985; Watts and Yancey 1994), or the characterization of isolates from specialized environmental niches, the current commercial systems may have limited utility. In addition, as anaerobic taxonomy changes through DNA and RNA studies, identification kits typically have more difficulty separating species, since taxonomy is no longer based solely on biochemical reactions.

Overnight systems such as the API 20A or Minitek systems perform fairly well with rapidly growing saccharolytic species. However, they frequently require supplementary tests (e.g., GLC) and usually have difficulty accurately identifying slow-growing, fastidious, or asaccharolytic species.

When the four-hour rapid systems were first introduced, the overnight systems usually performed equal or better to the rapid ones. Over time, however, the databases and the substrates used in the four-hour systems have been improved such that the rapid systems perform equal or better to the overnight ones. Therefore, the rapid systems are preferable because accuracy is equal or better, anaerobic incubation is not required, and results are available much sooner (Mangels 1992b; Phillips 1990).

Table 4.1 summarizes the major differences between the currently available anaerobic identification kits. The obvious difference between kits such as the API 20A and the rapid kits is the extended incubation under anaerobic conditions required by the growth dependent systems. Among the rapid systems one of the biggest differences is the inoculum concentration required. For many of the fastidious or slow-growing anaerobes, it can be very difficult to obtain a #4 or #5 McFarland turbidity without using growth from more than one plate. Thus, of the rapid systems the RapID ANA II is the best in this regard since it requires the lowest volume at a turbidity of only a #3 McFarland. Most of the systems have fairly similar databases with regard to the number and type of taxa. The only exception is the MicroScan Rapid Anaerobe database which contains only 58 taxa and is apparently overdue for an update. As discussed previously, the type of medium that is used can influence the enzymes that are detected. Thus, it is important to follow the manufacturer's instructions regarding the appropriate medium for preparing the inoculum. Although some of the systems can be read automatically, most can only be read manually. This is not necessarily a disadvantage since the human eye is extremely good at color discrimination and usually does as well, if not better, than the current automated systems. The key is that the analyst reading the panel should receive appropriate training on how to differentiate between weakly positive versus negative reactions. With appropriate training visual reading can be equal or superior to an automated reader (Arzese et al. 1994; Stoakes et al. 1990).

All of the rapid systems perform well when identifying the *Bacteroides fragilis* group, *Propionibacterium acnes*, and common clinical clostridia such as *Clostridium perfringens*. Performance with nonfermentative gram-positive cocci, *Fusobacterium* species, nonspore-forming gram-positive rods such as *Actinomyces* species, and the less common *Clostridium* species is more variable. Isolates from these groups may frequently require supplementary tests to obtain an accurate identification. Review of the available literature suggests that, overall, the RapID ANA II system performs the best of the rapid systems in terms of accuracy and usually has the lowest frequency of requiring supplementary tests. It is also the easiest and quickest to set up. Close in performance is the Rapid ID 32A system. For those laboratories interested in the semi-automated ATB identification instrument, the Rapid

Table 4.1. General Attributes of Anaerobe Identification Kits

Attribute	API 20A	Minitek Anaerobe II	AN-Ident	Rapid ID 32A	VITEK ANI	MicroScan	RapID ANA II
Number of Tests	20	18	21	29	28	24	18
Inoculum Concentration (McFarland Standard)	3	5	5	4	3	3	3
Incubation Condition	anaerobic	anaerobic	aerobic	aerobic	aerobic	aerobic	aerobic
Volume of Inoculum Required	5 ml	1.5 ml	3 ml	2 or 3 ml[4]	1.8 ml	3 ml	1 ml
Incubation Time	24 hours	48 hours	4 hours	4 hours	4 hours	4 hours	4 hours
Number of Taxa in Database[1]	77	114	84	81	84	58	95

Continued on next page.

Continued from previous page.

Attribute	API 20A	Minitek Anaerobe II	AN-Ident	Rapid ID 32A	VITEK ANI	MicroScan	RapID ANA II
Isolation Medium	non-selective agar	nonselective agar	enriched blood agar[2]	Columbia blood agar	no sugar- or bile-esculin–containing medium	non-selective agar	no sugar- or bile-esculin–containing medium
GLC in Supplementary Tests	Yes	Yes	Yes	No	No	Yes	No
Manual or Automatic Read	M	M	M	M or A	M[3]	M or A	M

[1]as of November 1994

[2]excluding TSA with 5 percent sheep blood and Schaedler's blood agar

[3]read manually, but requires VITEK filler and computer units

[4]3 ml required for inoculation by ATB system

ID 32A, as well as ID 32 strips for other microorganisms can be inoculated and read automatically. The other systems such as the VITEK ANI, AN-Ident, and the MicroScan Rapid Anaerobe panel, while accurate with some species, generally do not perform as well as the RapID ANA II or Rapid ID 32A.

REFERENCES

Adney, W. S., and R. L. Jones. 1985. Evaluation of the RapID-ANA system for identification of anaerobic bacteria of veterinary origin. *J. Clin. Microbiol.* 22:980–983.

Appelbaum, P. C., C. S. Kaufmann, and J. W. Depenbusch. 1985. Accuracy and reproducibility of a four-hour method for anaerobe identification. *J. Clin. Microbiol.* 21:894–898.

Appelbaum, P. C., C. S. Kaufmann, J. C. Keifer, and H. J. Venbrux. 1983. Comparison of three methods for anaerobe identification. *J. Clin. Microbiol.* 18:614–621.

Arzese, A., R. Minisini, and G. A. Botta. 1994. Evaluation of an automated system for identification of anaerobic bacteria. *Eur. J. Clin. Microbiol. Infect. Dis.* 13:135–141.

Bate, G. 1986. Comparison of Minitek Anaerobe II, API An-Ident, and RapID ANA systems for identification of *Clostridium difficile. Am. J. Clin. Pathol.* 85:716–718.

Burlage, R. S., and P. D. Ellner. 1985. Comparison of the PRAS II, AN-Ident, and RapID-ANA systems for identification of anaerobic bacteria. *J. Clin. Microbiol.* 22:32–35.

Celig, D. M., and P. C. Schreckenberger. 1991. Clinical evaluation of the RapID-ANA II panel for identification of anaerobic bacteria. *J. Clin. Microbiol.* 29:457–462.

Citron, D. M. 1984. Specimen collection and transport, anaerobic culture techniques, and identification of anaerobes. *Rev. Infect. Dis.* 6 (Suppl 1):S51–58.

Dowell, Jr., V. R., and T. M. Hawkins. 1974. Laboratory methods in anaerobic bacteriology. In *CDC Laboratory Manual*, DHEW Publication No. (CDC) 74-8272. Washington, DC: U.S. Government Printing Office.

Downes, J., and J. H. Andrew. 1988. Evaluation of the RapID ANA system as a four-hour method for anaerobe identification. *Pathology* 20:256–259.

Finegold, S. M., and W. L. George. 1989. *Anaerobic infections in humans.* New York: Academic Press, Inc.

Hansen, S. L., and B. J. Stewart. 1976. Comparison of API and Minitek to Center for Disease Control methods for the biochemical characterization of anaerobes. *J. Clin. Microbiol.* 4:227–231.

Hanson, C. W., R. Cassorla, and W. J. Martin. 1979. API and Minitek systems in identification of clinical isolates of anaerobic gram-negative bacilli and *Clostridium* species. *J. Clin. Microbiol.* 10:14–18.

Head, C. B., and S. Ratnam. 1988. Comparison of API ZYM system with API AN-Ident, API 20A, Minitek Anaerobe II, and RapID-ANA systems for identification of *Clostridium difficile*. *J. Clin. Microbiol.* 26:144–146.

Holdeman, L. V., E. P. Cato, and W. E. C. Moore, eds. 1977. *Anaerobe laboratory manual*, 4th ed. Blacksburg, VA: Virginia Polytechnic Institute and State University.

Holdeman, L. V., E. P. Cato, and W. E. C. Moore, eds. 1978. *Anaerobe laboratory manual update.* Blacksburg, VA: Virginia Polytechnic Institute and State University.

Hussain, H., R. Lannigan, B. C. Schieven, L. Stoakes, T. Kelly, and D. Groves. 1987. Comparison of RapID-ANA and Minitek with a conventional method for biochemical identification of anaerobes. *Diagn. Microbiol. Infect. Dis.* 6:69–72.

Karachewski, N. O., E. L. Busch, and C. L. Wells. 1985. Comparison of PRAS II, RapID ANA, and API 20A systems for identification of anaerobic bacteria. *J. Clin. Microbiol.* 21:122–126.

Kitch, T. T., and P. C. Appelbaum. 1989. Accuracy and reproducibility of the 4-hour ATB 32A method for anaerobe identification. *J. Clin. Microbiol.* 27:2509–2513.

Looney, W. J., A. J. C. Gallusser, and H. K. Modde. 1990. Evaluation of the ATB 32A system for identification of anaerobic bacteria isolated from clinical specimens. *J. Clin. Microbiol.* 28:1519–1524.

Mangels, J. I. 1992a. Microbiochemical systems for the identification of anaerobes. In *Clinical microbiology procedures handbook,* sec. 2.8., edited by H. D. Isenberg. Washington, DC: American Society for Microbiology.

Mangels, J. I. 1992b. Rapid enzymatic systems for the identification of anaerobes. In *Clinical microbiology procedures handbook,* sec. 2.9., edited by H. D. Isenberg. Washington, DC: American Society for Microbiology.

Marler, L. M., J. A. Siders, L. C. Wolters, Y. Pettigrew, B. L. Skitt, and S. D. Allen. 1991. Evaluation of the new RapID-ANA II system for the identification of clinical anaerobic isolates. *J. Clin. Microbiol.* 29:874–878.

Moore, H. B., V. L. Sutter, and S. M. Finegold. 1975. Comparison of three procedures for biochemical testing of anaerobic bacteria. *J. Clin Microbiol.* 1:15–24.

Murray, P. R., C. J. Weber, and A. C. Niles. 1985. Comparative evaluation of three identification systems for anaerobes. *J. Clin. Microbiol.* 22:52–55.

Pattyn, S. R., M. Leven, and L. Buffet. 1993. Comparative evaluation of the Rapid ID 32A kit system, miniaturized standard procedure, and a rapid fermentation procedure for the identification of anaerobic bacteria. *Acta Clin. Belg.* 48:81–85.

Phillips, I. 1990. New methods for identification of obligate anaerobes. *Rev. Infect. Dis.* 12 (Suppl 2):S127–132.

Quentin, C., M. Desailly-Chanson, and C. Bebear. 1991. Evaluation of AN-Ident. *J. Clin. Microbiol.* 29:231–235.

Rosenblatt, J. E. 1985. Anaerobic identification systems. *Clin. Lab. Med.* 5:59–65.

Schreckenberger, P. C., D. M. Celig, and W. M. Janda. 1988. Clinical evaluation of the VITEK ANI card for identification of anaerobic bacteria. *J. Clin. Microbiol.* 26:225–230.

Stager, C. E., and J. R. Davis. 1992. Automated systems for identification of microorganisms. *Clin. Microbiol. Rev.* 5:302–327.

Stargel, M. D., F. S. Thompson, S. E. Phillips, G. L. Lombard, and V. R. Dowell, Jr. 1976. Modification of the Minitek miniaturized differentiation system for characterization of anaerobic bacteria. *J. Clin. Microbiol.* 3:291–301.

Stenson, M. J., D. T. Lee, J. E. Rosenblatt, and J. M. Contezac. 1986. Evaluation of the AN-Ident system for the identification of anaerobic bacteria. *Diag. Microbiol. Infect. Dis.* 5:9–15.

Stoakes, L., T. Kelly, K. Manarin, B. Schieven, R. Lannigan, D. Groves, and Z. Hussain. 1990. Accuracy and reproducibility of the MicroScan Rapid Anaerobe identification system with an automated reader. *J. Clin. Microbiol.* 28:1135–1138.

Summanen, P., and H. Jousimies-Somer. 1988. Comparative evaluation of RapID ANA and API 20A for identification of anaerobic bacteria. *Eur. J. Clin. Microbiol. Infect. Dis.* 7:771–775.

Summanen, P., E. J. Baron, D. M. Citron, C. A. Strong, H. M. Wexler, and S. M. Finegold. 1993. *Wadsworth anaerobic bacteriology manual,* 5th ed. Belmont, CA: Star Publishing Co.

Watts, J .L., and R. J. Yancey, Jr. 1994. Identification of veterinary pathogens by use of commercial identification systems and new trends in antimicrobial susceptibility testing of veterinary pathogens. *Clin. Microbiol. Rev.* 7:346–356.

5

Gram-Positive Organisms

Loralyn Harris

Kraft Foods

Because of the rich diversity found in the group of microorganisms that are gram positive, the need for definitive identification of these organisms is quite varied. While some genera are easily differentiated with a few simple tests, many others contain species so closely related that a large series of tests is required to obtain identification to species level. Thus, commercial automated systems have concentrated on those organisms of major significance to clinical, pharmaceutical, and food-processing environments, and on those organisms most difficult to identify by conventional means.

The two commercial systems that will be discussed in this chapter for the identification of gram-positive organisms will be the AutoMicrobic System (AMS) and the Automated Tests for Bacteriology (ATB) system, both manufactured by bioMérieux Vitek, Inc., St. Louis, MO. The systems differ in the level of automation available and, in some cases, in the number and types of organisms in the database. The two systems can be used for the identification of other groups of microorganisms, and may also be discussed in other chapters.

AUTOMICROBIC SYSTEM

The AMS utilizes miniaturized test kits, also known as test cards, which are available for several different groups of microorganisms. The kits are about the size and shape of a credit card, and contain up to 30 different biochemical tests each. The Gram Positive Identification (GPI) test kit is used for the identification of staphylococci, streptococci, and micrococci, the *Bacillus* (BAC) test kit is used for *Bacillus* spp. identification, and a combination of GPI and Gram Negative Identification (GNI) test kits can be used for speciation of *Listeria*.

To use the system, the isolate in question is either taken from primary isolation media if suitable, or grown on appropriate media suggested by the manufacturer. A suspension of pure culture is prepared, and a corresponding test kit is coded by the analyst. The paired suspension/test kit is entered into the filler/sealer module of the system, which automatically fills and seals the test kit. The test card is then placed into the reader/incubator module of the system, and the remainder of the procedure is performed by the AMS. Cards are incubated at 35°C and scanned hourly for turbidity and/or color changes. Positive test results are analyzed by the system computer, compared to the database, and linked to an organism identification by the printed analyst's code. A complete report is printed out on the system printer, which includes the biochemical reactions, along with an organism "bionumber" and percentage confidence levels for the identification.

Due to the miniaturized nature of each of the biochemical reactions and the objective measurements taken by the instrument, genus and species identifications for gram-positive organisms can be obtained in 15–18 hours. The system is easy to inoculate, and only a small inoculum density is required, allowing for the identification of an organism from a primary isolation plate. No reagents are necessary, and the analyst does not even need to be present in the lab when the results become available. Technician variability is virtually eliminated, because the readings taken for each of the biochemical reactions are based on the percentage change of optical readings, not on a subjective assessment of color changes.

The microorganisms used to create the various AMS databases historically had been derived solely from clinical sources, as the instrument was originally designed for that purpose. As the market for the system increased, industrial users demanded more complete databases, using organisms obtained from processing environments. The system manufacturer responded by collecting a wide variety of microorganisms from its users, representing several industries, and adding them to the database population for each of the test kits. Users are still encouraged to submit new and unusual isolates to the manufacturer, to be added to the database if necessary. Software updates that include the expanded organism database are issued to all system users on a periodic basis.

If a user does not desire to submit microorganisms to the AMS manufacturer, a supplemental react file (SRF) is also available, which allows the user to create a proprietary database unique to a specific application. Multiple replicates of each isolate should be tested in the system, in order to generate all possible bionumbers that may be associated with that isolate. Once a database sufficient for the operator is established, future unknown cultures may be tested in the system, and matched to preestablished organisms. Culture collections unique to an industry or a particular processing location can by typed in this manner.

Tables 5.1 and 5.2 list the gram-positive organisms contained in the databases for the GPI and BAC test kits. Specific database limitations will be reviewed in more detail in the discussion of each microorganism. The capabilities of the AMS are different for each of the different subgroups of gram-positive microorganisms.

Table 5.1. Microorganisms Identified by the GPI Test Kit

Actinomyces pyogenes	*Staphylococcus saprophyticus*
Aerococcus sp.	*Staphylococcus sciuri*
Arcanobacterium haemolyticum	*Staphylococcus simulans*
Corynebacterium xerosis	*Staphylococcus warneri*
Corynebacterium spp.	*Staphylococcus xylosus*
Enterococcus avium	*Streptococcus acidominimus*
Enterococcus casseliflavus	*Streptococcus agalactiae*
Enterococcus durans	*Streptococcus anginosus*
Enterococcus faecalis	*Streptococcus bovis*
Enterococcus faecium	*Streptococcus bovis* other varieties
Enterococcus gallinarum	*Streptococcus constellatus*
Enterococcus hirae	*Streptococcus equi*
Erysipelothrix rhusiopathiae	*Streptococcus equinus*
Gemella morbillorum	*Streptococcus equisimilis*
Listeria spp.	*Streptococcus* Group G
Staphylococcus aureus	*Streptococcus intermedius*
Staphylococcus auricularis	*Streptococcus mitis*
Staphylococcus capitis	*Streptococcus mutans*
Staphylococcus cohnii	*Streptococcus pneumoniae*
Staphylococcus epidermidis	*Streptococcus pyogenes*
Staphylococcus haemolyticus	*Streptococcus salivarius* var. *salivarius*
Staphylococcus hominis	*Streptococcus sanguis* I and II
Staphylococcus hyicus	*Streptococcus uberis*
Staphylococcus intermedius	*Streptococcus zooepidemicus*
Staphylococcus lentus	

Table 5.2 Microorganisms identified by the BAC Test Kit

Bacillus alvei	*Bacillus megaterium*
Bacillus cereus	*Bacillus polymyxa*
Bacillus circulans	*Bacillus pumilis*
Bacillus coagulans	*Bacillus subtilis*
Bacillus firmus	*Bacillus subtilis (globigii)*
Bacillus laterosporus	*Bacillus sphaericus*
Bacillus lentus	*Bacillus stearothermophilus*
Bacillus licheniformis	*Bacillus thuringiensis*
Bacillus macerans	

Gram-Positive Cocci

The major groups of gram-positive cocci that can be identified by the GPI test kit belong to the genera *Streptococcus*, *Enterococcus*, and *Staphylococcus*. Other organisms in this group that are identified include *Gemella* spp. and *Aerococcus* spp. When using the GPI card to identify gram-positive cocci, the catalase, coagulase (when appropriate), and hemolytic activity of the organism must be noted on the test card in the appropriate designated areas. Proper identification of certain species cannot be made without these designations.

Because the environment of the GPI card is microaerophilic, strictly aerobic organisms will not grow in the test wells; thus, identification of these organisms cannot be made. As a result, members of the genera *Micrococcus*, *Planococcus*, and *Deinococcus* cannot be identified with the AMS. If a gram-positive, catalase-positive, coagulase-negative coccus is isolated, but does not give sufficient growth in the GPI card, one of these organisms should be suspected and confirmed with conventional tests (Schleifer et al. 1986, 1003–1013, 1035–1043). The AMS does have a test kit used for anaerobic organisms, the Anaerobe Identification (ANI) kit, which has recently been proposed for use in *Micrococcus* identification. This kit is an enzymatic kit, and does not depend on the growth of the microorganism to obtain biochemical reactions. Any bionumbers generated by the ANI test kit can be input into the SRF and a database for *Micrococcus* spp. can be created.

Strictly anaerobic organisms will not grow in the test wells of the card either, so members of the genera *Peptococcus*, *Peptostreptococcus*, *Ruminococcus*, *Coprococcus*, and *Sarcina* cannot be identified. The incubation of the test kits is performed at 35°C, so thermophilic organisms such as *Streptococcus salivarius* subsp.

thermophilus will not be identified. The manufacturer recommends propagating organisms to be placed into the test kits for 24 hours, aerobically at 35°C. Anaerobic and thermophilic cocci would not normally grow under such conditions and should be identified by other methods. There are a number of other organisms that may grow in the card, but are not in the database, most significantly the lactococci, *Leuconostoc* spp., *Pediococcus* spp., and other cocci important to food production. If identification of these organisms is desired, they may be tested in the system, and input into the SRF. Some work done in this area has demonstrated that these genera will typically get an identification from the GPI test kit of some other *Streptococcus* or *Staphylococcus* spp. With the SRF, the primary AMS identification will still be a misidentification, but SRF identification would be custom-designated to identify the correct organism.

Of the catalase-negative cocci that can be identified by the AMS, the pyogenic and viridans streptococci and the enterococci are best represented in the database (see Table 5.1), along with several other significant streptococci. *Stomatococcus* spp., which are very weak catalase producers that may be confused with the streptococci, are not identified by the AMS. These organisms may be distinguished from the streptococci by the very distinct capsule formation typical of the species.

Experience with the GPI card for identification of streptococci had demonstrated some early difficulties with identification of β-hemolytic streptococci, but improvements in the database for these organisms has led to improved rates of correct identifications. In several evaluations 86–97 percent of enterococcal strains, 80–100 percent of pneumococcal, and 100 percent of Group B non-hemolytic strains were correctly identified. Problems were still encountered with viridans and β-hemolytic streptococci. Only 57–79 percent of viridans streptococci and 50–57 percent of β-hemolytic strains were correctly identified (Ruoff et al. 1982; Applebaum et al. 1984; Facklam et al. 1985). Some strains in these groups were identified better than others, for example, *Streptococcus mitis* had identification rates of 79–100 percent, while *Streptococcus sanguis* I had rates of 29–43 percent correct. Streptococci identifications are usually completed in 8–12 hours.

Problems with clumping of the inoculum suspension can cause erroneous readings of the GPI card, and thus lead to misidentifications. Therefore, it is important to mix the colony in the saline suspension with a swab, and not rely on vortexing alone to minimize clumping. Addition of a small amount of sterile 10 percent glycerol to the saline can also help suspend clumping organisms. A list of recommended tests for separating species pairs with identical GPI test kit results (a report of "good confidence, marginal separation") is provided by the manufacturer, and can also be augmented with conventional methods (Hardie et al. 1986, 1043–1071; Facklam and Carey 1985, 154–175). Serogrouping is a valuable supplemental test to include for definitive identification in many cases.

The genus *Staphylococcus* is well represented in the GPI database, with only a few of the lesser-known species not available. As in the case of the streptococci, the SRF can be used with the other *Staphylococcus* spp., if known control organisms are used to create the file. Again, it is critical that the catalase and coagulase activities of the isolate be recorded on the test card, or misidentifications will result. The coagulase reaction is especially important for *Staphylococcus aureus* strains, as this organism will be misidentified without it. The AMS does have the capability of printing corrected reports, so that if any of the supplemental tests is inadvertently left off the test kit at the time of incubation, the result can be added back after the identification has been made. Most identifications for the staphylococci are completed within 10 hours.

Published reports of clinical evaluations (Almeida et al. 1983; Grasmick et al. 1983) give varying results of the accuracy of the GPI kit for *Staphylococcus* identification. Initial trials resulted in overall accuracy rates varying from 67.3 percent to 83.2 percent of isolates. The greatest accuracy rates achieved in these early evaluations were with *Staphylococcus epidermidis* (95.7–95.9 percent correct) and *Staphylococcus saprophyticus* (87.1–100 percent correct). Because of the limitations of the first GPI database (only 6 *Staphylococcus* spp. represented), difficulties were encountered with most of the other staphylococci. *Staphylococcus hominis* and *Staphylococcus warneri* were often misidentified as *Staphylococcus saprophyticus*, and *Staphylococcus haemolyticus* and *Staphylococcus simulans* were usually unidentified. With major improvements in the GPI database, most of these difficulties have been resolved, and these species are now easily identified. Many of the limitations in differentiating staphylococci with the AMS can be overcome with running select supplemental tests, most importantly novobiocin resistance. As with the streptococci, tests for separating isolates with "good confidence, marginal separation" reports are listed by the manufacturer, or other conventional methods (Kloos and Lambe 1991, 227–237) may be used.

Gram-Positive Nonspore-Forming Rods

Only a selected group of the gram-positive nonspore-forming organisms are included in the GPI database (Table 5.1). As a result, if identification of any other gram-positive nonspore-formers is desired, either conventional tests (Kandler and Weiss 1986, 1209–1234; Jones and Collins 1986, 1261–1434) or the SRF should be used. The identification of *Lactobacillus* spp. is particularly challenging, considering the large number of carbohydrate fermentation tests that are used to differentiate between species of lactobacilli. A suitable substitute for conventional tubed media is the API 50 CH, or Rapid CH test system, also manufactured by bioMérieux Vitek, Inc., which is a set of test strips containing cupules with 50 different carbohydrate fermentation tests. Inoculation of the test strips, while more cumbersome than automated methods, is still much easier than inoculating up to 50 different tubed media. Results of the tests can then be compared to a published reference (Kandler and Weiss 1986) for identification to species level.

Little published data are available on the performance of the GPI card for the identification of *Erysipelothrix* spp., *Actinomyces* spp., *Arcanobacterium* spp., or *Corynebacterium* spp. Quality control and American Type Culture Collection (ATCC) strains used on this system have performed satisfactorily, and anecdotal information has demonstrated the system to work well for these organisms.

The performance of the AMS for the identification of *Listeria* spp. has been more extensively studied (Heisick et al. 1989; Knight et al. 1988). Two of the test kits are required for the identification of *Listeria* to species level, the GPI and GNI test kits. The GPI card is used to obtain the genus identification and a mannitol reaction result, and the GNI card is used to obtain rhamnose and xylose reaction results. Along with Christie-Atkins-Munch-Peterson (CAMP) and nitrate reduction tests results (Lovett and Hitchins 1988, 29.01–29.14), the identification of *Listeria* to species level can be made, based on reactions in Table 5.3. This method is able to differentiate seven *Listeria* species: *monocytogenes, innocua, ivanovii, seeligeri, welshimeri, grayi,* and *murrayi.* Test results are available within 24 hours, compared to 7 days with the conventional method.

This method of identifying *Listeria* spp. has been collaboratively studied, and has obtained first action approval from the Association of Official Analytical Chemists International (AOACI) (Harris and Humber 1993). The system was able to correctly identify 90.8 percent of *Listeria* spp. isolates used in the study. It had the most difficulty identifying *Listeria ivanovii* on a consistent basis, with incorrect identifications resulting from false-negative salicin reactions. Misidentifications also resulted from insufficient growth in the GPI card by *Listeria grayi* and *Listeria murrayi* isolates. In most of these cases, retesting of the isolates after they had been propagated at 4°C for several days resulted in the correct identification.

Gram-Positive Spore-Forming Rods

The genera *Bacillus, Clostridium,* and *Sporolactobacillus* make up the group of gram-positive spore-forming rods, and can be easily distinguished from each other by various means. *Clostridium* spp. and *Sporolactobacillus* spp. are catalase negative; *Clostridium* spp. are strict anaerobes. Automated identification of these mircoorganisms will be discussed in other chapters. In this section only the gram-positive, catalase-positive spore-forming rods, namely *Bacillus* spp., will be discussed.

One of the first test kits designed for the AMS specifically for the industrial market was the BAC test kit. All of the organisms used to create the original BAC database were derived from industrial sources, primarily the pharmaceutical industry. Preliminary evaluations of the system concentrated on organisms used as biological indicators of sterilization processes (Odlaug et al. 1982), such as *Bacillus stearothermophilus, Bacillus subtilis,* and *Bacillus pumilis.* Some inadequacies were evident: Certain species would consistently be identified as other *Bacillus* spp., and many species could not be identified at all. As a result, industrial representatives were actively solicited for isolates to augment the database, and the present database has been expanded to include many other species of interest (Table 5.2). Given the size and complexity of the genus *Bacillus,* however, there are still many species that are not included, and these would need to be identified by conventional methods (Gordon et al. 1973; Claus and Berkeley 1986, 1105–1139). For those organisms included in the BAC database, an identification can usually be obtained within 6–15 hours, compared to 7–10 days with conventional biochemical testing.

The system works slightly differently for mesophilic and thermophilic species. For mesophilic species test kits are inoculated and incubated in the automated reader according to standard procedure. For thermophilic species test cards must be incubated off-line in a sealed, humidified container at 55°C for

Table 5.3. Listeria Species Determination with GPI and GNI Test Kits and Supplementary Tests

| | | GPI CARD | GNI CARD | | SUPPLEMENTAL TESTS | | |
| | | | | | | CAMP Test | |
Organism	Identification	Mannitol Result	Rhamnose Result	Xylose Result	Beta Hemolysis	*Staphylococcus aureus* Result	*Rhodococcus equi* Result
L. monocytogenes	Listeria spp.	−	+	−	+	+	−
L. innocua	Listeria spp.	−	+	−	−	−	−
L. ivanovii	Listeria spp.	−	−	+	+	−	+
L. seeligeri	Listeria spp.	−	−	+	V	V	−
L. welshimeri	Listeria spp.	−	V	+	−	−	−
L. grayi	Listeria spp.	+	−	−	−	−	−
L. murrayi	Listeria spp.	+	V	−	−	−	−

(−) Negative result, (+) Positive result, (V) Variable result

24 hours. After incubation, the cards, which are marked to indicate thermophilic incubation, are placed into the reader module for a single-time reading and analysis. The computer analysis and report generation are still the same as for mesophilic organisms. It is critical that the test cards remain in a sealed, humidified environment during 55°C incubation, otherwise bubbles will form in the test card that would interfere with the data analysis. There are some *Bacillus* strains that will not form smooth suspensions in the saline solution used to inoculate the test kits. For these isolates a heavier than normal inoculum should be suspended in the saline, left to settle out, and just the top portion used to inoculate another saline tube prior to submitting the card to the inoculation cycle.

As with the other AMS test kits, there are situations where additional conventional biochemical testing is necessary to complete the differentiation between two closely related species. In the case of *Bacillus subtilis,* pigment production on various media may be used to differentiate the various strains within this species (Gordon et al. 1973; Claus and Berkeley 1986, 1105–1139). Testing for the production of protein parasporal crystals is necessary to distinguish between the closely related species *Bacillus cereus* and *Bacillus thuringiensis.* Other tests necessary to distinguish between two species when the system gives a "good confidence, marginal separation" notation are referenced by the system manufacturer.

When the BAC test kit was originally evaluated, its performance relative to conventional testing was only marginally acceptable. The database was limited, and depending on the types of isolates tested, the ability of the system to identify organisms accurately to species level without extensive confirmatory testing range from 47 percent to 92 percent of isolates (Chadick 1993; Odlaug et al. 1982). With the improvement of the BAC database, and the addition of several more *Bacillus* spp., the overall rate of correct identifications to species level improved to 76 percent, along with a significant decrease in the number of additional confirmatory tests needed. The most valuable supplemental tests to use were the Vogues-Proskauer reaction and nitrate reduction (Chadick 1993). Due to the time-consuming and lengthy conventional procedures often necessary for the identification of *Bacillus* spp., the BAC test kit offers a practical alternative for more rapid screening of suspect *Bacillus* spp. isolates.

AUTOMATED TESTS FOR BACTERIOLOGY SYSTEM

The ATB system is a semiautomatic system that utilizes plastic test strips supporting a series of cupules that contain 32 biochemical tests, based either on assimilation of carbon substrates or the detection of chromogenic reaction compounds. The test strips are inoculated either manually or automatically from a suspension of the test organism, and are incubated off-line in a typical lab incubator. The test strips can be read manually and typed into the system computer, or they can be entered into an instrument that automatically reads each biochemical reaction and inputs the information to the system computer. The computer then analyzes the data and prints out a report of the organism identification. The ATB offers greater flexibility for the smaller laboratory over fully automatic systems, in that it is much cheaper to purchase and operate. Several of the different test strips available can be read within 4 hours, much more

rapidly than conventional methods, and even more rapidly than some automated methods.

The tests for the ATB do require more labor to set up and inoculate, and because the strips are incubated off-line, there is more labor involved in reading the
strips. Reagents must be added to some of the strips, and each strip must be manually placed into the instrument and removed prior to reading the next one. If the
test strips are to be read manually, the problems of subjectivity and ambiguous
test reactions also arise.

There are two test strips available for the identification of gram-positive organisms, the ID 32 STAPH, which can identify staphylococci, micrococci, and
Stomatococcus and *Aerococcus* spp., and the rapid ID 32 STREP, which identifies
streptococci, enterococci, lactococci, and several genera of gram-positive nonspore-forming rods. Tables 5.4 and 5.5 list the organisms that can be identified by
each of these test strips. As can be seen from the tables, the databases for these test
kits are much more extensive than those of the AMS, and include a much larger
variety of species for each of the genera. Several genera are included here that are
not available with the AMS, namely *Micrococcus* spp., *Stomatococcus* spp.,
Gardnerella vaginalis, *Lactococcus* spp., and *Leuconostoc* spp. The origins of the microorganisms, although again primarily derived from clinical sources, are also

Table 5.4. Microorganisms Identified by ID 32 STAPH Test Strip

Aerococcus viridans	*Staphylococcus equorum*
Micrococcus kristinae	*Staphylococcus gallinarum*
Micrococcus luteus	*Staphylococcus haemolyticus*
Micrococcus lylae	*Staphylococcus hominis* I and II
Micrococcus nishinomiyaensis	*Staphylococcus hyicus*
Micrococcus roseus	*Staphylococcus intermedius*
Micrococcus varians	*Staphylococcus kloosii*
Staphylococcus arlettae	*Staphylococcus lentus*
Staphylococcus aureus	*Staphylococcus lugdunensis*
Staphylococcus auricularis	*Staphylococcus saprophyticus*
Staphylococcus capitis	*Staphylococcus schleiferi*
Staphylococcus caprae	*Staphylococcus sciuri*
Staphylococcus carnosus	*Staphylococcus simulans*
Staphylococcus chromogenes	*Staphylococcus warneri*
Staphylococcus cohnii I and II	*Staphylococcus xylosus*
Staphylococcus epidermidis	*Stomatococcus mucilaginosus*

Table 5.5. Microorganisms Identified with the Rapid ID 32 STREP Test Strip

Aerococcus viridans	*Streptococcus cecorum*
Enterococcus avium	*Streptococcus constellatus*
Enterococcus casseliflavus	*Streptococcus defectivus*
Enterococcus durans	*Streptococcus downei*
Enterococcus faecalis	*Streptococcus dysgalactiae*
Enterococcus faecium I and II	*Streptococcus equi* var. *equi*
Enterococcus gallinarum	*Streptococcus equi* var. *zooepidemicus*
Enterococcus hirae	*Streptococcus equinus*
Erysipelothrix rhusiopathiae	*Streptococcus equisimilus*
Gardnerella vaginalis	*Streptococcus gordonii*
Gemella haemolysans	*Streptococcus* Group L
Gemella morbillorum	*Streptococcus intermedius*
Lactococcus garvieae	*Streptococcus mitis* I and II
Lactococcus lactis var. *cremoris*	*Streptococcus mutans*
Lactococcus lactis var. *lactis*	*Streptococcus oralis*
Lactococcus raffinolactis	*Streptococcus pneumoniae*
Leuconostoc spp.	*Streptococcus porcinus*
Listeria grayi	*Streptococcus pyogenes*
Listeria murrayi	*Streptococcus saccharolyticus*
Listeria spp.	*Streptococcus salivarius* var. *salivarius*
Streptococcus acidominimus	*Streptococcus salivarius* var. *thermophilus*
Streptococcus adjacens	*Streptococcus sanguis*
Streptococcus agalactiae	*Streptococcus sobrinus*
Streptococcus alactolyticus	*Streptococcus suis*
Streptococcus anginosus	*Streptococcus uberis* I and II
Streptococcus bovis I and II	*Streptococcus vestibularis*
Streptococcus canis	

more varied, with many species coming from pharmaceutical or food industry sources.

To date, assessment of the ATB system has been limited to clinical evaluation (Freney et al. 1992). Of 413 strains of streptococci, lactococci, enterococci, and various other gram-positive species, 95.3 percent were correctly identified, with 25.1 percent of these requiring extra supplemental tests. Only 4.7 percent of the isolates were either unidentified or misidentified. Little published information is available on the use of the system with the ID 32 STAPH test strip. The lower cost of this automated system, coupled with its relative ease of use, and the size and completeness of the gram-positive databases, make this system a viable alternative to conventional methods for the smaller laboratory, and a useful adjunct to other fully automated methods for the large laboratory.

REFERENCES

Almeida, R. J., J. H. Jorgensen, and J. E. Johnson. 1983. Evaluation of the AutoMicrobic system gram-positive identification card for species identification of coagulase-negative staphylococci. *J. Clin. Microbiol.* 18:438–439.

Applebaum, P. C., M. R. Jacobs, J. I. Heald, W. M. Palko, A. Duffett, R. Crist, and P. A. Naugle. 1984. Comparative evaluation of the API 20S system and the AutoMicrobic system gram-positive identification card for species identification of streptococci. *J. Clin Microbiol.* 19:164–168.

Chadick, D. 1993. Personal communication.

Claus, D., and R. C. W. Berkeley. 1986. Genus *Bacillus*. In *Bergey's manual of systematic bacteriology*, vol. 2, edited by P. H. A. Sneath. Baltimore: Williams & Wilkins.

Facklam, R., G. S. Bosley, D. Rhoden, A. R. Franklin, N. Weaver, and R. Schulman. 1985. Comparative evaluation of the API 20S and AutoMicrobic gram-positive identification systems for non-beta-hemolytic streptococci and aerococci. *J. Clin. Microbiol.* 21:535–541.

Facklam, R. R., and R. B. Carey. 1985. Streptococci and aerococci. In *Manual of clinical microbiology*, edited by E. H. Lennette, A. Balows, W. J. Hausler, H. J. Shadomy. Washington, DC: American Society for Microbiology.

Freney, J., S. Bland, J. Etienne, M. Desmonceaux, J. M. Boeufgras, and J. Fleurette. 1992. Description and evaluation of the semiautomated 4-hour rapid ID 32 STREP method for identification of streptococci and members of related genera. *J. Clin. Microbiol.* 30:2657–2661.

Gordon, R., W. C. Haynes, and C. H. Pang. 1973. The genus *Bacillus*. In *Agriculture Handbook No. 427*. Washington, DC: Agricultural Research Service.

Grasmick, A. E., N. Naito, and D. A. Bruckner. 1983. Clinical comparison of the AutoMicrobic system gram-positive identification card, API Staph-Ident, and conventional methods in the identification of coagulase-negative *Staphylococcus* spp. *J. Clin. Microbiol.* 18:1323–1328.

Hardie, J. M., J. Rotta, and J. O. Mundt. 1986. Genus *Streptococcus*. In *Bergey's manual of systematic bacteriology*, vol. 2, edited by P. H. A. Sneath. Baltimore: Williams & Wilkins.

Harris, L., and J. Humber. 1993. AutoMicrobic system for biochemical identification of *Listeria* species isolated from foods: Collaborative study. *J. AOAC Int.* 76:822–830.

Heisick, J. E., D. E. Wagner, M. L. Nierman, and J. T. Peeler. 1989. *Listeria* spp. found on fresh market produce. *Appl. Environ. Microbiol.* 55:1925–1927.

Jones, D., and M. D. Collins. 1986. Irregular, nonsporing gram-positive rods. In *Bergey's manual of systematic bacteriology* vol. 2, edited by P. H. A. Sneath. Baltimore: Williams & Wilkins.

Kandler, O., and N. Weiss. 1986. Genus *Lactobacillus*. In *Bergey's manual of systematic bacteriology*, vol. 2, edited by P. H. A. Sneath. Baltimore: Williams & Wilkins.

Kloos, W. E., and D. W. Lambe, Jr. 1991. *Staphylococcus*. In *Manual of clinical microbiology*, edited by A. Balows, W. J. Hausler, K. L. Herrmann, H. D. Isenberg, and H. J. Shadomy. Washington, DC: American Society for Microbiology.

Knight, M. L., J. F. Black, and D. W. Wood. 1988. Industry perspectives on *Listeria monocytogenes*. *J. Assoc. Off. Anal. Chem.* 71:682–683.

Lovett, J., and A. D. Hitchins. 1988. *Listeria* isolation. In *Bacteriological analytical manual*. Arlington, VA: AOAC International.

Odlaug, T. E., V. Jarzynski, R. A. Caputo, and C. C. Mascoli. 1982. Evaluation of an automated system for rapid identification of *Bacillus* biological indicators and other *Bacillus* species. *J. Parent. Sci. Technol.* 36:47-54.

Ruoff, K. L., M. J. Ferraro, M. E. Jerz, and J. Kissling. 1982. Automated identification of gram-positive bacteria. *J. Clin. Microbiol.* 16:1091–1095.

Schleifer, K. H., M. Kocur, T. Bergan, and R. G. E. Murray. 1986. The family Micrococcaceae and the family Deinococcaceae. In *Bergey's manual of systematic bacteriology*, vol. 2, edited by P. H. A. Sneath. Baltimore: Williams & Wilkins.

6
Yeasts

Gerard J. Osterhout
William G. Merz
Johns Hopkins Medical Institutions

The term *yeast* is used to describe unicellular fungi that reproduce principally by budding or fission. Colonies are smooth, moist, and creamy to pasty in texture, consisting of oval to spherical cells. Several may appear extremely mucoid owing to the production of a mucopolysaccharide capsule. They are taxonomically diverse, consisting of approximately 60 genera and over 600 species. Yeasts may also be classified on their ability to reproduce sexually. The true or perfect yeasts, which can reproduce sexually, are classified by their method of sexual reproduction as Ascomycetes and Basidiomycetes. The Fungi Imperfecti or Deuteromycetes only reproduce by asexual means. Most of the pathogenic yeasts are classified in this group.

Classically, the identification and speciation of yeast and yeast-like fungi has been based on a combination of morphologic, serological, and biochemical tests. Distinctive morphologic features include production of blastoconidia (buds), arthroconidia, germ tubes, chlamydospores, and pseudohyphae. Speciation is accomplished using biochemical tests, including carbohydrate fermentation and assimilation criteria. These methods are often time consuming and difficult to interpret due to overlapping color reactions or growth/no growth end points.

CHEMOTAXONOMY

The use of chemical analysis of microbial components (i.e., lipids, polysaccharides, proteins, and nucleic acids) has been increasingly applied to bacterial taxonomy (Brondz and Olsen 1986; Lechevalier 1982; Moss 1981, 1985).

Chemotaxonomic methods have also found increasing applicability to yeast detection and identification (Olsen 1990). Analytical methodologies utilized included gel electrophoresis, orthogonal-field-gel electrophoresis, spectrophotometry, proton magnetic resonance, high-performance liquid chromatography, gas chromatography, and combined gas chromatography-mass spectrometry.

Merz et al. (1988) used orthogonal-field-alternation gel electrophoresis to establish electrophoretic karyotypes for strains of *Candida albicans*. They detected much greater strain variation than revealed by existing biotyping techniques, thus expanding the scope of epidemiological studies.

An automated enzymatic method developed by Switchenko et al. (1994) spectrophotometrically determined D-arabinitol and used it as a potentially rapid method for detecting invasive candidiasis. This was developed as a more practical approach from the technically demanding methods of Roboz et al. (1990) and Wong and Castellanos (1989), using chiral stationary phase gas chromatography. Cellular carbohydrates (Brondz and Olsen 1990; Wejman 1979; Wejman and Rodriguez De Miranda 1988) have been increasingly utilized to differentiate yeast genera. Brondz and Olsen (1990) applied principal component analysis to glucose, mannose, and galactose components to distinguish *Saccharomyces cerevisiae* from *Torulopsis glabrata* and *Candida albicans*. When combined with cellular fatty acid (CFA) analysis (Brondz et al. 1989), all three of these organisms could be readily distinguished.

Gas chromatography has emerged as the predominant analytical technique in diagnostic microbiology (Morgan 1984; Moss 1981). Technical advances in recent years through the development of fused-silica capillary columns, automatic injection systems, digital integrators, and standardized calibrators have increased the applicability of this technology to the clinical microbiology laboratory and reference laboratories (Moss et al. 1980). Traditionally, it has been restricted to the analysis of end products of metabolism in the form of short-chain acids produced by anaerobic bacteria as an adjunct to biochemical analysis. Application to the end products of metabolism by yeasts, however, are limited (Lategan 1981). Considerable attention has been given to analysis of lipids. Bacterial lipid determination, particularly CFA analysis by gas-liquid chromatography, has long been recognized as a valuable chemotaxonomic tool for the classification and identification of gram-negative nonfermentative bacteria (Moss et al. 1988; Osterhout et al. 1991; Oyaizu and Komagata 1983; Veys et al. 1989). Compared with conventional and other chemical identification methods, the advantages of CFA analysis include its relative speed and simplicity and the lack of the need for recultivation of the organism after initial growth, or specialized and expensive reagents. Increasing attention has been given to the value of CFAs in yeast taxonomy (Chen 1981; Gangopadhyay et al. 1979; Kobayashi et al. 1987; Kock et al. 1985; Moss et al. 1982; Rattray et al. 1975).

Cellular Fatty Acids of Yeast

Yeasts contain mainly saturated and unsaturated fatty acids 14–18 carbons in length (Rattray 1988). The predominant acids, typified by most eukaryotic cells, are palmitate (16:0), palmitoleate (16:1), stearate (18:0), oleate (18:1), linoleic acid

(18:2), and linolenic acid (18:3). Minor amounts of myristic (14:0) and myristoleic (14:1) acids may also be present. By convention, the number in front of the colon represents the number of carbon atoms in the molecule, and the number after indicates the number of double bonds.

Yeasts possess a relatively limited spectra of CFAs compared to bacteria, which may also possess branched chain, hydroxy, and cyclopropane acids or combinations thereof. Polyunsaturated fatty acids tend to be absent in bacteria, while the predominant monounsaturated 18 carbon fatty acid is vaccenic acid. Since relatively few fatty acids are present in yeast, this suggests that the resolving power of this methodology will be very limited. Thus, distinctive CFA composition when present is accounted for primarily by quantitative as opposed to qualitative differences.

Early reports on CFA analysis discussed its potential usefulness as a taxonomic tool (Gangopadhyay et al. 1979; Gunasekaran and Hughes 1980; Rattray et al. 1975; Shaw 1974). Moss et al. (1982) analyzed a diverse group of yeasts using a fused-silica capillary column. The high resolving power of this column provided accurate quantitative determination of the CFAs, including complete separation of oleic and vaccenic acids that will coelute on packed columns. They concluded that the Basidiomycetous yeasts could be separated from the Ascomycetous species, although further classification was not possible due to overlapping profiles. Some conflicting results with Gangopadhyay were also noted. Using a variety of parameters, such as the presence of vaccenic acid and ratios of 16:1/16:0, they demonstrated that *Torulopsis glabrata* had a distinctive profile. The only yeast to possess a higher level of vaccenic acid is *Kluyveromyces polysporus* (Augustyn et al. 1992). It was also shown that *Saccharomyces cerevisiae*, like *Torulopsis glabrata*, lacks the polyunsaturated fatty acids characteristic of other yeasts. The lack of polyunsaturated fatty acids, however, is not valid for all *Saccharomyces* spp. The large amount of 16:1 in *Saccharomyces* spp., compared with the modest amounts of 16:1 and large amounts of 18:2 in *Candida*, separate these genera.

The earlier reports on CFA analysis did not utilize a uniform culture procedure, which has been shown to have a significant influence on fatty acid composition (Rattray et al. 1975). Thus, no direct comparison between the lipid composition and the taxonomic positions could be made. Kock et al. (1985) developed a standardized culture condition and used it in constructing a rapid yeast identification process. Using this method, 10 species of the genus *Kluyveromyces* (Cottrell et al. 1985) and five *Saccharomyces* species (Kock et al. 1986) could be differentiated. The method was also useful for determining relationships between anamorphic and teleomorphic forms (Viljoen et al. 1987, 1988, 1989). Data reported in these studies was compromised by the inefficiencies associated with the packed columns used to generate the results.

In a series of articles (Augustyn and Kock 1989; Augustyn 1989; Augustyn et al. 1990, 1991, 1992), promising results were obtained when CFA analysis was used to differentiate selected yeasts associated with the wine industry. The results were achieved using a polar capillary column that enabled optimal separation of CFAs, especially the polyunsaturated CFAs. *Saccharomyces cerevisiae* was differentiated from all other strains of the 8 genera and 36 species tested, except for one strain of *Pachytichospora transvaalensis*. However, it could not be distinguished from strains of *Saccharomyces bayanus* or *Saccharomyces pastorianus*. Recently, 123 representative yeasts associated with a Vienna sausage processing plant were identified using CFA analysis when combined with conventional techniques (Viljoen et al. 1993).

Augustyn et al. (1989) used a capillary column to differentiate 13 strains of *Saccharomyces cerevisiae*. However, they acknowledged that with the addition of more strains the method may be invalidated. Separation was only achieved by considering the contribution of minor acids. They also noted that the method would be less laborious if pattern recognition software was utilized in interpreting the data.

Viljoen and Kock (1989) concluded that the combination of CFA composition, electrophoretic karyotypes, coenzyme Q systems, guanine + cytosine contents, serological studies, and proton magnetic resonance spectra of mannans provided valuable criteria in determining relationships among 117 strains within the genus *Candida*.

Although prior studies indicated promising results for rapid speciation, Augustyn et al. (1992) concluded that the results previously observed were only valid for the strains used in that particular study. In most of their studies only a few strains were used to establish the basis for the conclusion. Subsequent analyses and integration of the data with previous reports revealed much more heterogeneity and overlapping CFA profiles between species and genera.

The standard method for the assessment of CFAs has been methanolysis, which generates the corresponding fatty acid methyl esters (FAMEs). Volatile fatty acids such as C12 can be lost in this derivatization procedure (Bryn and Jantzen 1982). Transesterification with ethanol, propanol, and butanol (Brondz et al. 1989) enabled 100 percent recovery of C10 and C12. The authors concluded that these acids could be used to further facilitate taxonomic differentiation between yeasts. Previously, these acids were not considered important in yeast chemotaxonomy (Rattray 1988).

Computerized multivariate analysis programs have enabled a tremendous amount of information to be derived from sugars and, especially, fatty acid analysis data (Brondz and Olsen 1992). Principal component analysis and partial least-squares discriminant analysis of both carbohydrate and fatty acid variables were used to evaluate the proposed merger of *Candida albicans* and *Torulopsis glabrata* into the same genus (Brondz and Olsen 1990). Recently, Marumo and Aoki (1990), using discriminant analysis of CFAs, concluded that *Torulopsis glabrata*, *Cryptococcus neoformans*, and most strains of the six species of *Candida* tested, could be differentiated. Overall, there was 95 percent agreement with conventional methods.

The data generated in all of the previous studies discussed relied on manual fatty acid identification and chromatogram interpretation, which requires considerable time and expertise on the part of the investigator.

Automated CFA Analysis

The only automated CFA identification system commercially available is the Microbial Identification System (MIS). It combines high resolution gas-liquid chromatography with a computer interface and software package for the identification of yeasts based exclusively on CFA composition. The system is capable of identifying over 100 fatty acids ranging in length from 10–20 carbons. No databases for the short chain acids (1–7 carbons) or long chain (24–90 carbons) mycolic acids are currently available. For reviews on applications of this system to a wide variety of microorganisms, see Welch (1991) and Stager and Davis (1992).

CFA Chromatography

Fatty acid methyl esters are analyzed on a Hewlett-Packard 5890A gas chromatograph equipped with a flame ionization detector, automatic sampler, integrator, and computer. Separation of the FAMEs is achieved with a fused-silica capillary column with cross-linked 5 percent phenylmethyl silicone. The specific operating parameters of the instrument are controlled and set automatically by the computer software. Chromatograms with peak retention times and areas are produced on a recording integrator and electronically transferred to the computer for analysis, storage, and report generation. Peak naming and column performance are achieved through the use of calibration standards containing straight-chain saturated and hydroxy fatty acids.

Data Analysis

Cellular fatty acids are identified on the basis of equivalent chain length data. This value is a representation of a fatty acid's retention times as it relates to a series of straight-chain saturated FAMEs found in the calibration mixture. Since equivalent chain length plots of fatty acids of a homologous series are linear, the computer automatically interpolates from a preestablished equivalent chain length plot the identity of each sample peak. Peak area values are expressed as a percentage of the total of the area of all peaks in the chromatogram, exclusive of the solvent front.

Following computer analysis, a hard-copy report is generated for each sample. In addition to sample identification, a variety of parameters concerning each chromatographic peak is printed, including retention time, area, equivalent chain length, peak name (specific fatty acid), and percentage of total fatty acids present.

Organism identification is based on computer comparison of an unknown organism's FAME profile with those of predetermined library profiles using a covariance matrix and principal component analysis that are subjected to pattern recognition algorithms. Library organism profiles are constructed on the basis of a multivariate Gaussian model of at least 10 strains. The correlation of an unknown organism's profile with a library entry is expressed as a similarity index (SI) on a numeric scale of 0.0 to 1.0, with 1.0 being considered identical. A library generation software (LGS) package is available as an accessory to the system. This enables the user to create customized library databases. Also included in the LGS are two cluster analysis programs, the dendrogram and two-dimensional plots, that are purported to be useful in tracking nosocomial infections or establishing the source of contamination in a pharmaceutical or food manufacturing process. One report in the literature using the MIS (Merz et al. 1992) concluded that while the CFA profile was helpful in supporting the identification of *Candida lusitaniae* strains, it could not be used for typing isolates within the species.

Culture Conditions and Fatty Acid Analysis

Valid comparison to library entries can only be accomplished when the requirements for growth media, incubation temperature, and time are strictly followed. Databases exist for isolates grown on Sabouraud dextrose agar for 24 hr at both 28°C and 35°C. These growth conditions approximate those utilized by Gangopadhyay et al. (1979), Kobayashi et al. (1987), Marumo and Aoki (1990), and Moss et al. (1982).

The extraction and derivitization method is described by Miller (1982). Briefly, approximately 40 mg (wet wt) of cells are gently scraped from the surface of the agar and then transferred to a screw-capped tube with a Teflon®-lined cap. Cells are saponified by heating them at 100°C for 30 minutes following the addition of 1 ml of 15 percent NaOH in 50 percent aqueous methanol. The hydrolysate is cooled to ambient temperature; 2 ml of methanolic HCl are added, and the mixture heated at 80°C for 10 min. The methylated fatty acids are quickly cooled to ambient temperature and extracted through the addition of 1.25 ml hexane-methyl-*tert*-butyl ether (1:1 vol/vol) with end-over-end mixing. The phases are allowed to separate, the lower aqueous layer is removed, and 3 ml of dilute NaOH saturated with NaCl is added to the remaining organic layer. After clarification of the phase interface, approximately two-thirds of the organic layer containing the FAMEs is transferred to a septum-capped vial for analysis. Typical chromatograms generated by the MIS are depicted in Figure 6.1 for *Candida albicans* and *Cryptococcus neoformans*.

System Evaluation

No evaluations of this methodology applied to yeasts are present in the literature. To assess the utility of this automated method, a database was created by the authors consisting of 13 genera and 30 species (see Table 6.1). The library entries were constructed using American Type Culture Collection (ATCC) isolates, well-characterized reference strains, and fresh clinical isolates. To limit the scope of the library, only isolates from clinical specimens were used. Additionally, only those strains capable of growth at 37°C were considered.

The results from an evaluation using fresh clinical isolates are summarized in Tables 6.2 and 6.3 (Osterhout and Merz 1990). An identification with an SI value ≥ 0.5 and no second choice is considered a good match. As can be seen, only *Candida krusei, Saccharomyces cerevisiae,* and *Torulopsis glabrata* could be consistently identified with confidence. The results paralleled those observed in a previous study with fewer library entries (Osterhout et al. 1989). Among the *Candida* spp. tested, <50 percent of the strains could be identified based exclusively on CFA analysis. Among the non-*Candida* species, the results are slightly better, although artificially inflated due to the greater percentage of *Saccharomyces cerevisiae* and *Torulopsis glabrata* strains. Rarely is only one choice supplied by the MIS. Normally, multiple identifications with very similar SI values are provided. Also, the correct identification is not necessarily listed as a possible choice, thus eliminating the prospect of merely confirming one of the listed choices as the definitive identification.

Since only a few strains were available to build some of the library entries, it is expected that the true heterogeneity of a particular species may not be

Figure 6.1. Chromatograms of CFA detected for a typical isolate of *Candida albicans* and *Cryptococcus neoformans*. Number before the colon indicates the number of carbon atoms; number after the colon indicates the number of double bonds.

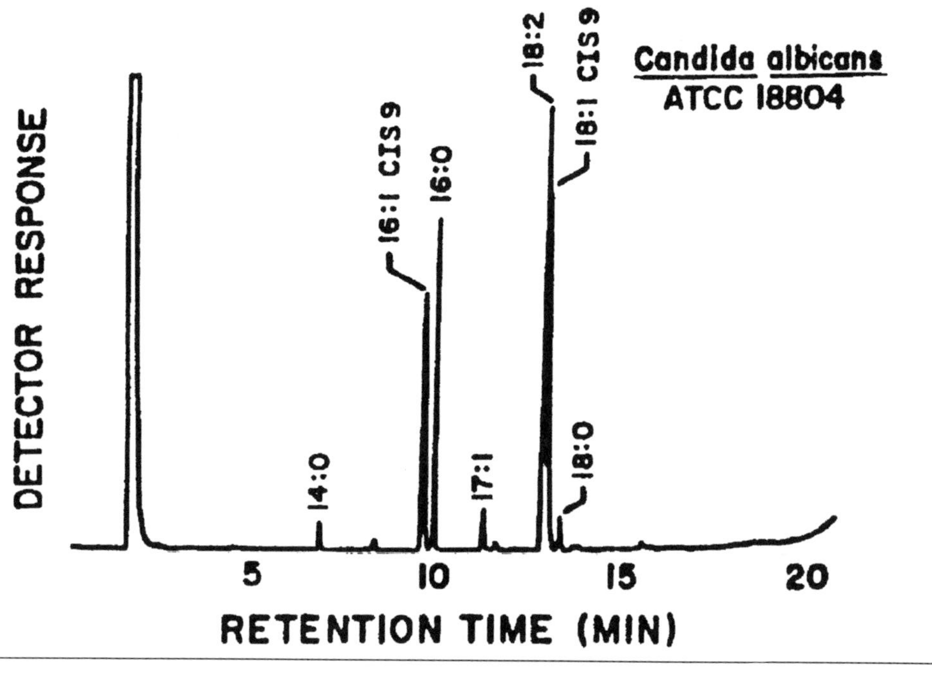

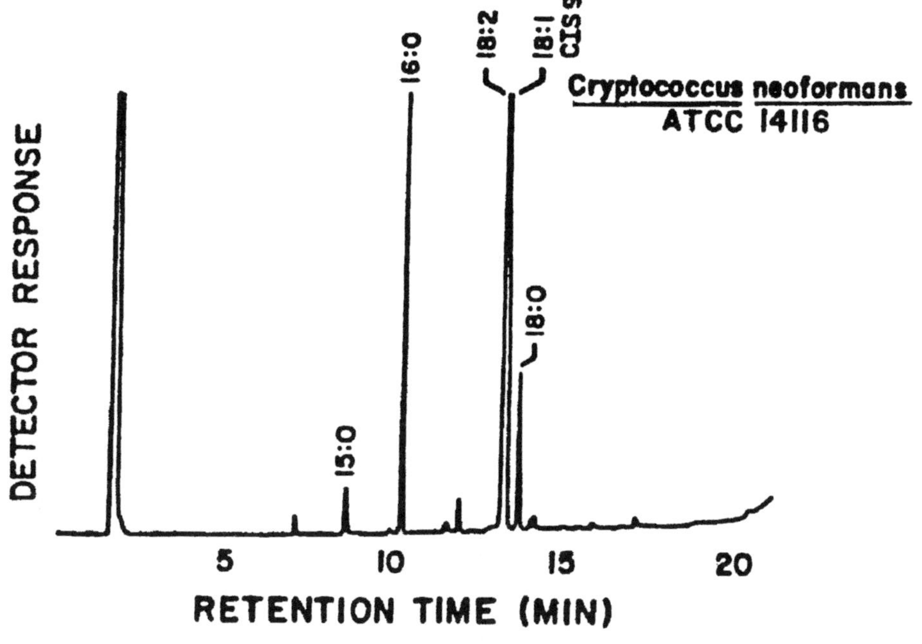

Table 6.1. Library Database Utilized in System Evaluation

Blastoschizomyces capitatus	*Candida* spp. (11)
Cryptococcus spp. (3)	*Geotrichum candidum*
Hanseniaspora uvarum	*Hansenula anomala*
Kluyveromyces marxianus	*Pichia* spp. (4)
Rhodotorula spp. (3)	*Saccharomyces cerevisiae*
Sporobolomyces salmonicolor	*Torulopsis glabrata*
Trichosporon beigelii	

Table 6.2. Summary of CFA Analysis for Yeast Identification: *Candida* Species

Candida Species	Tested	Overall	No 2nd Choice	w/Multi ID or SI < 0.5	Discrepant
			$SI^c > 0.5$		
albicans	21	19 (90)[b]	7 (33)	12 (57)	2 (10)
guilliermondii	2	2 (100)	0 (0)	2 (100)	0 (0)
kefyr	3	3 (100)	0 (0)	3 (100)	0 (0)
krusei	11	11 (100)	11 (100)	0 (0)	0 (0)
lambica	2	2 (100)	0 (0)	2 (100)	0 (0)
lusitaniae	16	16 (100)	12 (75)	4 (25)	0 (0)
parapsilosis	10	7 (70)	3 (30)	4 (40)	3 (30)
tropicalis[a]	29	24 (83)	12 (41)	12 (41)	5 (17)
Totals	94	84 (89)	45 (48)	39 (41)	10 (11)

[a]No ID given for 1 isolate

[b]Number of strains (%)

[c]Similarity index

Table 6.3. Summary of CFA Analysis for Yeast Identification: Non-*Candida* Species

Yeast Species	Tested	Overall	No 2nd Choice	w/Multi ID or SI < 0.5	Discrepant
			SI[c] > 0.5		
		Agreement			
Cryptococcus neoformans	19	14 (74)[b]	3 (16)	11 (58)	5 (26)
Rhodotorula spp.	2	2 (100)	0 (0)	2 (100)	0 (0)
Saccharomyces cerevisiae	12	12 (100)	11 (92)	1 (8)	0 (0)
Torulopsis glabrata	32	32 (100)	32 (100)	0 (0)	0 (0)
Trichosporon beigelii[a]	11	7 (64)	1 (9)	6 (55)	4 (36)
Totals	76	67 (88)	47 (62)	20 (26)	9 (12)

[a]No ID given for 1 isolate

[b]Number of strains (%)

[c]Similarity index

completely ascertained. The addition of more strains may invalidate the uniqueness previously described for a particular entry.

The expansion of the database to include more genera and species only further diminishes the utility of this method. To illustrate this point, a strain of *Candida krusei* was tested against the authors' narrowly defined database and the manufacturer's library that consists of 23 genera and 195 species or subspecies collected from a variety of environmental sources. In the former the isolate was exclusively identified as *Candida krusei* (SI 0.634) and no second choice. In the latter the following choices were listed as possible identifications: *Candida acidothermophilum* (SI 0.829), *Candida valida* (SI 0.526), and *Candida krusei* (SI 0.525).

APPLICATIONS

The identification of yeasts based exclusively on CFA analysis generally is not feasible. This is especially true when applied to a large number of phylogenetically diverse groups. These conclusions are consistent with opinions reached in previous reports (Augustyn et al. 1992; Lechevalier and Lechevalier 1988; Moss et al. 1982; Shaw 1974). When a narrowly defined group is considered, such as those from a particular environmental source, or capable of growth at 37°C, the utility of this technique is improved.

Although impractical as an independent identification technique, CFA analysis has been successively incorporated into industrial processes. Augustyn and Kock (1989) were able to distinguish between strains of *Saccharomyces cerevisiae*, and thus concluded that it is an easy and cheap method to determine the causes of stuck fermentations. Cellular fatty acid composition combined with morphologic data has been used as a rapid means to monitor fungal contamination in a pilot bioprotein plant (Botha and Kock 1993). Based on the absence of 18:3, 69 percent of the potential fungal contaminants could be eliminated. The remainder required morphologic investigation. The speed and simplicity inherent with the MIS system may prove to be instrumental in facilitating similar industrial applications, especially by increasing throughput.

As a routine identification tool, automated CFA analysis should be useful in providing preliminary taxonomic grouping. Large diagnostic centers handling a high volume of cultures may find this feature particularly attractive. A possible approach to yeast identification that incorporates the MIS in a clinical laboratory is described in Figure 6.2. Identifications based primarily on CFA analysis should be considered only if a single choice is listed with an SI ≥ 0.5.

Modification of the existing MIS procedure, such as alteration of the growth media, or substitution of the derivitization reagent, may improve the applicability to certain discrete taxonomic groups. The overall utility is not expected to significantly improve as a result of the inherent limitations previously discussed.

Figure 6.2. Possible application of CFA analysis.

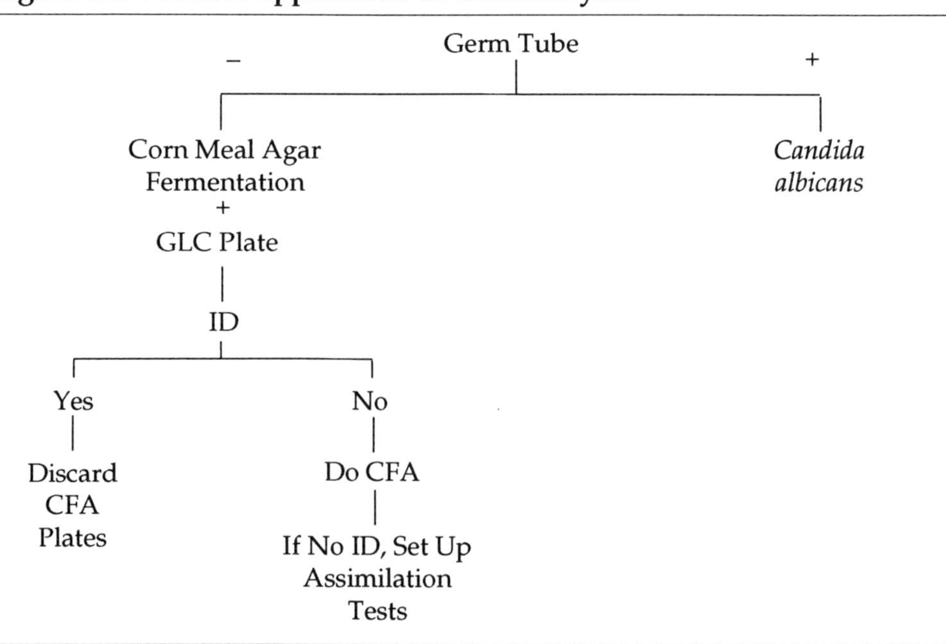

REFERENCES

Augustyn, O. P. H. 1989. Differentiation between yeast species, and strains within a species, by cellular fatty acid analysis. 2. *Saccharomyces cerevisiae. S. Afr. J. Enol. Vitic.* 10:8–17.

Augustyn, O. P. H, and J. L. F. Kock. 1989. Differentiation of yeast species, and strains within a species, by cellular fatty acid analysis. 1. Application of an adapted technique to differentiate between strains of *Saccharomyces cerevisiae. J. Microbiol. Methods.* 10:9–23.

Augustyn, O. P. H., D. Ferreira, and J. L. F. Kock. 1991. Differentiation between yeast species, and strains within a species by cellular fatty acid analysis. 4. *Saccharomyces* sensu stricto, *Hanseniaspora, Saccharomycetes,* and *Wickerhamiella. System. Appl. Microbiol.* 14:324–334.

Augustyn, O. P. H., J. L. F. Kock, and D. Ferreira. 1990. Differentiation between yeast species, and strains within a species, by cellular fatty acid analysis. 3. *Saccharomyces* sensu lato, *Arxiozyma,* and *Pachytichospora. System. Appl. Microbiol.* 13:44–55.

Augustyn, O. P. H., J. L. F. Kock, and D. Ferreira. 1992. Differentiation between yeast species, and strains within a species, by cellular fatty acid analysis. 5. A feasible technique? *System. Appl. Microbiol.* 15:105–115.

Botha, A., and J. L. F. Kock. 1993. Application of fatty acid profiles in the identification of yeasts. *Int. J. Food Microbiol.* 19:39–51.

Brondz, I., and I. Olsen. 1986. Review. Microbial chemotaxonomy. Chromatography, electrophoresis, and relevant profiling techniques. *J. Chromatogr.* 379:367–411.

Brondz, I., and I. Olsen. 1990. Multivariate analyses of cellular carbohydrates and fatty acids of *Candida albicans, Torulopsis glabrata,* and *Saccharomyces cerevisiae. J. Clin. Microbiol.* 28:1854–1857.

Brondz, I., and I. Olsen. 1992. Review of chemosystematics: Multivariate approaches to oral bacteria and yeasts. *Acta. Odontol. Scand.* 50:321–336.

Brondz, I., I. Olsen, and M. Sjöström. 1989. Gas chromatographic assessment of alcoholyzed fatty acids from yeasts: A new chemotaxonomic method. *J. Clin. Microbiol.* 27:2815–2819.

Bryn, K., and E. Jantzen. 1982. Analysis of lipopolysaccharides by methanolysis, trifluoroacetylation, and gas chromatography on a fused-silica capillary column. *J. Chromatogr.* 240:405–413.

Chen, E. C. H. 1981. Fatty acid profiles of some cultured and wild yeasts in brewery. *J. ASBC* 39:117–124.

Cottrell, M., J. L. F. Kock, P. M. Lategan, P. J. Botes, and T. J. Britz. 1985. The long-chain fatty acid compositions of species representing the genus *Kluyveromyces. FEMS Microbiology Letters.* 30:373–376.

Gangopadhyay, P. K., H. Thadepalli, I. Roy, and A. Ansari. 1979. Identification of species of *Candida, Cryptococcus* and *Torulopsis* by gas-liquid chromatography. *J. Infect. Dis.* 140:952–958.

Gunasekaran, M., and W. T. Hughes. 1980. Gas-liquid chromatography: A rapid method for identification of different species of *Candida*. *Mycologia* 72:505–511.

Kobayashi, K., H. Suginaka, and I. Yano. 1987. Analysis of fatty acid composition of *Candida* species by gas-liquid chromatography using a polar column. *Microbios* 51:37–42.

Kock, J. L. F., M. Cottrell, and P. M. Lategan. 1986. A rapid method to differentiate between five species of the genus *Saccharomyces*. *Appl. Microbiol. Biotechnol.* 23:499–501.

Kock, J. L. F., P. M. Lategan, P. J. Bates, and B. C. Viljoen. 1985. Developing a rapid statistical identification process for different yeast species. *Microbiol. Methods* 4:147–154.

Lategan, P. M. 1981. Characterization of pathogenic species of *Candida* by gas chromatography: Preliminary findings. *J. Med. Microbiol.* 14:219–222.

Lechevalier, M. P. 1982. Lipids in bacterial taxonomy. *In CRC handbook of microbiology,* edited by A. I. Laskin, and H. A. Lechevalier. Boca Raton, FL: CRC Press Inc.

Lechevalier, H., and M. P. Lechevalier. 1988. Chemotaxonomic use of lipids—an overview. *In Microbial lipids,* vol. 1, edited by C. Ratledge, and S. G. Wilkinson, London: Academic Press Inc.

Marumo, K., and Y. Aoki. 1990. Discriminant analysis of cellular fatty acids of *Candida* species, *Torulopsis glabrata,* and *Cryptococcus neoformans* determined by gas-liquid chromatography. *J. Clin. Microbiol.* 28:1509–1513.

Merz, W. G., C. Connelly, and P. Hieter. 1988. Variation of electrophoretic karyotypes among clinical isolates of *Candida albicans. J. Clin. Microbiol.* 26:842–845.

Merz, W. G., U. Khazan, M. A. Jabra-Rizk, L. Wu, G. J. Osterhout, and P. F. Lehmann. 1992. Strain delineation and epidemiology of *Candida (Clavispora) lusitaniae. J. Clin. Microbiol.* 30:449–454.

Miller, L. T. 1982. Single derivitization method for routine analysis of bacterial whole-cell fatty acid methyl esters, including hydroxy acids. *J. Clin. Microbiol.* 16:584–586.

Morgan, M. A. 1984. Gas-liquid chromatography in the clinical microbiology laboratory. *Lab. Med.* 15:544–550.

Moss, C. W. 1981. Gas-liquid chromatography as an analytical tool in microbiology. *J. Chromatogr.* 203:337–347.

Moss, C. W. 1985. Uses of gas-liquid chromatography and high-pressure liquid chromatography in clinical microbiology. *In Manual of clinical microbiology,* 4th ed., edited by E. H. Lennette, A. Balows, W. J. Hausler, Jr., and H. F. Shadomy. Washington, DC: American Society for Microbiology.

Moss, C. W., S. B. Dees, and G. O. Guerrant. 1980. Gas-liquid chromatography of bacterial fatty acids with a fused-silica capillary column. *J. Clin. Microbiol.* 12:127–130.

Moss, C. W., T. Shinoda, and J. W. Samuels. 1982. Determination of cellular fatty acid composition of various yeasts by gas-liquid chromatography. *J. Clin. Microbiol.* 16:1073–1079.

Moss, C. W., P. L. Wallace, D. G. Hollis, and R. E. Weaver. 1988. Cultural and chemical characterization of CDC groups EO-2, M-5, and M-6, *Moraxella (Moraxella)* species, *Oligella urethralis, Acinetobacter* species, and *Psychrobacter immobilis. J. Clin. Microbiol.* 26:484–492.

Olsen, I. 1990. Chemotaxonomy of yeasts. *Acta. Odontol. Scand.* 48:19–25.

Osterhout, G. J., and W. G. Merz. 1990. Role of cellular fatty acids (CFA) in the identification of yeasts in a clinical mycology laboratory, abstract F-8. *Abstr. 90th Annu. Meet. Am. Soc. Microbiol.* Washington, DC: American Society for Microbiology.

Osterhout, G. J., V. H. Shull, and J. D. Dick. 1991. Identification of clinical isolates of Gram-negative nonfermentative bacteria by an automated cellular fatty acid identification system. *J. Clin. Microbiol.* 29:1822–1830.

Osterhout, G. J., L. Wood, and W. G. Merz. 1989. Cellular fatty acids (CFA) analysis for identification of medically important yeast, abstract F-49. *Abstr. 89th Annu. Meet. Am. Soc. Microbiol.* Washington, DC: American Society for Microbiology.

Oyaizu, H., and K. Komagata. 1983. Grouping of *Pseudomonas* species on the basis of cellular fatty acid composition and the quinone system with special reference to the existence of 3-hydroxy fatty acids. *J. Gen. Appl. Microbiol.* 29:17–40.

Rattray, J. B. M. 1988. Yeasts. *In Microbial lipids*, vol. 1, edited by C. Ratledge, and S. G. Wilkinson. London: Academic Press, Inc.

Rattray, J. B. M., A. Schibeci, and D. K. Kidby. 1975. Lipids in yeasts. *Bact. Rev.* 39:197–231.

Roboz, J., E. Nieves, and J. F. Holland. 1990. Separation and quantification by gas chromatography-mass spectrometry of arabinitol enantiomers to aid the differential diagnosis of disseminated candidiasis. *J. Chromatogr.* 500:413–426.

Shaw, N. 1974. Lipid composition as a guide to the classification of bacteria. *Adv. Appl. Microbiol.* 17:63–108.

Stager, C. E., and J. R. Davis. 1992. Automated systems for identification of microorganisms. *Clin. Microbiol. Rev.* 5:302–327.

Switchenko, A. C., C. G. Miyada, T. C. Goodman, T. J. Walsh, B. Wong, M. J. Becker, and E. F. Ullman. 1994. An automated enzymatic method for measurement of D-arabinitol, a metabolite of pathogenic *Candida* species. *J. Clin. Microbiol.* 32:92–97.

Veys, A. W., W. Callewaert, E. Waelkens, and K. V. D. Abbeele. 1989. Application of gas-liquid chromatography to the routine identification of nonfermenting gram-negative bacteria in clinical specimens. *J. Clin. Microbiol.* 27:1538–1542.

Viljoen, B. C., and J. L. F. Kock. 1989. A taxonomic study of the yeast genus *Candida* Berkhout. *Syst. Appl. Microbiol.* 12:91–102.

Viljoen, B. C., J. L. F. Kock, and K. Thoupou. 1989. The significance of cellular long-chain fatty acid compositions and other criteria in the study of the relationship between sporogenous Ascomycetes species and asporogenous *Candida* species. *Syst. Appl. Microbiol.* 12:80–90.

Viljoen, B. C., J. L. F. Kock, and T. J. Britz. 1988. The significance of long-chain fatty acid composition and other phenotypic characteristics in determining relationships among some *Pichia* and *Candida* species. *J. Gen. Microbiol.* 134:1893–1899.

Viljoen, B. C., G. A. Dykes, M. Callis, and A. vonHolly. 1993. Yeasts associated with Vienna sausage packaging. *Int. J. Food Microbiol.* 18:53–62.

Viljoen, B. C., J. L. F. Kock, H. B. Muller, and P. M. Lategan. 1987. Long-chain fatty acid compositions of some asporogenous yeasts and their respective ascosporogenous states. *J. Gen. Microbiol.* 133:1019–1022.

Wejman, A. C. M. 1979. Carbohydrate composition and taxonomy of *Geotrichum, Trichosporon,* and applied genera. *Antonie van Leeuwenhoek J. Microbiol.* 45:119–127.

Wejman, A. C. M., and L. Rodrigues de Miranda. 1988. Carbohydrate patterns of *Candida, Cryptococcus,* and *Rhodotorula* species. *Antonie van Leeuwenhoek J. Microbiol.* 54:535–543.

Welch, D. F. 1991. Applications of cellular fatty acid analysis. *Clin. Microbiol. Rev.* 4:422–438.

Wong, B., and M. Castellanos. 1989. Enantioselective measurement of the *Candida* metabolite D-arabinitol in human serum using multidimensional gas chromatography and a new chiral phase. *J. Chromatogr.* 495:21–30.

7

Probe Technology and Automation

Joseph O. Falkinham, III
Virginia Polytechnic Institute and State University

Probes are molecules that are added to samples for either identification, detection, or enumeration of microorganisms or viruses. Usually, the target is a component of a microorganism or virus, such as DNA, RNA, protein, polysaccharide, or lipid. The probe is capable of specifically reacting with the target molecule. Following the probe-target reaction, the complex is detected, usually by labeling the probe, and the amount measured by some means. Probes can be used following isolation of a microorganism, virus, or component from a sample or can be used directly with the sample without some intervening isolation step.

A variety of molecules have been used as probes, including nucleic acids (RNA or DNA) and antibodies. Nucleic acid probes are used for the detection and enumeration of DNA or RNA sequences that are components of microorganisms or viruses. DNA probe technology is illustrated in Figure 7.1. Targets for antibody probes can be whole microorganisms or viruses or their unique protein, carbohydrate, polysaccharide, or lipid components. The basis for the formation of the probe-target complexes is specific binding. Single-stranded nucleic acid probes are capable of forming stable hybrids with single-stranded target molecules through hydrogen bonds between complementary bases (i.e., adenine and thymine [or uracil] and guanine and cytosine). Antibody molecules can also bind specifically to target compounds to form stable complexes.

The major focus of this review of probe technology shall be on nucleic acid probes because that technology can be combined with nucleic acid amplification using the polymerase chain reaction (PCR) technology (Mullis 1987; Mullis et al. 1987). This technology can result in million- to billionfold increases (i.e., amplification) of individual target sequences or probe-target complexes, resulting in

Figure 7.1. DNA probe technology.

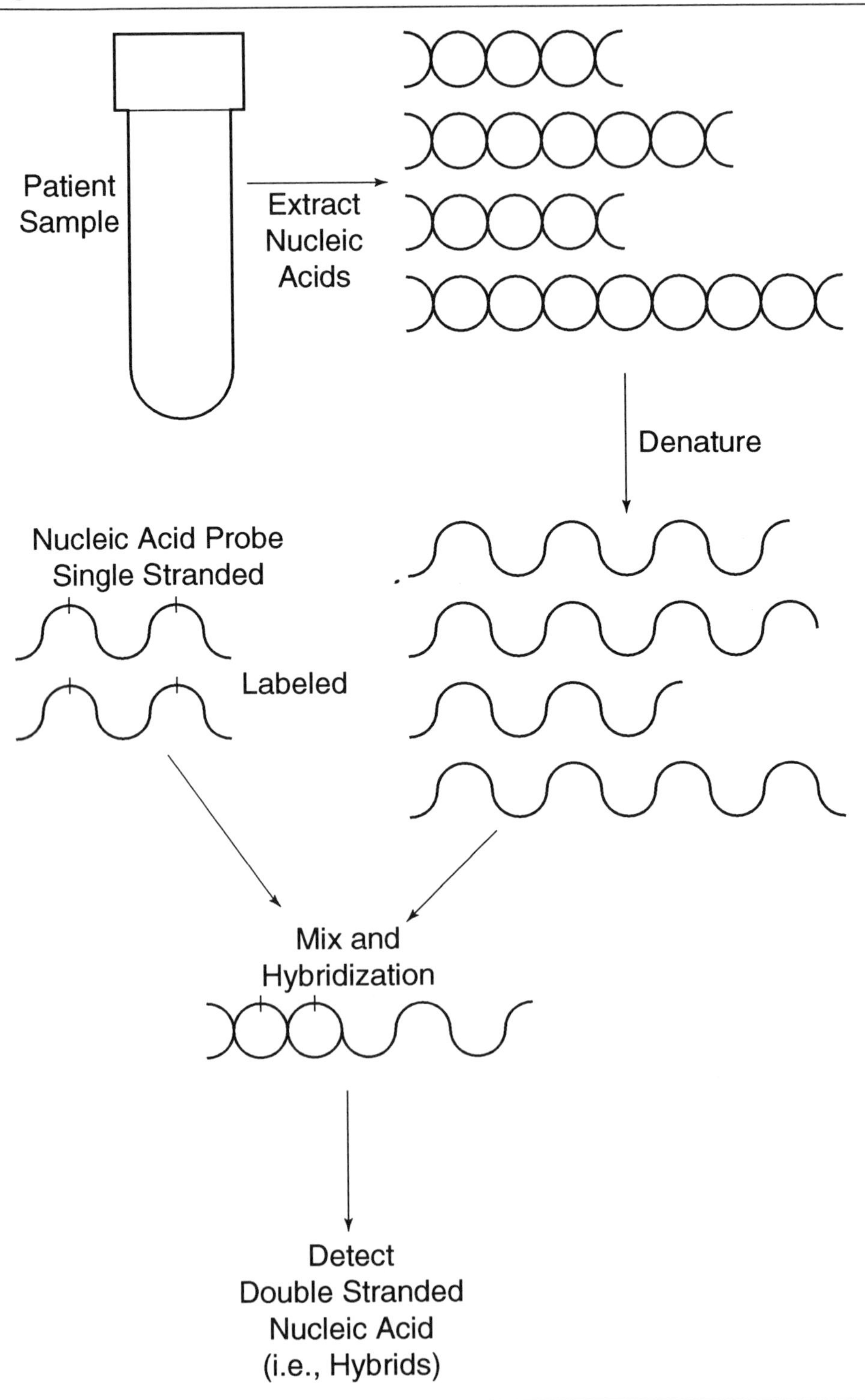

significant increases in sensitivity of detection. There is no analogous method to amplify protein, polysaccharide, or lipid components of target microorganisms or viruses, with the exception of growth.

ADVANTAGES AND DISADVANTAGES OF PROBE TECHNOLOGY

There are a number of clear advantages to the employment of probe technology for the detection or enumeration of specific microorganisms or viruses. First, sensitive and specific probes can be developed for the identification, detection, or enumeration of microorganisms, viruses, or their components. Even in the presence of other materials, probes can detect picogram levels of specific targets. Further, probes have been developed that are capable of distinguishing between closely related targets (Macario and deMacario 1990).

Probe technology may result in significant time savings. Normally, identification, detection, or enumeration of microorganisms or viruses is based on their ability to replicate under suitable conditions. This requires additional time and can be quite expensive for microbes or viruses with fastidious growth requirements. By contrast, because probe-based identification, detection, or enumeration is based on the presence of a microorganism, virus, or particular component, the time required for growth is not required. The advantage of probe-based systems is most clearly seen in the use of probes for the detection of *Mycobacterium tuberculosis*. Normally, 4–8 weeks of incubation of culture media is required for detection of *Mycobacterium tuberculosis* colonies. By contrast, these colonies can be detected by nucleic acid probes in as short as 4 hours. If the detection of a microorganism involves concentration and microscopic examination (e.g., protozoal parasites), probe technology may shorten the time required by eliminating the concentration steps.

Probe technology offers clear advantages for the detection of microorganisms that cannot be cultivated conveniently in the laboratory (e.g., protozoal parasites). Probes do not require growth of the target for identification, detection, or enumeration. Further, probe technology can identify, detect, or enumerate all microbial cells, whether they are capable of growth or not. Thus, viable but unculturable bacterial pathogens (e.g., stressed fecal coliforms) can be identified, detected, or enumerated by probe technology. However, this aspect of probe technology may, in some instances, cause a problem. Dead organisms or viruses can be identified, detected, or enumerated by probe technologies. These dead or inactive organisms or viruses may not be of significance for an ongoing surveillance or quality assurance (QA) program.

Probe technology can have other disadvantages.

1. Probe technology raises the required level of sophistication for the detection of microorganisms, viruses, or their components. Consequently, employment of probe technology may require the purchase of additional equipment and training for personnel.

2. Probe technology may not provide the required level of sensitivity without some method of amplifying either the target or probe-target complex.

3. Probe-target reactions may be subject to interference. Probes may detect nonviable or inactive microorganisms or viruses whose presence is unimportant.

PROBE TECHNOLOGY BASICS

Probe technology is based on the binding of a molecule—the probe—with a particular microorganism, virus, or an individual and unique component, and the detection of the probe-target complex. Most commonly, probes are nucleic acid molecules for the detection of either DNA or RNA, or are antibody molecules for the detection of proteins, carbohydrates, polysaccharides, or lipids.

Sensitivity of probe technology is based on the strength of the signal provided by either the bound probe itself or the method of detection of the probe-target complex. Sensitivity is also directly influenced by the recovery of the target from a sample, the ability of the recovered target to react with the probe, and whether the probe-target reactions or the probe-target complex detection system are inhibited by components of the reaction mixture (Hames and Higgins 1985; Harlow and Lane 1988). Low recovery, target modification, and inhibition either of the probe-target reaction or of the generation of the signal from the detection system would result in false-negative reactions.

Specificity of probe technology is based on the ability of the probe to react with the unique target and to no other targets. The presence of related targets or nonspecific binding of the probe to components in the reaction mixture are expected to directly influence the specificity of a particular probe (Hames and Higgins 1985). Binding to related targets or nonspecific binding would result in misidentification, false-positive detection, or overestimates of numbers of target microorganisms or viruses.

NUCLEIC ACID PROBES

Nucleic acid probes are single-stranded DNA or RNA molecules. Detection is based on the formation of a "hybrid" between the nucleic acid probe and single-stranded DNA or RNA recovered from a microorganism or virus or a sample containing both (Macario and deMacario 1990). Hybridization is due to the ability of the single-stranded nucleic acid sequence (i.e., the probe) to form a stable, hydrogen-bonded, double-stranded molecule (i.e., the hybrid) with a single-stranded target sequence, because the probe's sequence is complementary to one strand of a portion of that DNA or RNA (i.e., the target). Nucleic acid probes can be long or short; short probes are preferred because of their rapid diffusion and low cost of synthesis (Anderson and Young 1985). However, long probes may provide more specificity of reaction. Because the conditions of the reaction between the nucleic acid probe and the target can be adjusted to require an exact match for hybrid formation, the reaction can be quite specific. Alternatively, conditions can be chosen to allow some degree of mismatches (i.e., not complementary) between the probe and the target nucleic acid sequences (Anderson and Young 1985; Young and Anderson 1985).

ANTIBODY PROBES

Antibodies are proteins produced by mammals that are capable of binding and forming complexes with different molecules, called antigens. Antigens can be proteins, polysaccharides, or lipids. Even molecules that are normally unable to elicit antibody formation in mammals can be made antigenic by coupling with another molecule, called a hapten. A variety of different antibodies, each able to bind and form a complex with a specific antigen, can be produced by a mammal. Since these antibody-forming cells cannot be propagated in a culture medium and a single reactive antibody molecule is preferred as a probe, a specific type of antibody-forming cell, a hybridoma, is employed for both the selection and production of the desired antigen-specific antibody probe (Harlow and Lane 1988).

CRITERIA FOR PROBE SELECTION

Whether one isolates a nucleic acid or antibody probe, the probe must meet certain criteria. The following are major criteria for probe selection.

- Sensitivity

- Specificity

- Interference

- Stability

Interference includes modification of either the probe or target, masking either the probe or target, inhibition of the probe-target reaction, and inhibition of the signal from the detection system. Stability includes both the stability of the probe before reaction (e.g., shelf life) and the stability of the probe, target, and probe-target complex in the reaction. If any are unstable, false-negative identification, detection, or underestimates of numbers of microorganisms or viruses will result.

The process of developing a probe involves not only defining the target of the probe, but identifying the conditions under which the probe-target reaction is to be performed and the complex detected. These conditions include the actual composition of the reaction mixture, the temperature of reaction, and the presence or absence of interfering microorganisms, viruses, their components, or other compounds in the sample or sample extract (Anderson and Young 1985; Young and Anderson 1985). Identification of the desired probe should be conducted under conditions that will mimic exactly the conditions under which the probe will be employed in practice. Otherwise, the sensitivity and specificity of the probe, measured during its identification and development, will be different, since the conditions directly influence the probe-target reaction.

ISOLATION OF PROBES

Isolation of Nucleic Acid Probes (DNA or RNA)

Nucleic acid probes, specific for a particular sequence, can be isolated or constructed in a variety of ways. Probes can be isolated by cloning and screening DNA sequences (Gopo et al. 1988), by hybrid selection (Kohne 1989), and by organic synthesis from sequence data (Falkinham 1994). In addition, probes are available commercially.

Nucleic acid probes should be tested to ensure that there are no cross-reactions with other target-related nucleic acid sequences that might be present in the reaction mixture or sample extract. Self-complementary regions within the probe should be avoided because they would prevent hybridization to the target. The length of nucleic acid probes influences sensitivity and specificity of detection. Assuming that the reaction conditions remain constant, long probes are usually more specific than short probes, because they require a longer complementary sequence for hybrid formation. Employment of long probes may result in more sensitive detection, compared to short probes, because a long probe may carry a greater signal for detection or offer more sequences for binding of a detection molecule. However, the advantage of long probes in carrying a stronger signal is balanced by the fact that long probes have a slower rate of diffusion to the target, especially a problem in filter hybridization (Anderson and Young 1985). Consequently, the rate of hybrid formation is lower and longer incubation is required. Finally, larger probes are more expensive to synthesize as compared to short probes.

Isolation of Antibody Probes

Upon the injection of a purified antigen into a mammal, white blood cells capable of forming an antibody that reacts with the antigen are stimulated to proliferate. Since each antibody-forming cell produces a single antibody and most antigens are complex, a variety of different antibodies are produced, each capable of binding to the antigen. This is called a polyclonal antibody. These antibody-forming cells are incapable of being cloned and grown in a culture medium. Consequently, rather than using a polyclonal antibody as a probe, the antibody-forming cells are fused with a type of malignant cell (i.e., multiple myeloma). The products of these fusions are cells, called hybridomas, which can be grown continuously in a culture medium and are capable of producing a single antibody, called a monoclonal antibody. Methods for immunizing animals with antigen, developing methods for screening for antibody production, producing and growing hybridomas, and producing monoclonal antibodies are provided in Harlow and Lane (1988). Following the production of monoclonal antibody-forming hybridomas, individual hybridomas are screened for both the level of production and specificity of reaction with antigen. Following identification of a hybridoma meeting the desired criteria, the monoclonal antibody can be purified (e.g., protein A or affinity column) and labeled with either I^{125} or enzymes (e.g., peroxidase or alkaline phosphatase). Methods for performing these manipulations are provided in Harlow and Lane (1988).

ISOLATION OF PROBE TARGET

Probes can be used directly on samples or can be employed following the recovery of some fraction. One method of fraction recovery is the growth of microorganisms (e.g., bacteria, yeast, or fungi) on a culture medium. Viruses can be grown on suitable host cells. The resulting microbial colonies or purified viruses can then be subjected to steps necessary for the isolation of the probe's target, free from interfering compounds. If the target is a surface molecule (e.g., protein, polysaccharide, or lipid), whole cells or virus may be a suitable substrate for the reaction with the probe. If the probe-detection system is sensitive enough, the microorganism or virus can be identified directly using the probe, without having to go through a growth stage. If the target (e.g., DNA or RNA) was not a surface accessible molecule, the target would require isolation.

Major reasons for subjecting the sample to some isolation step include (1) releasing the target molecule so it can react with the probe, (2) separating the target from molecules that might modify or mask the target, (3) separating the target from molecules that might inhibit or interfere with the probe-target reaction, and (4) separating the probe-target complex from molecules that might interfere with the probe-target complex detection system.

Target Nucleic Acid (DNA or RNA) Isolation

There are many approaches for the isolation of nucleic acids from different starting materials. These methods are outlined in Falkinham (1994) and details have been published for the isolation of nucleic acids from microbial cultures (Atlas and Bej 1994), patient specimens (Boom et al. 1990), food samples (Andersen and Omiecinski 1992), water samples (Holben et al. 1988; Knight et al. 1990; Somerville et al. 1989), and soil samples (Atlas et al. 1992). Primary choice of one method of isolation over another for a particular sample must be based principally on the yield of nucleic acid and the absence of interfering (i.e., false-positive) or inhibiting (i.e., false-negative) components. Care must be taken to avoid degradation of nucleic acids (either DNA or RNA) during the isolation. Because lysis of cells results in the release of DNases and RNases, it is common to include inhibitors of these enzymes in the lysis solutions. Inclusion of detergents, chaotropic agents, and protein-denaturing agents are other ways of reducing nucleic acid degradation. Two known inhibitors of nucleic acid isolation and hybridization reactions are humic acid (Tebbe and Vahjen 1993) and polysaccharides (Do and Adams 1991; Fang et al. 1992). Methods to remove both classes of compounds are provided by these authors. Finally, nucleic acid isolation methods should be rather simple and short, especially if a large number of samples is to be processed in either a stepwise or automated fashion.

Target Protein, Polysaccharide, Carbohydrate, and Lipid Isolation

Methods are quite varied for the isolation of proteins, polysaccharides, or lipids from microorganisms or viruses. A number of general approaches for isolation of these fractions are provided in Gerhardt et al. (1994). Development of methods for the isolation of any of these components should be guided by the original literature describing their isolation and characterization. The same primary considerations as were listed for nucleic acid isolation methods must be included in any decisions concerning the isolation of proteins and other possible target molecules of microbes and viruses. Like nucleic acid isolation, care must be taken to avoid degradation of proteins, polysaccharides, and lipids during the isolation. Because lysis of cells results in the release of proteases and other polymer-degrading enzymes, it is necessary to include protease, polysaccharidase, or lipase inhibitors. If a large number of samples is to be processed, the isolation methods should be rather simple and short.

PROBE–TARGET REACTION

For some samples it is possible to perform the probe-target reaction without any separation or isolation of the target (e.g., whole microbial cells or virus). Alternatively, a simple release and recovery step (e.g., alkaline lysis of cells and denaturation of DNA) may be sufficient if no inhibitory compounds are present in the sample extract. Whether the target is directly available for reaction with the probe or the sample requires treatment before the probe-target reaction, the conditions must be adjusted to permit the probe to react with the target. Further, conditions must be suitable for the probe-target detection signal.

Nucleic Acid Hybridization

Following isolation of DNA, it must be made single stranded (i.e., denaturation) in order to permit hybridization with a nucleic acid probe. In many instances (alkaline lysis of cells), isolation and denaturation of DNA to single strands can occur in a single step. Ribosomal RNA (rRNA) is often preferred as target because it is single stranded and does not require denaturation (Lane and Collins 1991). Further, there are thousands of copies of the large and small rRNA molecules in cells (Kohne 1989). Thus, there is a natural amplification of the target nucleic acid sequences.

Stringency
Stringency refers to the extent of the hydrogen-bonded pairing between nucleic acid bases in the probe and target. The stability of double-stranded nucleic acid hybrids is influenced, in part, by the length of the duplex, the percentage of guanine + cytosine (G+C) pairs within the duplex, and the number of mismatches. Further, the stability of duplexes is influenced by the ionic conditions and temperature of the reaction (Anderson and Young 1985; Young and Anderson 1985). As a consequence, conditions can be chosen that require exact complementarity between bases in the probe and target for the probe's entire length. This would be a highly stringent reaction. By contrast, the conditions may be altered (e.g., lower temperature) that allow for a certain degree of mismatching between the probe and the target. Stringency may be important in samples that contain closely

related targets. Reactions of high stringency may be required to avoid hybridization of the probe to related target sequences (i.e., false-positive reactions).

Liquid and Filter Hybridization

One of the factors that is important to consider in developing a probe-based detection system is whether the probe-target reaction is to occur in solution (Young and Anderson 1985) or on a solid matrix (e.g., nitrocellulose paper; Twomey and Krawetz 1990; Anderson and Young 1985). For nucleic acid probes, there are a number of factors to consider (Table 7.1). For liquid hybridization reactions between the probe and the target under conditions of excess probe, the formation of a complex is proportional to the concentration of the target (Kohne 1989). Thus, liquid hybridization can be used for enumeration of microorganisms or viruses in a sample (Kohne 1989; Young and Anderson 1985). Unfortunately, there is no strict proportionality between complex formation and target numbers for solid phase hybridization (Anderson and Young 1985).

In cases where the probe is labeled for detection, one disadvantage of liquid hybridization is that either the unbound probe must be destroyed or removed before detection of the probe-target complexes, adding an additional step to the process. By contrast, probe removal is rather straightforward if the target nucleic acid is immobilized on a solid matrix. Another advantage of solid phase hybridization reactions is that the reaction product is available as a record. Accordingly, many solid phase reactions are used where presence/absence answers are required; liquid phase reactions are employed in instances where a specific numerical value is required.

Inhibition of Hybridization

Nucleic acid hybridization reactions are susceptible to inhibition due to the factors listed below.

- Degradation or destruction of target

- Modification of target

- Masking of target

- Modification of conditions for probe-target reaction

Table 7.1. Liquid and Solid Phase Probe-Target Reactions

Liquid	Solid
Proportional to Number	Not Proportional to Number
Separate Unbound Probe	Unbound Probe Easily Separated
Reaction Record Separate	Reaction is the Record

- Degradation or destruction of probe
- Modification of probe
- Masking of probe
- Prevention of probe-target reaction (false negative)
- Nonspecific probe binding (false positive)
- Inhibition of probe detection

Target or probe nucleic acid molecules could be destroyed, modified, or masked such that they would be unable to react with each other. Further, inhibitors of nucleic acid hybridization, including humic acid and polysaccharides, have been reported. Inhibition can be detected by the employment of positive controls. If inhibitors are present, an additional step of separating the target from the inhibitors is required.

Antibody-Antigen Reactions

Antibody-antigen complexes are held together by noncovalent bonds, with the contribution of the individual associations each contributing to the stability of the complex (Harlow and Lane 1988). Because the stability of these complexes, like nucleic acid hybrids, directly influences the ability to detect them, initial experiments must establish the parameters for stable antibody-antigen complex formation. As is the case for the employment of nucleic acid probes, an initial decision must be made concerning the format of the assay. For antibody-antigen reactions these include the following:

1. Antibody capture assays (under conditions of either antigen or antibody excess or under conditions of either antibody or antigen competition)
2. Antibody sandwich assays
3. Antigen capture assays using competition reactions

In addition, as was the case for nucleic acid probe-based systems, liquid and solid phase assay systems may be used (Harlow and Lane 1988).

DETECTION OF PROBE–TARGET COMPLEX FORMATION

Criteria to be considered for the detection of probe-target complexes are listed below.

- Equipment required
- Shelf life
- Sensitivity of detection
- Safety
- Ease of disposal of reagents

- Simplicity of use
- Desired result
- Necessary records
- Cost
- Technical skills available

The major decision concerns the desired level of sensitivity. Though radioisotopes offer a strong signal, their employment requires adequate provision for safe handling and disposal. Luminescent reactions can be employed to generate data, which can be saved using photographic film as a record. One limitation of luminescent reactions is that light may be generated in a flash, requiring sophisticated equipment. Shelf life is another consideration for selection of a method of detection. Some radioisotopes have half-lives that may limit their utility.

Problem of the Unreacted Probe

The major problem for the detection of any probe-target complex occurs when a labeled probe is employed. Provision must be made for the removal or destruction of unreacted, yet labeled, probe. Any probe that remains in the liquid reaction mixture or in the solid matrix will provide a signal, resulting in mistaken identification, false-positive detection, or an overestimate of the number of targets.

Detection of Nucleic Acid Hybridization

Both isotopic and nonisotopic methods for detection of hybridization between probe and target are available. A number of examples are listed below.

- Radioactive isotopes
- Colorimetric reactions
- Fluorescence
- Chemiluminescence
- Bioluminescence

Methods for isotopic labeling of both DNA and RNA nucleic acid probes (e.g., iodination, end labeling, and nick translation) are provided in Gerhardt (1994). An excellent review of nonisotopic labeling techniques is provided by Kricka (1992a) with detailed discussions in Kessler (1992) and Kricka (1992b). Nonisotopic labeling methods, including colorimetric (Leary et al. 1983), fluorescent (Woolford and Dale 1992), chemiluminescent (Arnold et al. 1989), and bioluminescent (Geiger 1992) techniques, are now available. Any of these techniques offers the same level of sensitivity of detection as do radioisotopes, without the hazards and requirements attendant on the employment of radioisotopes. Nonisotopic detection techniques can be employed with either liquid or solid phase reactions (e.g., Dubitsky et al. 1992). Also, there are two other advantages to the employment of nonisotopic labeled probes.

1. They are more stable than radioactively-labeled probes (i.e., no decay).

2. Evidence of hybridization can be gained rapidly, without the need for long incubations to generate autoradiographs.

Detection of Antibody-Antigen Formation

Antibody-antigen formation can be measured in a variety of ways. Antibodies can be labeled with radioisotopes (i.e., I^{125}) or nonisotopically with biotin or enzymes (e.g., peroxidase or alkaline phosphatase) (Harlow and Lane 1988). Some type of method must be employed to permit the separation of labeled antibody bound to antigen from unreacted, labeled antibody. One format that is employed widely is the 96-well, microtiter dish. One embodiment has antibody bound to the walls of each well, where it is free to react with antigen. Any antigen in the sample is bound to antibody and remains attached following washing. Enzyme-labeled antibody is next added, which binds to bound antigen. Any unreacted antibody can be washed away and labeled antibody probe can be measured by its ability to catalyze a reaction. This method is called enzyme-linked immunosorbent assay (ELISA).

SENSITIVITY AND SPECIFICITY OF PROBE TECHNOLOGY

Sensitivity of probe identification, detection, and enumeration systems is based on several factors including recovery of target, presence of interfering agents, and conditions of the probe-target reaction. Low recovery or masking of the target and the presence of agents preventing or reducing probe-target complex formation will reduce sensitivity (Tebbe and Vahjen 1993; Do and Adams 1991; Fang et al. 1992). One of the advantages of the detection of rRNA sequences is their enormous number in individual cells (at least 10,000 per individual *Escherichia coli* cell; Kohne 1989; Lane and Collins 1991). To date, sensitivity of either nucleic acid- or antibody-based probe technologies have still not reached the levels of sensitivity achieved by culture methods for microorganisms (Falkinham 1994).

Specificity of probe-based identification, detection, and enumeration systems is not expected to be influenced by low target recovery. However, probe-target complex formation will be directly influenced by the presence of target-related microbial cells, viruses, or components. This is particularly a problem for nucleic acid-based probe systems if the related sequences can form duplexes, because conditions for the hybridization reaction lack a high degree of stringency (Anderson and Young 1985; Young and Anderson 1985). The presence of such hybrids would result in incorrect identification, false-positive detection, or overestimates of the number of target microorganisms or viruses. Consequently, employment of probe-based identification, detection, and enumeration systems requires appropriate controls.

Although it may be difficult to isolate or construct specific nucleic acid probes, there are a wide variety of species-specific probes (Macario and deMacario 1990). For example, of 200 clones of a genomic library screened, 8 probes specific for *Salmonella typhimurium* were isolated by Gopo et al. (1988).

However, if a probe specific for a species and other closely related species exist and may be within samples, the isolation of a specific probe may be more difficult. For example, Ambrosio et al. (1991) were able to isolate only a single *Mycobacterium paratuberculosis*-specific probe, which failed to react with representatives of the closely related species, *Mycobacterium avium*, only after screening 650 clones of a genomic library.

QUANTITATION OF PROBE DATA

As noted above, a quantitative relationship between probe-target complex formation and target number can be attained in liquid hybridization (Young and Anderson 1985), and can be achieved under conditions of probe excess over a rather wide range. This requires that there be a linear relationship for the detection system. Though quantitation of radioactivity and luminescence meets that standard, other methods of labeling, such as enzyme conjugation, are semiquantitative at best (Harlow and Lane 1988; Kricka 1992b).

Quantitative relationships are more difficult to attain, employing systems in which the probe-target complex is immobilized. While such systems may be convenient and permit easy separation of unbound probe from probe-target complexes, the solid matrix may mask the signals generated by the detection system. Further, it is not clear that probe-target complex formation kinetics are strictly linear (Anderson and Young 1985; Hames and Higgins 1985). Target bound in the matrix may be less available to form a complex with the probe and there may be masking of immobilized targets by other target molecules at high target concentrations. Because of the need to identify nonlinearity in either liquid or solid phase reactions, one is well advised to carry out a number of probe-target reactions at different concentrations.

QUALITY ASSURANCE CONSIDERATIONS

Probe technology lends itself well to a QA program. A list of positive and negative controls, necessary for monitoring any probe-based detection system, is presented in Table 7.2. In addition, inclusion of probe-target dose-response samples provides for QA data and a standard curve for quantitation. In instances where a presence/absence result is generated, the QA program will monitor the highest dilution (i.e., lowest number or concentration) of target yielding a positive result. For those probe-based systems in which a continuous variable numerical result is generated, standard QA statistical methods can be employed.

AMPLIFICATION TECHNOLOGIES

Amplification methods have been added to nucleic acid-based probe detection systems, in order to increase sensitivity detection or enumeration. Either the target alone or the probe-target complex can be amplified. Target amplification

Table 7.2. Controls for Probe-Based Detection

Positive	Control Description	Negative
	Probe Absent	–
	Target Absent	–
+	Target-Probe Present	
+	Target-Probe Dose Response	

includes the growth of microorganisms, viruses, or their components under conditions suitable for their replication. Though amplification of proteins, polysaccharides, or lipids independently of the host cell or virus is not possible, it is possible to amplify nucleic acid sequences using the PCR. Either DNA (Fiss et al. 1992; Starnbach et al. 1989) or RNA (usually rRNA; Boddinghaus et al. 1990) can be amplified; the latter requires a step of reverse transcription to produce a double-stranded DNA from the RNA. Amplification of the probe-target complex can also be performed by either employing enzyme cycling or the PCR.

Amplification Methods

Target Amplification

The most common method for target amplification, which is not confined to its employment in probe-based detection and enumeration systems, is growth of microorganisms or viruses. Colony formation is an example of such amplification. The sample may be plated directly or can be subjected to some extraction or enrichment technique to yield cells, or virus, free from other materials (e.g., inhibitors of growth) before inoculation into a selective or nonselective medium. Nonselective media are preferred because there is less chance that the growth of the desired, target microorganism will be inhibited. Most selective media are inhibitory, even toward the selected microbe. However, with growth there is a chance that not only the target microorganism or virus will be amplified, but other, related microbes or viruses will grow. Thus, one problem with amplification methods is the attendant increase in sensitivity that can result in a decrease in specificity.

Target nucleic acid sequences can also be amplified employing the PCR (Mullis 1987; Mullis et al. 1987; Atlas and Bej 1994). DNA or RNA sequences that are unique to the target microorganism can be amplified using the cyclic PCR reaction. Amplification of the unique sequence is accomplished by the use of two primers for DNA replication, complementary to opposite strands of the sequence and flanking the ends of the sequence. PCR amplification of RNA (reverse transcription PCR or RT–PCR) requires an initial reverse transcription of a specific, unique RNA sequence with the enzyme reverse transcriptase.

PCR or RT–PCR amplification can be performed alone without a probe. In this application detection and enumeration of a microorganism is based on the appearance and yield of the amplified, unique double-stranded product. Alternatively, detection of the amplified PCR product can be performed with the aid of a nucleic acid probe, capable of hybridizing with a sequence within the amplified product. Detection in that case requires denaturation of double-stranded PCR product, but sensitivity of detection is increased.

Probe-Target Amplification

Enzyme amplification is initiated with the binding of an enzyme-linked probe to the target. Following removal of unbound probe, substrate for the enzyme is added and enzyme cycling can be carried out until the desired level of sensitivity is attained (Atlas and Bej 1994).

PCR amplification of a probe-target hybrid is performed as is PCR amplification of target DNA. The double-stranded probe-target nucleic acid duplex is used as the substrate for a PCR reaction. As was the case for PCR amplification of target DNA or RNA, direct detection of the amplified probe-target complex is possible. Alternatively, the amplified probe-target product can be detected by employing a nucleic acid probe.

Disadvantages of Amplification

Inclusion of an amplification cycle in a probe-based detection system has disadvantages that should be considered. In spite of the power of increased sensitivity as a consequence of amplification, loss of specificity can occur (Haff 1993). This can occur with any of the methods described above. Related microorganisms or viruses can grow under conditions selective for the growth of a particular target microbe or virus. Because probe-target reactions are, like all chemical reactions, subject to mass action principles, amplification of a related target can result in a false-positive reaction (i.e., 1 percent in the study of Clarridge et al. 1993), or a falsely positive value (Haff 1993). Inhibitors of the PCR reaction were found in 7.3 percent of samples in one study (Clarridge et al. 1993). In addition to possible loss of specificity, amplification leads to increased complexity of the measurement, time increases because of the length of the amplification cycle (i.e., as long as overnight), and new equipment requirements and costs. The increased cost or time requirements may argue against inclusion of an amplification cycle in nucleic acid–based probe identification, detection, or enumeration systems.

APPLICATIONS OF PROBE TECHNOLOGY

Probe technology has a wide range of applications for the continual QA of pharmaceutical and food products and to monitor water quality (Table 7.3). Probe technology has several advantages over the conventional culture of microorganisms and viruses, including the speed of obtaining a result, the identification, detection, or enumeration of damaged microorganisms or viruses that cannot grow (i.e., damaged), yet could grow under conditions achieved in product use.

Pharmaceutical and food products require testing for the presence of contaminating microorganisms or viruses. Both nucleic acid and antibody probes could be employed for identification, detection, or enumeration. In addition to whole

Table 7.3. Applications of Probe Technology

Pharmaceutical and Food Products
    Presence of Contaminating Microorganism(s)
    Presence of Contaminating Virus
    Presence of Contaminating Microbial or Viral Component(s)
    Presence of Toxin(s)
    Quality Assurance

Process Water Quality
    Presence and Enumeration of Microbial Pathogen(s)
        Fecal Coliforms
        Salmonella, Shigella, Vibrio
        Legionella, Mycobacterium

    Presence and Enumeration of Viral Pathogen(s)
        Hepatitis Virus(s)
        Enteric Virus(s)

    Presence of Toxins

microorganisms or viruses, a unique component of a microorganism or virus could be detected by probe technology. Components of particular interest includes toxins produced by bacteria (e.g., botulism), fungi (e.g., aflatoxin), or as part of a virus. For the direct detection of a component, the probe would necessarily be an antibody. For indirect detection the probe could be a nucleic acid that was specific for the gene encoding that toxin. Not only can the data concerning the presence or absence of a contaminating microorganism or virus be generated by probe technology, but that data can be employed in a QA program to document good manufacturing practices.

Water quality monitoring is another process where probe technology has been applied (Schmidt et al. 1991). Nucleic acid or antibody probes specific for those microorganisms routinely identified and enumerated (i.e., fecal coliforms) could be employed with significant savings in time before detection. This may be of particular advantage under circumstances where large rainwater flows may prevent a water treatment system from producing effluent of adequate quality. Rather than wait the required 24–48 hours incubation for detection of fecal coliforms, probes could be employed to provide results within hours of sample collection. The same type of probe-based identification, detection, and enumeration system could be employed for other water-borne human pathogens, including the causative agents of diarrheal diseases (e.g., *Salmonella*; Gopo et al. 1988) and emerging water-borne pathogens (e.g., *Legionella*; Starnbach et al. 1989). Probe-based systems for the detection and enumeration of water-borne viral pathogens (e.g., hepatitis virus) are particularly attractive targets for development. Mammalian enteric viruses require sophisticated culture systems for their detection, identification, and enumeration. Further, incubation periods are rather long. As was pointed out for fecal coliforms, rapid recognition of the presence of enteric viruses in water may prevent widespread instances of water-borne disease outbreaks.

AUTOMATION OF PROBE TECHNOLOGY

Probe technology, whether involving nucleic acid or antibody probes, lends itself to automation, because the probes can be treated as individual reactants. Thus, addition of the probe is analogous to the addition of a particular catalyst or component of a reaction. An automated probe-based system could be developed for continuous, and perhaps even real-time, identification, detection, or enumeration of microorganisms, viruses, or their components as part of pharmaceutical or food production or water quality monitoring. Incidently, an instrument for automated probe identification, detection, or enumeration could be developed for the market as well.

Criteria for Automation of Probe Technology

Criteria for automation of probe technology are listed below.

- Frequency of sampling

- Required data

- Required data format

- Required sensitivity

- Required specificity

- Possible interference

- Cost constraints

These are not a great deal different from those considered for development of a probe-based system. In fact, before development of a probe-based identification, detection, or enumeration system, consideration of its ultimate employment, including whether the probe will be used in an automated or manual system, must be made.

Automation would be desirable in cases where a high rate of material flow (e.g., pharmaceutical or food product, water, or effluent) requires a high frequency of sample collection and analysis. These situations would be met where a process is operated continuously over time. Automated probe technology is advantageous in cases where rapid responses to the presence of contaminants in a product or water are required, especially if real-time data for the presence of a contaminant could be provided, like pH measurements.

In addition to considering the frequency of sampling required, automation of probe technology requires knowledge of the required data and its format. For example, a probe-based system for detection would have fewer requirements than one that was needed to provide exact numbers of contaminants. Knowledge of the desired sensitivity and specificity of measurement is required. The former requires knowledge of the composition of the sample and the presence or absence in interfering agents. Further, it may require knowledge of regulatory guidelines (e.g., fecal coliforms per 100 ml water). Consideration of specificity targets requires knowledge of the presence of cross-reactive microorganisms, viruses, or their components. Finally, as is always the case, cost is an important factor to consider. Savings of technical time must be more than the cost of operating the automated system and the amortization of the development costs over some reasonable period of time.

Steps of Automation

The steps of an automated probe-based identification, detection, or enumeration system are outlined below.

- Sample removal
- Amplification (if required)
- Cell lysis and target release
- Target recovery
- Target amplification (if required)
- Probe-target reaction
- Removal of unbound probe
- Detection of probe-target complexes
- Data analysis (including controls)

The entire process can include steps for amplification. As was noted above, amplification of the target or of the probe-target complex can be performed. However, amplification requires time and that time may prevent the generation of results in the desired time frame.

Automation of a probe-based system could involve either liquid probe-target reactions or solid-phase reactions. The former system could be similar to autoanalyzers used in clinical laboratories for the measurement of serum components (e.g., glucose). Solid phase reactions could be performed by immobilizing a sample on a paperlike endless belt that could be drawn through different reaction vessels. Separation of a labeled probe from the probe-target complex would still be required.

A future direction of probe-based systems would be to construct a sensor in which the probe and the reagents for detection of the probe-target complex would be present. Such a probe would provide continuous, real-time monitoring as is possible for pH and glucose.

ACKNOWLEDGMENTS

The author acknowledges the support of the National Institutes of Allergy and Infectious Disease for research conducted in his laboratory and to the employees of Gen-Probe, Inc., for their willingness to educate the author about the practical aspects of nucleic acid probe development.

REFERENCES

Andersen, M. R., and C. J. Omiecinski. 1992. Direct extraction of bacterial plasmids from food for polymerase chain reaction amplification. *Appl. Environ. Microbiol.* 58:4080–4082.

Anderson, M. L. M., and B. D. Young. 1985. Quantitative filter hybridization. In *Nucleic acid hybridization*, edited by B. D. Hames, and S. J. Higgins. Oxford: IRL Press.

Arnold, L. J., Jr., P. W. Hammond, W. A. Wiese, and N. C. Nelson. 1989. Assay formats involving acridinium-ester-labeled DNA probes. *Clin. Chem.* 35:1588–1594.

Atlas, R. M., and A. K. Bej. 1994. Polymerase chain reaction. In *Methods for general and molecular bacteriology*, edited by P. Gerhardt. Washington, DC: American Society for Microbiology.

Atlas, R. M., G. Sayler, R. S. Burlage, and A. K. Bej. 1992. Molecular approaches for environmental monitoring of microorganisms. *BioTechniques* 12:706–717.

Boddinghaus, B., R. Rogall, R. Flohr, H. Blocker, E. C. Bottger. 1990. Detection and identification of mycobacteria by amplification of rRNA. *J. Clin. Microbiol.* 28:1751–1759.

Boom, R., C. J. A. Sol, M. M. M. Salimans, C. L. Jansen, P. M. E. Wertheim-vanDillen, and J. van der Noordaa. 1990. Rapid and simple method for purification of nucleic acids. *J. Clin. Microbiol.* 28:495–503.

Clarridge, III, J. E., R. M. Shawar, T. M. Shinnick, and B. B. Plikaytis. 1993. Large-scale use of polymerase chain reaction for detection of *Mycobacterium tuberculosis* in a routine mycobacteriology laboratory. *J. Clin. Microbiol.* 31:2049–2056.

Do, N., and R. P. Adams. 1991. A simple technique for removing plant polysaccharide contaminants from DNA. *BioTechniques* 10:1331–1334.

Dubitsky, A., J. Brown, and H. Brandwein. 1992. Chemiluminescent detection of DNA on nylon membranes. *BioTechniques* 13:392–400.

Falkinham, III, J. O. 1994. Nucleic acid probes. In *Methods for general and molecular bacteriology*, edited by P. Gerhardt. Washington, DC: American Society for Microbiology.

Fang, G., S. Hammar, and R. Grumet. 1992. A quick and inexpensive method for removing polysaccharides from plant genomic DNA. *BioTechniques* 13:52–56.

Fiss, E. H., F. F. Chehab, and G. F. Brooks. 1992. DNA amplification and reverse dot blot hybridization for detection and identification of mycobacteria to the species level in the clinical laboratory. *J. Clin. Microbiol.* 30:1220–1224.

Geiger, R. E. 1992. Detection of alkaline phosphatase by bioluminescence. In *Nonisotopic DNA probe techniques*, edited by L. J. Kricka. San Diego: Academic Press, Inc.

Gerhardt, P. 1994. *Methods for general and molecular bacteriology.* Washington, DC: American Society for Microbiology.

Gopo, J. M., R. Mellis, E. Filipska, R. Memeveri, J. Filipski. 1988. Development of a *Salmonella*-specific biotinylated DNA probe for rapid routine identification of *Salmonella. Mol. Cell Probes* 2:271–279.

Haff, L. 1993. PCR: applications and alternative technologies. *BioTechnology* 11:938–939.

Hames, B. D., and S. J. Higgins. 1985. *Nucleic acid hybridization.* Oxford: IRL Press.

Harlow, E., and D. Lane. 1988. *Antibodies. A laboratory manual.* New York: Cold Spring Harbor Laboratory.

Holben, W. E., J. K. Jansson, B. K. Chelm, and J. M. Tiedje. 1988. DNA probe method for detection of specific microorganisms in the soil bacterial community. *Appl. Environ. Microbiol.* 54:703–711.

Kessler, C. 1992. Nonradioactive labeling methods for nucleic acids. In *Nonisotopic DNA probe techniques*, edited by L. J. Kricka. San Diego: Academic Press, Inc.

Knight, I. T., S. Shults, C. W. Kaspar, and R. R. Colwell. 1990. Direct detection of *Salmonella* spp. in estuaries by using a DNA probe. *Appl. Environ. Microbiol.* 56:1059–1066.

Kohne, D. E. 1989. U.S. Patent 4,851,330.

Kricka, L. J. 1992. *Nonisotopic DNA probe techniques.* San Diego: Academic Press, Inc.

Kricka, L. J. 1992. Nucleic acid hybridization test formats: strategies and applications. In *Nonisotopic DNA probe techniques*, edited by L. J. Kricka. San Diego: Academic Press, Inc.

Lane, D. J., and M. L. Collins. 1991. Current methods for detection of DNA/ribosomal RNA hybrids. In *Rapid methods and automation in microbiology and immunology*, edited by A. Vaheri, R. C. Tilton, and A. Balows. New York: Springer-Verlag.

Leary, J. J., D. J. Brigati, and D. C. Ward. 1983. Rapid and sensitive colorimetric method for visualizing biotin-labeled DNA probes to DNA or RNA immobilized to nitrocellulose bioblots. *Proc. Natl. Acad. Sci. USA* 80:4045–4049.

Macario, A. J. L., and E. C. deMacario. 1990. *Gene probes for bacteria.* New York: Academic Press, Inc.

Mullis, K. B. 1987. U.S. Patent 4,683,202.

Mullis, K. B., H. A. Erlich, N. Arnheim, G. T. Horn, R. K. Saiki, and S. J. Scharf. 1987. U.S. Patent 4,683,195.

Schmidt, T. M., E. F. DeLong, and N. R. Pace. 1991. Phylogenetic identification of uncultivated microorganisms in natural habitats. In *Rapid methods and automation in microbiology and immunology,* edited by A. Vaheri, R. C. Tilton, and A. Balows. New York: Springer-Verlag.

Somerville, C. C., I. T. Knight, W. L. Straube, and R. R. Colwell. 1989. Simple, rapid method for direct isolation of nucleic acids from aquatic environments. *Appl. Environ. Microbiol.* 55:548–554.

Starnbach, M. N., S. Falkow, and L. S. Tompkins. 1989. Species-specific detection of *Legionella pneumophila* in water by DNA amplification and hybridization. *J. Clin. Microbiol.* 27:1257–1261.

Tebbe, C. C., and W. Vahjen. 1993. Interference of humic acids and DNA extracted directly from soil in detection and transformation of recombinant DNA from bacteria and a yeast. *Appl. Environ. Microbiol.* 59:2657–2665.

Twomey, T. A., and S. A. Krawetz. 1990. Parameters affecting hybridization of nucleic acids blotted onto nylon or nitrocellulose membranes. *BioTechniques* 8:478–481.

Woolford, A. J., and J. W. Dale. 1992. Simplified procedures for detection of amplified DNA using fluorescent label incorporation and reverse probing. *FEMS Microbiology Letters* 99:3211–316.

Young, B. D., and M. L. M. Anderson. 1985. Quantitative analysis of solution hybridization. In *Nucleic acid hybridization,* edited by B. D. Hames, and S. J. Higgins. Oxford: IRL Press.

8

Methodology for Detection of Mycoplasmas in Cell Cultures Used to Produce Human Pharmaceuticals

Gerald K. Masover
Frances A. Becker

Genentech, Inc.

The first report of a mycoplasma contaminant in a cell culture was made by Robinson et al. (1956). Soon after that report, many workers using cell cultures in various types of research found mycoplasma contaminants in their cultures. At about that time the poliomyelitis vaccine was being produced in primary cell cultures, and there was concern regarding the safety of vaccines produced in this manner. As a result, in 1962 the U.S. Public Health Service established a requirement for mycoplasma testing of viral vaccines produced in cell cultures.

The requirement for mycoplasma testing has since been extended to all biologicals produced in cell cultures, including pure protein pharmaceuticals. The testing requirement for viral vaccines is given in the *Code of Federal Regulations* (21 CFR 610.30) and for all other biologicals made in cell lines in a document entitled *Points to Consider in the Characterization of Cell Lines Used to Produce Biologicals* (1993), which was issued by the FDA's Center for Biologics Evaluation and Research (CBER). This latter document provides some guidance about which tests for mycoplasma should be done and some minimal requirements for performance of the testing.

This chapter describes in detail two methods for detecting mycoplasmas in cell cultures used to produce human pharmaceuticals: a cultural method and a fluorescence staining method, referred to as the DNAF method because of its use of DNA-binding fluorochromes (DNAFs). These two methods constitute a mycoplasma testing program that satisfies the FDA's *Points to Consider*.

BASIC BIOLOGY OF MYCOPLASMAS

The trivial term *mycoplasma* is commonly used to describe microorganisms that make up a large group of prokaryotic cells that are phenotypically distinguishable from other bacteria by both the total lack of a bacterial cell wall and by their very small size (Razin and Freundt 1984, 740–742). The absence of a cell wall contributes to some of the unique properties of mycoplasmas, such as sensitivity to osmotic shock and detergents, resistance to cell wall-active antibiotics such as penicillin, and the formation of very small colonies that can have the appearance of a fried egg when viewed microscopically. The very small size of individual cells places mycoplasmas among the smallest free-living prokaryotes. However, they are sufficiently large and complex to be infected themselves by mycoplasma viruses (Maniloff 1992).

Recognition of the tiny characteristic colony that is made by mycoplasmas on agar is important for the diagnosis of mycoplasma infection, and the mechanism of colony formation should be understood. Individual mycoplasma cells are small enough (≈ 0.3 mm-diameter spheres) to penetrate the interstices of the fibrils in the agar preparation on which they are plated. They begin dividing within the agar, and the nascent colony grows as a spherical mass embedded in and among the agar fibrils just below the agar surface. When the edge of the spherical colony reaches the surface, it grows along the free water layer that exists on properly hydrated agar gels and creates an area of surface growth both wider and thinner than the subsurface growth. When the colony is viewed from either above or below, the observer sees a deep central area surrounded by a wider surface area. The deep growth and surface growth combined create what is frequently referred to as the fried egg appearance. Mycoplasma colonies range in size from approximately 50 to 500 µm in diameter (Razin and Oliver 1961; Razin 1983, 83–88). Colonies are best viewed with the aid of a microscope, and it is helpful to continually focus the microscope above and below the agar surface in order to appreciate the surface and subsurface character of the typical colony. Mycoplasma colony formation is represented diagrammatically in Figure 8.1 as originally reported by Razin and Oliver (1961). Typical colonies are shown in Figure 8.2.

Numerous external factors can alter the classical appearance of mycoplasma colonies, such as the gel strength of the agar, the crowding of colonies, and the moisture level during incubation of the plates. Some common artifacts, referred to as "pseudocolonies" (Hayflick 1965; Masover and Hayflick 1981), may be observed, depending on the inoculum composition and the agar formulation. Pseudocolonies are often comprised of crystalline materials deposited on the agar; occasionally, however, this artifact can be confused with atypical

Figure 8.1. The development of mycoplasma colonies on agar.
(A) Vertical section through the agar before inoculation: (a) free water film; (b) network of agar fibrils. (B) A drop containing a viable mycoplasma particle is placed on the agar. (C) Approximately 15 minutes after inoculation, the drop is absorbed by the agar, forming a slight swelling. (D) Approximately 3–6 hours after inoculation, the viable particle has penetrated into the agar. (E) Approximately 18 hours after inoculation, a small spherical colony of organisms intertwined in the agar fibrils has been formed below the agar surface. (F) Approximately 24 hours after inoculation, the colony approaches the agar surface. (G) Approximately 24–48 hours after inoculation, the growth spreads into the free water film forming the peripheral zone; (c) central zone; (d) peripheral zone. (Source: Razin, S., and O. Oliver. 1961. Morphogenesis of mycoplasma and bacterial L-form colonies. *J. Gen. Microbiol.* 24:225–237. Reprinted with permission.)

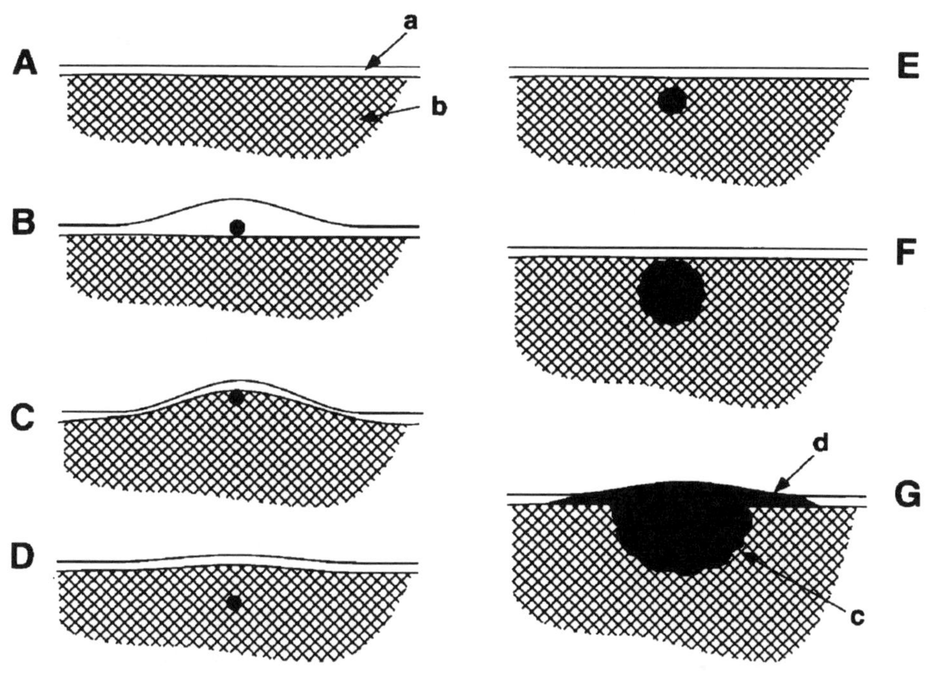

mycoplasma colonies. To prepare for performing mycoplasma testing, analysts should become familiar with the appearance of mycoplasma colonies made by different genera and species, as well as under different conditions, and develop a degree of expertise at the recognition of artifacts.

Since mycoplasmas have no cell wall, they are gram negative. Taxonomically, the absence of a cell wall separates mycoplasmas from other bacteria and puts them into a distinct class named Mollicutes (Latin, *mollis* = soft, *cutes* = skin). This

Figure 8.2. Typical mycoplasma colonies: (a) Colonies of *Mycoplasma hyorhinis*; (b) colonies of *Mycoplasma hominis*; (c) colonies of *Mycoplasma pneumoniae*. These colonies were growing on Modified Edward agar and were photographed using Kodak Tri-X pan 400 ASA film, a 2.5× Zeiss objective lens and a 10× Zeiss kpl eyepiece. Bar = 300 μm.

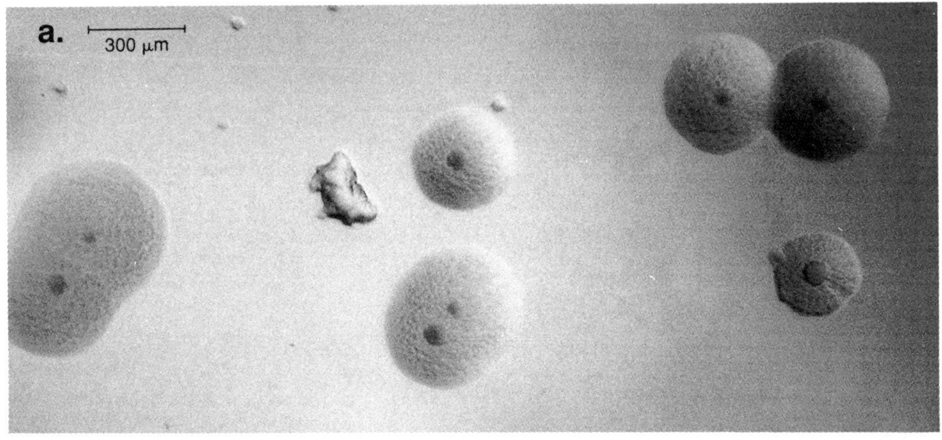

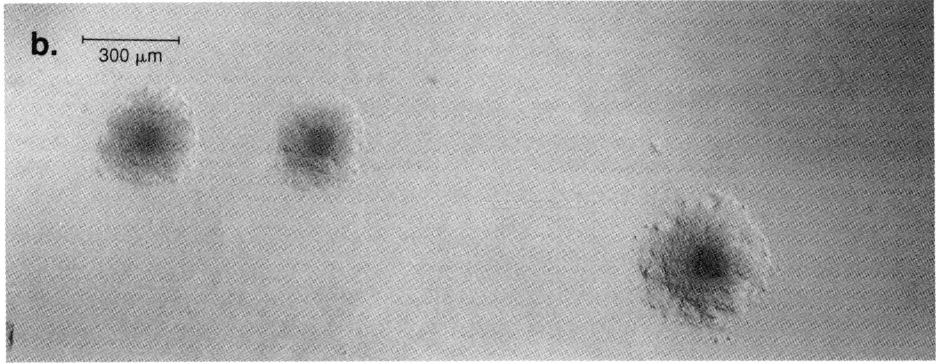

class is the only one in Division Tenericutes—one of four divisions in the kingdom Prokaryotae (Gibbons and Murray 1978).

The mycoplasma genome is a single, circular, double-stranded DNA molecule, which replicates in the same manner as the DNA of other bacteria. It is unique among prokaryotes in that it is very small and has a very low molecular percent of guanine + cytosine (G+C). Mycoplasma ribosomes and mechanisms for protein synthesis resemble those of other bacteria, and the ribosomes are susceptible to antibiotics that work at the 16S ribosomal subunit. A more thorough discussion of the molecular biology and genetics of these organisms is given by Razin (1985) and Maniloff et al. (1992).

The mycoplasmas that have been cultivated and identified are parasites of humans, animals, arthropods, and plants. The primary habitats of human and animal mycoplasmas are the mucous surfaces of the respiratory and genital tracts. In some animals mycoplasmas are found in the eyes, alimentary canal, mammary glands, and joints. The widespread use of cell cultures for biomedical research and other applications stimulated increased interest in the occurrence of mycoplasmas in cell cultures as well. The epidemiology of cell culture contaminant mycoplasmas was recently reviewed by Barile and Rottem (1993, 155–193). More thorough discussions on the biology of mycoplasmas can be found in comprehensive microbiological handbooks (Masover and Hayflick 1981; Razin 1991) or in special publications on the subject (Maniloff et al. 1992; Razin and Tully 1983; Tully and Razin 1983; Barile and Razin 1979; Tully and Razin 1995).

CULTURAL METHOD

Basis for the Method

Broth enrichment and demonstration of typical colonies on agar is a very powerful classical microbiological method because it takes advantage of the capacity of the subject microbial cells to replicate. In a broth culture individual cells may replicate to a high concentration that, alone, often provides diagnostic information by production of a pH change, seen as a medium color change or slight turbidity. More importantly, the increased concentration also increases the probability of selecting viable cells with a given sample volume. When a broth sample is subcultured onto agar and incubated, the mycoplasmas replicate to form a visible entity—a colony—from an invisible entity—a single viable cell or small clump of them (colony forming unit, CFU). This technique allows a much larger volume to be tested by inoculation into broth and later subculture to agar than would be feasible by direct inoculation onto agar alone. The subcultures must be performed at various times to allow amplification and recovery of viable cells of both rapid- and slow-growing organisms in the broth.

Preparation of Media

There is agreement in the mycoplasma literature that there is no single universal culture medium that will grow all species of Mollicutes well. Having stated this, it seems that the difficulties encountered with the growth of Mollicutes on agar relate most often to attempts at isolating them from natural hosts. However, tissue cultures are not natural hosts. Approximately 20 mycoplasma species have been reported as cell culture contaminants (Barile and Rottem 1993; 155–193), and all of them, with the exception of *Mycoplasma hyorhinis* DBS 1050, grow on the more commonly used mycoplasma medium formulations. The DBS 1050 strain of *Mycoplasma hyorhinis* is readily detected by the DNAF procedure and has been recently reported to be cultivable on agar by modification of the usual media and growth conditions (Del Giudice et al. 1980; Kotani et al. 1990; Gardella and Del Giudice 1995).

With respect to the use of antimicrobics in media, FDA guidance permits the use of penicillin. We have chosen to omit it in the interest of using the mycoplasma test as an adjunctive test for bacteria and fungi as well. Other antimicrobics such as thallium acetate, which often is included in media used to detect clinical mycoplasmas, are not permitted in media used for testing pharmaceutical cell cultures.

Two widely used mycoplasma media were recently compared using a quantitative growth promotion test. The results of this comparison showed that all strains tested, except *Mycoplasma hyorhinis* DBS 1050, grew on both Modified Edward and SP4 media, but that the SP4 medium appeared to give somewhat better growth for the common cell culture contaminant strains (Masover et al. 1994). The preparation of these two media, which are representative of the general types of media in use, is described here.

Modified Edward Medium

The majority of species classified as *Mycoplasmas* or *Acholeplasmas* will grow well on the type of medium originally devised by Edward (1947) and modified by Hayflick (1965). The essential and major nutrient constituents are beef heart infusion, peptone, fresh yeast extract, and unheated horse serum. The heart infusion, peptone, and a small amount of sodium chloride are generally supplied in one of the commercial PPLO (pleuropneumonia-like organisms) broth or agar preparations. Various minor constituents, such as specific sugars, selected amino acids, and pH indicators, are generally added to this formulation. Because mycoplasmas do not synthesize nucleic acid precursors, many medium formulations contain either native DNA or its precursors.

Formulations for unsupplemented and supplemented Modified Edward broth and agar media are given in Tables 8.1–8.4.

SP4 Medium

SP4 medium was originally developed as a culture medium for members of the genus *Spiroplasma* (spiral-shaped mollicutes that infect plants and insects) that could not be grown on either conventional mycoplasma medium or on conventional spiroplasma medium (Tully et al. 1977). It was subsequently shown to be useful in isolating other mollicutes from both clinical and cell culture samples. It differs from the Modified Edward type of medium mostly by (1) inclusion of one of the defined cell culture medium formulations, (2) reduction of the mycoplasma broth base preparation, (3) use of fetal bovine serum rather than horse serum, and (4) use of noble agar as the solidifying agent.

Table 8.1. Modified Edward Mycoplasma Broth Medium, Unsupplemented (for 1 liter supplemented medium)

PPLO broth without crystal violet (Difco 0554-01 or equivalent)	21 g
Phenol red 0.5% w/v aqueous stock solution	10 ml
Calf thymus DNA (Sigma D-1501 or equivalent)	0.02 g

Distilled water, sufficient quantity to make 692 ml

— Adjust pH to 7.4 ± 0.1 with 1 N NaOH or 1 N HCl.

— Sterilize by autoclaving at 121°C for 15 minutes.

— Store at 2–30°C.

— Expiration is 6 months from date of preparation.

Table 8.2 Modified Edward Mycoplasma Broth Medium, Supplemented (1 liter)

The following sterile components are combined aseptically:

Unsupplemented mycoplasma broth	692 ml
Glucose-arginine stock solution*	8 ml
Fresh yeast extract**	100 ml
Unheated horse serum	200 ml

— Aseptically adjust the pH of the complete broth to 7.4 ± 0.1, if necessary, by dropwise addition of sterile 1 N NaOH or 1 N HCl.

— Store at 2–8°C.

— Expiration is 90 days from the date of preparation.

*This is a filter-sterilized aqueous solution containing 25 percent w/v of glucose and 25 percent w/v of arginine HCl. It gives a final concentration of 0.2 percent w/v in the supplemented broth.

**Fresh yeast extract refers to a recently prepared hot water extract of live yeast cells. It has been found to be important for the isolation of some clinical specimens and is routinely used in many mycoplasma medium formulations (Hayflick 1965; Freundt 1983). Fresh yeast extract is prepared by slowly adding 250 grams of active dry baker's yeast to 1 liter of boiling water and then allowing the mixture to boil slowly for about 15 minutes. The boiled preparation is clarified by centrifugation (8000 × G) or filtration. The yield from the clarification step is about 50 percent of the starting volume. The extract may be sterilized by autoclaving or filtration. We prefer stepdown filtration with a 0.2 μm filter as the final step. The final extract is brought to pH 7.4 ± 0.1 and stored frozen (≤ -20°C) for not more than 6 months. It may be reclarified after thawing, if necessary, by a second filtration through a 0.2 μm filter.

Table 8.3. Modified Edward Mycoplasma Agar Medium, Unsupplemented (for 1 liter supplemented medium)

The formulation for unsupplemented agar medium is identical to that for the unsupplemented broth except that 14 g of purified agar (Difco 0560-01 or equivalent) is added for each 692 ml portion.

PPLO broth without crystal violet (Difco 0554-01 or equivalent)	21 g
Phenol red 0.5% w/v aqueous stock solution	10 ml
Calf thymus DNA (Sigma D-1501 or equivalent)	0.02 g
Purified agar (Difco 0560-01, or equivalent)	14 g

Distilled water, sufficient quantity to make 692 ml

— Adjust pH to 7.4 ± 0.1 with 1 N HCl or 1 N NaOH.

— Sterilize by autoclaving at 121°C for 15 minutes.

— Store in convenient aliquots at room temperature.

— Expiration is 6 months from date of preparation.

Table 8.4. Modified Edward Mycoplasma Agar Medium, Supplemented (1 liter)

The sterile supplements used are the same as for the broth medium.

Unsupplemented mycoplasma agar	692 ml
Glucose-arginine stock solution*	8 ml
Fresh yeast extract	100 ml
Unheated horse serum	200 ml

— Aseptically adjust the pH of the complete broth to 7.4 ± 0.1, if necessary, by dropwise addition of sterile 1 N NaOH or 1 N HCl.

— Store at 2–8°C.

— Expiration is 90 days from the date of preparation.

*This is a filter-sterilized aqueous solution containing 25 percent w/v of glucose and 25 percent w/v of arginine HCl. It gives a final concentration of 0.2 percent w/v in the supplemented broth.

Current formulations for unsupplemented and supplemented SP4 broth and agar media, which are slightly modified from the original, are given in Tables 8.5–8.8.

Table 8.5. SP4 Mycoplasma Broth Medium, Unsupplemented (for 1 liter supplemented medium)

Mycoplasma broth base (BBL, 11458)	3.5 g
Tryptone (Difco)	10 g
Peptone (Difco)	5.3 g
Deionized water	655 ml
Phenol red (0.1% w/v aqueous solution)	20 ml

— Heat to dissolve and adjust to pH 7.0–7.2 if necessary.

— Sterilize by autoclaving at 121°C for 15 minutes.

Table 8.6. SP4 Mycoplasma Broth Medium, Supplemented (1 liter)

The following sterile components are combined aseptically:

Unsupplemented SP4 broth medium (for 1 liter)	
Arginine HCl (42% w/v aqueous solution)	5 ml
Glucose (50% w/v aqueous solution)	10 ml
CMRL 1066 tissue culture medium with glutamine (10x, Gibco)	50 ml
Fresh Yeast Extract (15% w/v aqueous solution, Gibco)	35 ml
Yeastolate (Bacto, Difco, 4% w/v aqueous solution)	50 ml
Fetal bovine serum (heated at 56°C, 30 minutes)	170 ml

— Aseptically adjust the pH of the complete broth to 7.0–7.2, if necessary, by dropwise addition of sterile 1 N NaOH or 1 N HCl.

— Store at 2–8°C.

Table 8.7. SP4 Mycoplasma Agar Medium, Unsupplemented (for 1 liter supplemented medium)

The formulation for unsupplemented SP4 agar is the same as that for the unsupplemented broth medium with two exceptions: (1) 0.8 g of noble agar is added to the unsupplemented broth formulation for each 100 ml volume of finished agar desired and (2) the phenol red is omitted.

Mycoplasma broth base (BBL, 11458)	3.5 g
Tryptone (Difco)	10 g
Peptone (Difco)	5.3 g
Noble agar	8 g
Deionized water	675 ml

Table 8.8 SP4 Mycoplasma Agar Medium, Supplemented (1 liter)

The same sterile components are aseptically mixed with the unsupplemented agar medium as are used to make the supplemented broth.

Unsupplemented SP4 agar (for 1 liter)

Arginine HCl (42% w/v aqueous solution)	5 ml
Glucose (50% w/v aqueous solution)	10 ml
CMRL 1066 tissue culture medium with glutamine (10x, Gibco)	50 ml
Fresh yeast extract 15% w/v aqueous solution, Gibco)	35 ml
Yeastolate (Bacto, Difco, 4% w/v aqueous solution)	50 ml
Fetal bovine serum (heated at 56°C, 30 minutes)	170 ml

    — Store at 2–8°C.

General Considerations for Medium Preparation

When preparing supplemented agar medium, caution must be used to mix components at the appropriate temperatures. Since agar solidifies at approximately 44°C and the serum protein is denatured and will precipitate at approximately 60°C, the medium components must be combined at an intermediate temperature. Sterile unsupplemented agar may be held at 50–55°C in a water bath after it has been liquefied in a microwave oven, an autoclave, or boiling water. The other sterile components are best warmed to avoid local solidification of the agar when adding components. Best mixing occurs when the supplements are heated briefly to near the agar hold temperature (50–55°C). The complete agar medium is mixed well without excessive agitation and should be dispensed (approximately 8.0 ml per 60 mm petri plate) immediately after it is prepared. Bubbles, if any, that form on the agar surface may be removed by "brushing" lightly with a bunsen burner flame. It is important to avoid subsurface bubbles by careful dispensing, because large numbers of subsurface bubbles will render the agar unfit for use. The final product should be clear. The finished plates should be packaged and stored in a manner that maintains their sterility and cleanliness and prevents excessive water loss.

Use of the highest grades of agar is frequently recommended for mycoplasma agar media, because lesser grades are occasionally toxic to some potential isolates (Barile and Rottem 1993; Olson and Barile 1992). The better quality agars also result in very clear agar media with fewer artifacts that might be confused with mycoplasma colonies.

For all medium formulations the entire lot of supplemented mycoplasma broth or agar is incubated at 35–37°C for 2 days as a sterility check. All units that are observed to be microbially contaminated are discarded. The entire lot is discarded if more than 10 percent of the units are contaminated.

Each lot is tested for growth promotion as described for positive controls under Inoculation and Incubation below. A lot of supplemented medium meets the growth promotion requirements if each of the positive control organisms is positive for growth. For agar each of the organisms must form at least one colony on at least one plate. For broth each of the organisms must form at least one colony on at least one subculture plate. Medium lots that fail to meet this criterion may be retested once. The growth promotion test may be performed concurrently with the use of the medium in mycoplasma detection tests. However, the mycoplasma detection test results will be considered invalid if the medium used fails to meet the requirements of the growth promotion test.

The 90-day expiration was conservatively assigned for supplemented Modified Edward media based on rigorous quantitative growth promotion testing of numerous lots of media against a number of different mycoplasma strains in our laboratory. We have not performed the same testing for determination of SP4 media expiration dating, but expect it would be similar.

Preparation and Maintenance of Mycoplasma Stocks for Positive Controls

To carry out mycoplasma testing in a manner consistent with other microbiological procedures used in the pharmaceutical industry, the methods used must ensure that appropriately characterized and accurately titered positive control mycoplasmas are available for use in the tests. The following methodology for growing, quantitating, and storing mycoplasma stocks is designed to make available growing log phase mycoplasma cultures that can be frozen at high titer for subsequent dilution and use as positive controls for mycoplasma testing.

Mycoplasmas do not generally make turbid broth cultures or grossly visible colonies on agar. Therefore, log phase cultures must be prepared and/or selected for freezing by indirect methods and shown retrospectively to contain a suitable titer of viable organisms. Even though mycoplasmas are not easily visualized, their metabolic activity is generally sufficient to cause discernible modification of the pH of their environment. The pH modification of broth medium during mycoplasma growth has been studied sufficiently to relate color change in an appropriate broth medium to the approximate titer of organisms (Masover et al. 1977).

3 × 5 Dilution Scheme

The 3 × 5 dilution scheme greatly enhances the chance of selecting a culture near the optimal time for freezing physiologically active organisms at an adequate titer for the use intended.

1. Arrange 15 containers of complete mycoplasma broth (up to 100 ml/container) in 3 columns of 5 containers each (refer to Figure 8.3). Label the columns A, B, and C. Label the rows 10^{-1}, 10^{-2}, 10^{-3}, 10^{-4}, and 10^{-5}.

2. Inoculate the first container in column A (10^{-1}) from frozen or reconstituted lyophilized stock and mix well (1 part stock plus medium to make 10 parts A).

3. Inoculate the first container in column B from the first container in column A and mix well (1 part A plus medium to make 3 parts B).

4. Inoculate the first container in column C from the first container in column B and mix well (1 part B plus medium to make 2 parts C).

5. Perform five serial tenfold dilutions starting at the first container in each column.

6. Incubate all the containers at 35–37°C.

7. Observe the containers daily for color and pH change.

8. At the appropriate time select and pool several dilutions with color and pH change typical of a log-phase culture for the species being grown. (Refer to color change criteria below.)

9. Aseptically dispense 1.0 ml aliquots of the pool into sterile freezing vials. Label the vials (organism and lot number) and store at –60°C or below.

10. Determine the titer of the frozen culture by colony counting of serial tenfold dilutions. (Refer to quantitation by colony counting below.)

Color Change Criteria

Species such as *Mycoplasma pneumoniae*, *Mycoplasma hyorhinis*, and *Acholesplasma laidlawii*, which are generally referred to as "fermenters," cause an acid color

Figure 8.3. Dilution scheme for preparation of mycoplasma stocks.

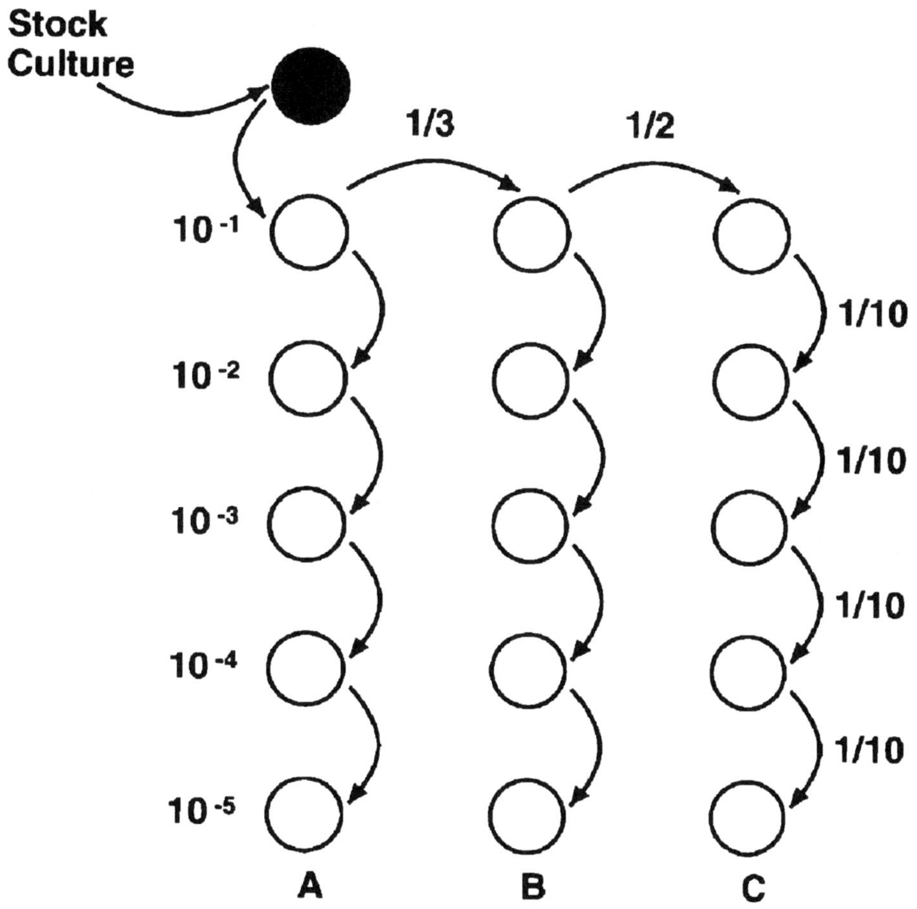

change—that is, from cherry red to orange to yellow in a medium containing phenol red pH indicator and starting at a pH between 7.0 and 7.4. For freezing choose cultures having the orange color to ensure viable log-phase mycoplasmas. Any unequivocal progressive color change of an inoculated container compared to an uninoculated control container is acceptable. Cultures that have progressed to the yellow color are in stationary or death phase and are not optimal.

Species such as *Mycoplasma orale*, *Mycoplasma hominis*, and *Mycoplasma arginini*, which are generally referred to as "nonfermenters" or "arginine hydrolizers," cause an alkaline color change—that is, from cherry red to deep red in a medium containing phenol red pH indicator and starting at a pH of 7.0 to 7.4. As with the fermenters, any unequivocal progressive color change of an inoculated container compared to an uninoculated control container is acceptable, and cultures having a slight but discernible color change are optimal.

Quantitation of Mycoplasma Stocks by Colony Counting

Mycoplasma stocks are quantitated by the classical dilution, plating, and counting methodology used for bacteria—a difference being the need to observe mycoplasma colonies with a microscope.

1. Quick-thaw one vial of the frozen stock in a 30–35°C water bath.

2. Prepare serial tenfold dilutions by adding 0.5 ml of stock to 4.5 ml of complete mycoplasma broth, mix thoroughly by vortexing, and use this dilution as inoculum for the next dilution. Continue this process until at least eight tenfold dilutions have been made.

3. Inoculate 0.2 ml of selected dilutions onto each of two 60 mm petri plates of complete mycoplasma agar. Allow the inoculum to absorb into the agar surface before incubating.

4. Incubate all plates at 35–37°C in closed chambers with the gaseous environment and for the period indicated for the species in Table 8.9.

5. Count all colonies on plates containing approximately 30 to 300 colonies using a microscope and final magnification of 25× to 100×. This process is greatly aided by use of a grid—either gridded plates or a gridded acrylic plate holder on the microscope stage.

6. Calculate the titer per ml of stock as follows. Express the result as CFU/ml.

$$\frac{\text{Average number of colonies/plate}}{0.2\ \text{ml/plate}} \times \frac{1}{\text{dilution}} = \text{number of CFU/ml}$$

Example: If two plates inoculated with the 10^{-4} broth dilution have 160 and 200 colonies, respectively, then

$$\frac{180\ \text{colonies/plate}}{0.2\ \text{ml/plate}} \times \frac{1}{10^{-4}} = 9 \times 10^6\ \text{CFU/ml}$$

7. Determine the initial titer of a new stock by performing steps 1 through 6 three separate times (preferably by a different analyst each time). If the difference between the highest and lowest results is not more than threefold, calculate the mean titer. If the difference is more than threefold, perform three additional titrations and calculate the mean titer using all six results.

Expiration Dating

Initially, expiration of mycoplasma stocks for positive controls is three years from the date of preparation (based on the experience of the authors). The expiration date of a stock may later be extended by demonstrating that the titer remains within $0.5\ \log_{10}$ CFU of the original titer. A new expiration date one year from the date of testing is then established. The expiration date may be extended an

Table 8.9. Incubation Conditions for Optimal Colony Formation

Mycoplasma Species	Optimal Gaseous Environment	Minimum Incubation Period (Days)
pneumoniae	aerobic	10
orale	anaerobic*	5
hominis	aerobic	5
hyorhinis	aerobic	5

*Generate anaerobic environment with a GasPak Plus (BBL, Becton Dickenson Microbiology Systems, Cockeysville, Maryland) system or equivalent.

unlimited number of times. We currently have active stocks that have been shown to be stable in storage for more than eight years.

Sampling and Sample Storage

Testing is done on a suspension of cells in the culture fluid. For production cell cultures that are grown in suspension, the sample to be tested is taken just prior to harvesting the cell culture fluid because that is the time at which contaminant mycoplasmas, if present, would be expected to be at the highest titer. If an adherent cell culture is to be tested, the analyst must assure that an appropriate number of cells is resuspended in the culture medium by scraping them off their growth surface. The cells should not be trypsinized in preparation for mycoplasma testing. Upon receipt by a testing laboratory, samples are stored at 2–8°C until frozen at –60°C or below. Samples must be frozen if they are not tested within 24 hours of sampling. The sample size should be sufficient to freeze a number of aliquots in the event it is necessary to perform a retest.

Consideration should be given to secondary samples taken from an entirely separate sampling port in the event there is a positive result and a need for a second sample from a process that has been long since concluded. The secondary sample provides a possible means for distinguishing between contamination during sampling and true contamination in the production cell culture.

Inoculation and Incubation

In the following procedure Day 0 refers to the day the test begins. Subsequent days refer to the number of elapsed days in the test—that is, Day 1 is the day after inoculation, when the inoculated media has been incubated for 1 day.

Inocula Preparation

- *Test samples*—Thaw frozen samples in a water bath (33–37°C) shortly before inoculation into test media.

- *Positive controls*—Thaw a vial of each positive control stock in a water bath (33–37°C). Dilute stocks in complete mycoplasma broth to 100 or less CFU per 0.2 ml for agar inocula and to 100 or less CFU per 10 ml for broth inocula.

Inoculation

On Day 0 inoculate mycoplasma broth and agar using a single lot of each medium according to the values given in Table 8.10.

Incubation

Incubate half of the plates in each group aerobically in sealed (watertight) chambers at 35–37°C. The sealed containers are necessary to prevent agar dessication during prolonged incubation. Incubate the remaining plates anaerobically at 35–37°C. The anaerobic environment may be generated by use of the GasPak Plus system or equivalent. All broths are incubated aerobically at 35–37°C. Since the plates contain rich medium and no antimicrobics, containers used to hold plates

Table 8.10. Inoculation Requirements for Day 0

Inocula	Agar		Broth (100 ml)	
	ml per plate	No. of plates	ml per bottle	No. of bottles
Test sample	0.2	4	10	1
No inoculum (negative control)	—	4	—	1
Mycoplasma hominis or *Mycoplasma orale* positive control				
a) diluted to ≤100 CFU/0.2 ml	0.2	2	—	—
b) diluted to ≤100 CFU/10 ml	—	—	10	1
Mycoplasma hyorhinis positive control				
a) diluted to ≤100 CFU/0.2 ml	0.2	2	—	—
b) diluted to ≤100 CFU/10 ml	—	—	10	1
Mycoplasma pneumoniae positive control				
a) diluted to ≤100 CFU/0.2 ml	0.2	2	—	—
b) diluted to ≤100 CFU/10 ml	—	—	10	1

Note: The inoculum should not be allowed to run to the edge of the plate where observation of colonies is difficult. The volume is such that it does not cover the entire plate and the inoculum should be allowed to absorb into the agar while the plate is level prior to being transported to the incubator.

should be disinfected prior to use to minimize microbial contamination. For the same reason containers of plates should not be opened for any reason, including interim observations, until the conclusion of the incubation period.

Note: The FDA's current *Points to Consider* (1993) does not require aerobic incubation. However, we have chosen to perform aerobic incubation because *Mycoplasma hyorhinis*, one of the more common cell culture contaminant mycoplasmas, is markedly inhibited by the GasPak environment, which is generally used to achieve the suggested anaerobic conditions (Polak-Vogelzang et al. 1983). As noted earlier, the DBS 1050 strain of *Mycoplasma hyorhinis* is not readily cultivable on any of the standard mycoplasma media under standard conditions. Other strains of *Mycoplasma hyorhinis*, however, grow very well on standard mycoplasma media under aerobic conditions.

Subculture 1

On Day 3 inoculate the broth cultures of the test sample, positive controls, and negative control onto additional plates of complete mycoplasma agar according to the values given in Table 8.11.

Subcultures 2 and 3

On days 7 and 14 repeat the subculture 1 procedure.

Observation and Identification of Mycoplasmas

Observation of Agar Plates

Observe mycoplasma agar plates microscopically after at least 14 days of incubation. Observe a minimum of 10 fields within the inoculated area of each 60 mm

Table 8.11. Inoculation Requirements for Day 3

Inocula	ml per plate	No. of plates
Broth culture of test sample	0.2	4
Broth culture of negative control	0.2	4
Broth culture of *Mycoplasma hominis* or *Mycoplasma orale* positive control	0.2	2
Broth culture of *Mycoplasma hyorhinis* positive control	0.2	2
Broth culture of *Mycoplasma pneumoniae* positive control	0.2	2

Notes:

1. Allow the inocula to be absorbed into the agar prior to incubation.

2. Incubate the Day 3 subculture plates at 35–37°C, half aerobically and half anaerobically, as was done for the direct inoculation plates.

plate at a total magnification of 40×. The appearance of mycoplasma colonies may be confirmed at 100× total magnification or higher. Observations are recorded as "positive," meaning that mycoplasma colonies were observed, or "negative," meaning that no mycoplasma colonies were observed. An ambiguous result should be checked by a second observer. If the result cannot be resolved as either positive or negative, a retest should be performed and the questionable colony(s) should be subcultured to a new agar plate. Agar subculture is accomplished by aseptically excising a block of agar containing the suspected colony, inverting it on a fresh plate, and pushing it across the fresh agar surface to transfer organisms from the old agar surface to the new one.

Observation of Positive Controls

The various mycoplasma species that may be used as positive controls differ in their growth and metabolic rates. Table 8.12 shows the typical growth pattern for each of the positive control organisms in their optimal gaseous environment. The Day 0 direct inoculation plates are inoculated with ≤100 CFU/0.2 ml/plate and typically contain 10 to 100 colonies. The subculture plates, however, originate from a less concentrated inoculum of 100 CFU/110 ml final volume of inoculated broth culture. *Mycoplasma hyorhinis*, *Mycoplasma hominis* and *Mycoplasma orale*, which grow fairly rapidly, all consistently grow to a concentration that is readily detectable in the Day 3 subculture plates, reach their maximum concentration in the broth on or before Day 7, and often are dead by Day 14. By contrast, the slow growing *Mycoplasma pneumoniae* generally does not reach detectable levels at the Day 3 subculture and often is still below detection limits at the Day 7 subculture. By the Day 14 subculture, however, it consistently results in large numbers of colonies on the plate.

Table 8.12. Typical Results for Positive Control Agar Plates

Positive Control	Day 0	Day 3	Day 7	Day 14[4]
Mycoplasma hominis[1]	+	+++	++	—
Mycoplasma orale[2]	+	++	+++	—
Mycoplasma hyorhinis[3]	+	++	+++	—
Mycoplasma pneumoniae[1]	+	—	+ or —	++

Legend: + = low # CFU, ++ = intermediate # CFU, +++ = maximal # CFU, — = no CFU.

Notes:

[1]*Mycoplasma pneumoniae* and *Mycoplasma hominis* generally grow either aerobically or anaerobically.

[2]*Mycoplasma orale* strains generally grow poorly or not at all in an aerobic environment.

[3]*Mycoplasma hyorhinis* strains grow poorly or not at all in an anaerobic environment.

[4]Day 3, Day 7, and Day 14 refer to day of subculture from broth.

Reaction with Dienes Stain (optional)

Dienes stain (Dienes 1939) was developed to distinguish mycoplasma colonies from bacterial and L-form colonies in clinical specimens. While the need to make this distinction is not anticipated with any frequency in testing pharmaceutical production cultures, the method is given here in the interest of completeness. The stain is prepared by dissolving 2.5 g of methylene blue, 1.25 g of azure II, 10 g of maltose, and 0.25 g of sodium carbonate in 100 ml of water. The stain is diluted between 1:10 and 1:30 just prior to use. It is applied by stroking an area adjacent to the suspected colony with a cotton swab that has been moistened with the stain. The stain will diffuse to the colony, which can then be examined under a microscope. Alternatively, the stain may be allowed to dry on a coverslip, which is then placed over the suspected colony(s), and is observed microscopically through the coverslip. The mycoplasma colonies stand out distinctly with dense blue staining centers and light blue peripheries. Viable bacterial colonies and colonies of bacterial L-forms will also stain initially, but they will decolorize in about 30 minutes. Mycoplasma colonies do not decolorize the stain. A source of error could be a colony of stained microorganisms composed of dead bacteria or L-forms.

Interpretation of Results

Valid Test

For a test to be considered valid, all of the following conditions must be met:

- All negative control plates must be negative for mycoplasma.

- At least one of the positive control plates for each positive control mycoplasma species must be positive.

- For a negative result no more than one plate in any group may be discarded for accidental causes.

Microbially contaminated plates are discarded if the contamination interferes with the observation of mycoplasma colonies. A "group" in this context is a set of agar plates that is inoculated at the same time with the same inoculum and incubated under the same conditions.

Negative Test

For a test sample to be considered negative, all agar cultures must be negative for mycoplasma after at least 14 days of incubation in a valid test.

Positive Test

Positive test results must be confirmed. If any culture is positive for the presence of mycoplasma in a valid test, the *aliquot* is considered positive for the presence of mycoplasma and another aliquot of the same sample is tested for the presence of mycoplasma. If the second aliquot is positive, the *sample* is considered to be positive for the presence of mycoplasma. An investigation is then performed, and an aliquot from a second sample from the same lot is tested as part of the investigation.

We consider the issue of the validity of the second sample to be sufficiently important to require an entirely separate sampling port system on every production vessel so that there is adequate assurance that the primary and secondary samples are independent. If the initial positive results cannot be accounted for based on determinate error (via the investigation and testing results on the second sample), the *lot* shall be considered to fail the requirements for absence of mycoplasma. If the investigation does not support the conclusion that mycoplasma contamination exists in the production lot and the test of the second sample is negative, the lot shall be considered to meet the requirements for absence of mycoplasma. This latter situation does not preclude rejection of the lot based on very conservative quality standards or dispositioning it to other than human use.

Retest Criteria

A retest should be performed under any one of the following circumstances:

- A test is invalid.

- Results are ambiguous.

- A determinate error is suspected in the assay.

DNAF METHOD

Basis for the Method

Because there is a strain of mycoplasma (*Mycoplasma hyorhinis* DBS 1050), which commonly infects cell cultures but is not readily grown on the commonly used agar preparations (Hopps et al. 1973), it is necessary to complement the cultural method with a method that will detect this particular contaminant. All *Mycoplasma hyorhinis* species cytadsorb strongly to cells in culture and are readily viewed by the use of DNAFs. Therefore, a method taking advantage of these properties is included in the mycoplasma detection program.

Fluorescence staining permits direct viewing of an individual mycoplasma organism. However, an individual mycoplasma is so small and so simple that even when being observed as a fluorescing body, one's confidence that what is being seen is actually a mycoplasma requires enhancement by the context in which it is being viewed. Thus, the ability to diagnose mycoplasma depends on the ability to see associations between mycoplasmas and their host cells as well as on the numbers of such associations one is able to observe.

The DNAF procedure is not specific for mycoplasmas. DNA-binding fluorochromes will stain any body that contains DNA, and even the more general diagnosis of microbial contamination using DNAF is dependent on the context in which it is seen. A DNAF stain can be applied directly to the cell culture sample to be tested for mycoplasma. Direct staining is rapid but has the disadvantages of a large number of artifacts and a lack of standardization. Alternatively, the DNAF stain can be applied to a test cell culture sample that has been inoculated into a known mycoplasma-free indicator cell culture.

Historically, the use of fluorochromes that bind to DNA, permitting direct viewing of individual cell-associated mycoplasmas, was reported by Russel et al. (1975) and by Chen (1977). Prior to this work, Fogh (1973, 65–106) had advocated the use of a cytochemical method that employed uninfected indicator cell cultures. Del Giudice and Hopps (1978, 57–69) combined the use of DNAF with the use of uninfected indicator cells to create the method generally used today.

The indicator cell culture method has the advantage of being more conducive to standardization by the inclusion of positive and negative controls and by the use of a more uniform cell substrate for the test. It also has the advantage of being a biological system in which low-level contaminant mycoplasmas can be enriched in number to make them more readily discernible and more confidently diagnosed. This method, which is the one outlined in the FDA's *Points to Consider* guideline, is described here.

Reagents

Reagents required for the DNAF procedure include fixative, stain, buffer, and mounting solution, prepared as described below.

Fixative

Use 1 part glacial acetic acid plus 2 parts absolute methanol prepared fresh for each use. Prepare at least 10 ml for each coverslip to be stained.

Stain

Use of bisbenzimide as the DNAF stain is accepted by the FDA. *Bisbenzimide 1000× concentrated stock solution* contains 5.0 mg of bisbenzimide (Hoechst 33258, 2'-[4-Hydroxyphenyl]-5-[4-methyl-1-piperazinyl]-2,5'-bi-1H-benzimidazole) per 100 ml of distilled water (50 mg/ml). Each 100 ml is preserved with 0.01 g of thimerosal. Store this solution at 2–8°C in the dark (foil wrapped). *Bisbenzimide stain working solution* (0.05 µg/ml) contains 0.1 ml of the concentrated stock solution plus 99.9 ml of citrate/phosphate buffer, pH 5.5. It is prepared fresh for each use.

Buffer

For each 100 ml of *citrate/phosphate buffer*, pH 5.5, mix 44 ml of 0.1 *M* citric acid with 56 ml of 0.2 *M* sodium phosphate dibasic.

Mounting Solution

The mounting solution consists of equal portions of glycerol and citrate/phosphate buffer, pH 5.5, to make the desired final volume (2 or 3 drops per slide are required). The addition of n-propyl gallate at a final concentration of 0.2 percent w/v, reduces photobleaching substantially (Giloh and Sedat 1982), allowing a particular area of a slide to be viewed for a much longer period or at several different times. It also enhances maintenance of fluorescence on storage and facilitates photography.

Preparation of Indicator Cells

Cells chosen as indicator cells should grow attached to a substrate and have a large ratio of cytoplasmic area to nuclear area. Vero and 3T6 cells are frequently used. Cells are best seeded the day prior to inoculation of a test sample and in a density that will be subconfluent at the end of the 3- to 5-day culture period. Prepare 6-well cell culture plates containing 1 presterilized 22 mm^2 number one coverslip per well.

Trypsinize an antibiotic-free stock culture of Vero cells and prepare a cell suspension containing 5×10^4 viable cells per ml. Dispense 4.0 ml of cell suspension per well ($\approx$ 3.8 cm in diameter) or adjust the volume of cell suspension proportionately to the size of culture vessels. Incubate the coverslip-cell cultures at 35–37°C in a humidified 4–6% CO_2 in air atmosphere.

Inoculation of Indicator Cells

Indicator cell cultures are inoculated 1 day after the cells are plated by adding the following to two wells each:

- Test sample, 0.5 ml. Multiple samples may be tested concurrently.

- Negative control, uninoculated wells.

- 100 CFU or less of positive control *Mycoplasma hominis* (derived from ATCC #23114) in 0.5 ml or less of complete mycoplasma broth. The intent of this positive control is to test a low number of poorly cytadsorbing mycoplasmas. Therefore, *Mycoplasma orale* strains or other poorly cytadsorbing mycoplasma species may also be used, but they may not give reliably positive results when inoculated in low numbers.

- 100 CFU or less of positive control *Mycoplasma hyorhinis*, strain BTS-7 (derived from ATCC #17981) in 0.5 ml or less of complete mycoplasma broth. The intent of this positive control is to test a low number of strongly cytadsorbing mycoplasmas. Therefore, other strains of *Mycoplasma hyorhinis* or other cytadsorbing mycoplasmas may be used.

DNAF Staining

After a three- to five-day incubation period, the coverslip cell cultures are stained as described below.

1. Remove medium from the well (do not allow the coverslip to dry).

2. Add fixative in sufficient volume (2–5 ml) to completely immerse the coverslip and let stand for 2 minutes.

3. Aspirate the fixative and then apply fresh fixative. Let stand for 5 minutes.

4. Aspirate the fixative and air dry for 30–60 minutes, or aspirate the fixative and rinse twice with citrate/phosphate buffer, aspirating after each rinse. The objective, in either case, is to remove the fixative prior to staining. Dried fixed slides may be stored for future staining.

5. Apply working solution of bisbenzimide stain (sufficient quantity to immerse the coverslip) and stain for 10–30 minutes in the dark.

6. Aspirate the stain and wash twice with distilled water.

7. Aspirate the distilled water and either mount or store if desired. Stained coverslips may be stored refrigerated for up to 7 days, either dry or immersed in citrate/phosphate buffer.

8. Mount the coverslips on a slide, cell side down on a drop of mounting solution. A more permanent mount can be made by sealing the mounted coverslip with petroleum jelly, nail polish, or an equivalent liquid barrier. Coverslips mounted in this manner and stored refrigerated in the dark are suitable for viewing even after storage for several weeks.

9. Observe microscopically (minimum 630×; oil immersion), using a microscope equipped with a fluorescence light source, an epi-illumination system, and fluorite objectives. The following excitation, barrier, and beam splitting filters are suitable for bisbenzimide:

Excitation	Band pass 365 nm; 11 nm band width
Barrier	Long wave pass 397 nm
Beam splitter	Dichromatic beam filter 395 nm

Interpretation of Results

The bisbenzimide-DNA complex emits 475 nm wavelength light, which is light blue. The indicator cell nuclei are intensely stained, generally ovoid, and have unbroken margins. Mitochondrial DNA and, therefore, cell cytoplasmic areas, cannot be seen with this stain. The background on a good preparation is black. Occasionally, micronuclei formed by anomalous mitotic events are seen, but they are usually much larger than mycoplasmas. Occasional mitotic figures will also be observed. The nuclei are approximately 10–20 μm in diameter (long axis), and the mycoplasmas are seen as small round bodies, approximately 0.3 μm in diameter. The mycoplasmas are generally not intracellular, but are adherent to the cell membrane and are seen in the areas where the cytoplasm is expected to be. When the indicator cells are well separated, heavily contaminated cells have a blue fluorescent nucleus with the cytoplasmic region virtually covered with discrete, small, blue fluorescent bodies; the space between the cells is generally black.

Mycoplasmas also cover the membrane above the nucleus, but these mycoplasmas are not distinguishable from the very brightly fluorescing nucleus. The extent of cytadsorption is variable with different mycoplasma species and is very important with respect to ease of diagnosis using this method. All of the *Mycoplasma hyorhinis* species, including the "noncultivable" strain DBS 1050, cytadsorb well and are readily observed by this method. All of the other common cell culture contaminants are expected to be detectable by the cultural methodology as well as by the DNAF method.

Valid Test

For a test to be valid, both negative control slides must be negative for mycoplasma and at least one positive control slide for each of the positive control mycoplasma species must be positive.

Negative Test

For a test sample to be negative, both coverslip-cell-culture DNAF stains must be negative for mycoplasma in a valid test.

Positive Test

In our laboratory we have established that for a test sample to be considered positive for mycoplasma, at least 5 percent of not less than 200 cells in a valid test must have multiple (more than five) blue fluorescent bodies the size of mycoplasmas associated with the indicator cells. In the event of an ambiguous result or a determinate error, a retest should be performed. Some samples, notably hybridoma cultures, contain particulate debris that react with the stain and renders the test result ambiguous. In such cases a retest is done in which 0.5 ml of supernatant medium and indicator cells resuspended by scraping from the original culture is subcultured into fresh indicator cultures after 3–5 days of incubation. The subculture is stained and observed after an additional 3–5 days of incubation.

Positive test results must be confirmed. If any cultures are positive for the presence of mycoplasma in a valid test, the *aliquot* is considered positive for the presence of mycoplasma and another aliquot of the same sample is tested for the presence of mycoplasma. If the second aliquot is positive, the *sample* is considered to be positive for the presence of mycoplasma. An investigation is then performed, and an aliquot from a second sample from the same lot is performed as part of the investigation. If the initial positive results cannot be accounted for based on determinate error (via the investigation and testing results on the second sample), the *lot* shall be considered to fail the requirements for absence of mycoplasma. If the investigation does not support the conclusion that mycoplasma contamination exists in the production lot and the test of the second sample is negative, the lot shall be considered to meet the requirements for absence of mycoplasma.

Retest Criteria

A retest should be performed in the event of an invalid test, an ambiguous result, or a determinate error.

IDENTIFICATION OF MYCOPLASMAS

The identification of mycoplasmas using the aforementioned methods depends on the ability to observe a typical colony on agar or to see sufficient numbers of round, blue, cell-associated, fluorescent bodies of the appropriate size in the DNAF test. When colonies are present on agar in sufficient number and have the classic fried egg morphology, appropriate size, texture, and interaction with the agar, one has a high confidence level that mycoplasmas are present. When, in the DNAF test, one sees a classical presentation of large numbers of strongly cell-associated bodies that have bound the stain and, in addition, the positive and negative controls produce expected results, then the identification of a microbial contaminant that is a mycoplasma is probable. Unfortunately, it is not reasonable to expect that a classical presentation will occur in all cases of true contamination. In cases where there is any ambiguity about a positive result, the microbiologist is obliged to exercise due diligence in resolution of the ambiguity.

Resolution of ambiguity about a positive result may necessitate identifying the suspected mycoplasma to the species level, which can be done based on antigenic analysis. The easiest and most direct way to accomplish it is by direct immunofluorescence on agar using incident illumination. This technique was first described by Del Giudice et al. (1967) and has been used virtually unchanged since that time. A detailed description of the method including preparation and fluoresceination of antisera, is given by Gardella et al. (1983, 431–439). The fluoresceinated antisera against specific mycoplasma species are not generally commercially available, but they are not difficult to prepare.

Although more than 20 mycoplasma species have been reported as contaminants of cell cultures, 95 percent of the isolates were one of 5 species: *Mycoplasma orale*, *Mycoplasma arginini*, *Mycoplasma hyorhinis*, *Mycoplasma fermentans*, and *Acholeplasma laidlawii* (Barile and Rottem 1993, 155–193). Thus, fluoresceinated antisera to these 5 species would be expected to identify 95 percent of the positive results. Recent reports indicate that both DNAF and fluorescent antibodies can be applied to the same cell culture preparation and that a particular cell or fluorescing body can be viewed for both stains, thereby enhancing confidence in the diagnosis, as well as providing species level identification (Freiberg and Masover 1990; Masover and Pleibel 1994; Masover and Becker 1995).

DISCUSSION

The cultural and the DNAF staining methods used for the detection of mycoplasmas are not new techniques. However, the application of these methodologies in the highly controlled and documented manner required by the GMP posture of today's pharmaceutical manufacturers is relatively recent. The two methods, as described above, were submitted to the FDA as part of a Product License Application in 1987 and were found acceptable for use in lot release testing of the first recombinant human pharmaceutical to be prepared in a eukaryotic cell culture (Activase®, recombinant human tissue-type plasminogen activator). Since then, these tests have been in continuous use in essentially the same format, with acceptable results. When properly performed by well-trained analysts, the methods are very reliable.

The expectation of the FDA is that there be no adventitious contaminants in the cell cultures used to produce pharmaceuticals. Each production culture is expected to be a pure culture of only the production cell. This means that detection of any quantity of any adventitious agent renders the production culture unfit for use and, as in the case of sterility testing of product, raises questions of sensitivity, specificity, and appropriateness of the tests employed.

In general, the sensitivity of tests for adventitious biological agents in cultures is a function of the sample size, so that the result "none detected" means less than one of whatever the assay measures per volume tested. This is the reason for testing a 10 ml sample rather than some lesser quantity in the mycoplasma cultural test, which is still considered to be the primary method for detecting mycoplasmas in pharmaceutical production cultures. Inasmuch as the purpose of these procedures is to detect a biologically active contaminant, it seems that a biological test will always be appropriate.

Other established methods for mycoplasma testing are reviewed by Barile and Rottem (1993, 155–193). Newer methodology is available but has not yet been validated for use in pharmaceutical applications. For example, the polymerase chain reaction (PCR) is now widely used with great success for the detection of mycoplasma DNA sequences in clinical and research samples. This method is currently being considered for the testing of pharmaceutical production cultures as well. The question of whether a method such as PCR might replace or be used adjunctively with a biological assay will have to be assessed. Concomitant application of a DNAF and a fluoresceinated antibody or combination of antibodies (Masover and Becker 1995) might also be considered for rapid screening, which is adjunctive to the classical biological methodology.

REFERENCES

Barile, M. F., and S. Razin, eds. 1979. *The mycoplasmas*, vol. 1: *Cell biology*. New York: Academic Press.

Barile, M. F., and S. Rottem. 1993. Mycoplasmas in cell culture. In *Rapid diagnosis of mycoplasmas*, edited by I. Kahane, and A. Adoni. New York: Plenum Press.

Center for Biologics Evaluation and Research. 1993. Points to consider in the characterization of cell lines used to produce biologicals. Rockville, MD: Department of Health and Human Services, U.S. Food and Drug Administration.

Chen, T. R. 1977. In situ detection of mycoplasma contamination in cell cultures by fluorescent Hoechst 33258 stain. *Exp. Cell Res.* 104:255–262.

Del Giudice, R. A., and H. E. Hopps. 1978. Microbiological methods and fluorescent microscopy for the direct demonstration of mycoplasma infection of cell cultures. In *Mycoplasma infection in cell cultures*, edited by G. J. McGarrity, D. G. Murphy, and W. W. Nichols. New York: Plenum Press.

Del Giudice, R. A., R. S. Gardella, and H. E. Hopps. 1980. Cultivation of formerly noncultivable strains of *Mycoplasma hyorhinis Current Microbiology.* 4:75–80.

Del Giudice, R. A., N. F. Robillard, and T. R. Carski. 1967. Immunofluorescence identification of *Mycoplasma* on agar by use of incident illumination. *J. Bacteriol.* 93:1205–1209.

Dienes, L. 1939. L organism of Kleineberger and *Streptobacillus moniliformis*. *J. Infect. Dis.* 65:24–42.

Edward, D. G. 1947. A selective medium for pleuropneumonia-like organisms. *J. Gen. Microbiol.* 1:238–243.

Fogh, J. 1973. Contaminants demonstrated by microscopy of living tissue cultures or of fixed and stained tissue culture preparations. In *Contamination in tissue culture*, edited by J. Fogh. New York: Academic Press.

Freiberg, E. F., and G. K. Masover. 1990. Mycoplasma detection in cell culture by concomitant use of bisbenzimide and fluoresceinated antibody. *In Vitro Cell Dev. Biol.* 26:585–588.

Freundt, E. A. 1983. Culture media for classic mycoplasmas. In *Methods in mycoplasmology*, vol. I, edited by S. Razin, and J. G. Tully. New York: Academic Press.

Gardella, R. S., R. A. Del Giudice, and J. G. Tully. 1983. Immunofluorescence. In *Methods in mycoplasmology*, edited by S. Razin, and J. G. Tully. New York: Academic Press.

Gardella, R. S., and R. A. Del Giudice. 1995. Growth of *Mycoplasma hyorhinis* Cultivar a on semisynthetic medium. *Appl. Environ. Microbiol.* 61:1976–1979.

Gibbons, N. E., and R. G. E. Murray. 1978. Proposals concerning the higher taxa of bacteria. *Int. J. Syst. Bacteriol.* 28:1–6.

Giloh, H., and J. W. Sedat. 1982. Fluorescence microscopy: Reduced photobleaching of rhodamine and fluorescein protein conjugate by n-propyl gallate. *Science* 217:1252–1255.

Hayflick, L. 1965. Tissue cultures and mycoplasmas. *Tex. Rep. Biol. Med.* 23 (suppl. 1):285–303.

Hopps, H. E., B. C. Meyer, M. F. Barile, and R. A. Del Giudice. 1973. Problems concerning "noncultivable" mycoplasma contaminants in tissue cultures. *Annals N. Y. Acad. Sci.* 225:265–276.

Kotani, H., G. H. Butler, D. Tallarida, C. Cody, and G. J. McGarrity. 1990. Microbiological cultivation of *Mycoplasma hyorhinis* from cell cultures. *In Vitro Cell. Dev. Biol.* 26:91–96.

Maniloff, J. 1992. Mycoplasma Viruses. In *Mycoplasmas: Molecular biology and pathogenesis*, edited by J. Maniloff, R. N. McElhaney, L. R. Finch, and J. B. Baseman. Washington, DC: American Society for Microbiology.

Maniloff, J., R. N. McElhaney, L. R. Finch, and J. B. Baseman, eds. 1992. *Mycoplasmas: Molecular biology and pathogenesis*. Washington, DC: American Society for Microbiology.

Masover, G. K., and F. A. Becker. 1995. Detection of mycoplasmas by DNA staining and fluorescent antibody methodology. In *Molecular and diagnostic procedures in mycoplasmology*, vol. II, edited by J. G. Tully, and S. Razin. New York: Academic Press.

Masover, G., and L. Hayflick. 1981. The genera *Mycoplasma, Ureaplasma,* and *Acholeplasma,* and associated organisms (thermoplasmas and anaeroplasmas). In *The procaryotes*, edited by M. P. Starr, H. Stolp, H. G. Truper, A. Balows, and H. G. Schlegel. Berlin: Springer-Verlag.

Masover, G. K., and N. Pleibel. 1994. Rapid definitive diagnosis of cell culture contaminant mycoplasmas by concomitant use of two fluorescent stains. *IOM Letters Vol. 3*, Program and Abstracts of the 10th International Congress of the International Organization for Mycoplasmology. 19–26 July, Bordeaux, France.

Masover, G. K., S. Razin, and L. Hayflick. 1977. Effects of carbon dioxide, urea and ammonia on growth of *Ureaplasma urealyticum* (T-strain mycoplasma). *J. Bacteriol.* 130:292–296.

Masover, G. K., F. A. Becker, G. Hawker, N. Pleibel, and S. Razin. 1994. Evaluation of two media widely used for detection of mycoplasmas in cell cultures. *IOM Letters Vol. 3*, Program and Abstracts of the 10th International Congress of the International Organization for Mycoplasmology. 19–26 July, Bordeaux France.

Olson, L. D., and M. F. Barile. 1988. Mycoplasma infection of cell cultures: Isolation and detection. *J. Tissue Cult. Meth.* 11:175–179.

Polak-Vogelzang, A. A., H. H. DeHaan, and J. Borst. 1983. Comparison of various atmospheric conditions for isolation and subcultivation of *Mycoplasma hyorhinis* from cell cultures. *Antonie van Leeuwenhoek* 49:31–40.

Razin, S. 1983. Identification of mycoplasma colonies. In *Methods in mycoplasmology*, vol. I, edited by S. Razin, and J. G. Tully. New York: Academic Press.

Razin, S. 1985. Molecular biology and genetics of mycoplasmas (Mollicutes). *Microbiol. Rev.* 49:419–455.

Razin, S. 1991. The Genera *Mycoplasma, Ureaplasma, Acholeplasma, Anaeroplasma,* and *Asteroleplasma.* In *The procaryotes*, edited by A. Balows, H. G. Scheifer, M. Dworkin, W. Harder, and K. H. Schleifer. New York: Springer-Verlag.

Razin, S., and E. A. Freundt. 1984. The Mollicutes, Mycoplasmatales and Mycoplasmataceae. In *Bergey's manual of systematic bacteriology*, vol. I, edited by N. R. Krieg, and J. G. Holt. Baltimore: Williams and Wilkins.

Razin, S., and O. Oliver. 1961. Morphogenesis of mycoplasma and bacterial L-form colonies. *J. Gen. Microbiol.* 24:225–237.

Razin, S., and J. G.Tully, eds. 1983. *Methods in mycoplasmology*, vol. I: *Mycoplasma characterization*. New York: Academic Press.

Robinson, L. B., R. H. Wichelhausen, and B. Roizman. 1956. Contamination of human cell cultures by pleuropneumonia-like organisms. *Science* 124:1147–1148.

Russel, C. W., C. Newman, and D. H. Williamson. 1975. A simple cytochemical technique for demonstration of DNA in cells infected with mycoplasmas and viruses. *Nature* 253:461–462.

Tully, J. G., and S. Razin, eds. 1983. *Methods in mycoplasmology*, vol. II: *Diagnostic mycoplasmology*. New York: Academic Press.

Tully, J. G., and S. Razin, eds. 1995. *Molecular and diagnostic procedures in Mycoplasmology*, vol I and II: Orlando: Academic Press.

Tully, J. G., R. F. Whitcomb, H. F. Clark, and D. L. Williamson. 1977. Pathogenic mycoplasmas: Cultivation and vertebrate pathogenicity of a new spiroplasma. *Science* 195:892–894.

9

Cellular Fatty Acid Analysis for the Classification and Identification of Bacteria

A. Ray Smith

University of Illinois at Urbana-Champaign
College of Veterinary Medicine

Joel P. Siegel

University of Illinois at Urbana-Champaign
Illinois Natural History Survey

There are many different ways to identify biologically and medically important organisms, but this chapter concentrates on the use of cellular fatty acid (CFA) analysis for the classification and identification of bacteria. The recent literature (1987–1995) is emphasized due to the development of computerized, automated gas chromatographs and the availability of commercial databases for bacterial identification.

The first to suggest that CFA analysis by gas-liquid chromatography (GLC) could be used to identify bacteria successfully was Abel et al. (1963). They identified different CFA patterns within various members of the family Enterobacteriaceae and in some gram-positive bacteria, but made no attempt to identify the specific CFAs. Their report established the potential usefulness of this technique and also established the basis for additional work.

Analysis using CFA is chemotaxonomy, since the identification and quantitation of fatty acids (FAs) are highly discriminatory properties (Tornabene 1985); the main requirement for performing CFA analysis is proper instrumentation (Welch 1991). Our experience is with the MIDI microbial identification system (MIS, Microbial ID, Inc., Newark, DE). This system is a fully automated, computerized, high resolution GLC unit. The computer-aided interpretation of data through the use of the MIDI libraries and library generation software (LGS) have greatly facilitated successful utilization of this procedure.

For one to use the MIS, the microorganisms must be grown on a standardized medium using prescribed temperature and time of incubation for results from different laboratories to be comparable. This is true for all CFA data because medium and temperature influence the CFA composition of the bacterial cell envelope. The FAs play a role in regulating membrane fluidity in response to temperature; therefore, comparing FA profiles from bacteria incubated at the same temperature is essential. In practice, the CFA process involves extraction (saponification), methylation, and esterification of the FAs comprising the cell walls according to standard chemical protocols (Miller and Berger 1985). Fatty acid methyl esters (FAMEs) are then analyzed by GLC. This analysis reveals both the identity of the FAs and their percent composition. With proper databases, the data can then be used to distinguish between species as well as identify strains within a species using principal component analysis (PCA) and the hierarchical unweighted pair group method arithmetic–averages-algorithm (UPGMA) clustering techniques (Siegel et al. 1995; Suzuki et al. 1993). The groupings are presented as dendrograms based on Euclidean distance.

NOMENCLATURE USED

Fatty acids are named according to the number of carbon atoms, type of functional groups, and location of the double bonds (Table 9.1). The convention of writing C (followed by the number of carbon atoms):(followed by the number of double bonds) will be used. For example, the saturated hexadecanoic FA (palmitic acid) is written $C_{16:0}$. Composition differences between gram-positive (*Staphylococcus aureus*) and gram-negative (*Pseudomonas aeruginosa*) organisms are also illustrated.

MIDI SYSTEM

The MIDI MIS was developed by Hewlett-Packard Co. (Avondale, PA) and is presently marketed by MIDI (Microbial ID, Inc.). The system consists of a gas chromatograph (GC) equipped with a flame ionization detector (FID) and a 5 percent methylphenyl silicon fused-silica capillary column (25 m by 0.2 mm); and an automatic sampler, an integrator, a computer, and a printer (White et al. 1988). Gases used at specified flow rates are hydrogen (as the carrier), air, and nitrogen. Ultrahigh purity grade hydrogen and nitrogen are used and zero-grade air is specified. Software libraries for the identification of aerobes, clinical isolates,

Table 9.1. Mean percent composition of cellular fatty acids comprising >1 percent of the cell wall for *Staphylococcus aureus* and *Pseudomonas aeruginosa*. The purpose of this table is to illustrate the nomenclature used and show composition differences between gram-positive and gram-negative organisms.

Simplified Name	Systematic Name	*Staphylococcus aureus*	*Pseudomonas aeruginosa*
Saturated			
$C_{12:0}$	Dodecanoic		5.0
$C_{14:0}$	Tetradecanoic	2.8	1.0
$C_{16:0}$	Hexadecanoic	5.6	23.7
$C_{18:0}$	Octodecanoic	16.1	tr.
$C_{20:0}$	Eicosanoic	15.3	
Unsaturated			
$C_{16:1}$	9-Hexadecanoic		25.8
$C_{18:1}$	9-Octadecanoic	1.4	30.1
Branched			
iso-$C_{14:0}$	12-Methyl tridecanoic	5.8	
iso-$C_{15:0}$	13-Methyl tetradecanoic	9.2	
anteiso-$C_{15:0}$	12-Methyl tetradecanoic	30.8	
iso-$C_{17:0}$	15-Methyl hexadecanoic	1.6	
anteiso-$C_{17:0}$	14-Methyl hexadecanoic	3.4	
Complex— Hydroxy			
3-OH-$C_{10:0}$	3-Hydroxy decanoic		3.8
2-OH-$C_{12:0}$	2-Hydroxy dodecanoic		4.6
3-OH-$C_{12:0}$	3-Hydroxy dodecanoic		4.5

anaerobes, mycobacteria, and yeasts are available from MIDI. In addition, LGS can be purchased from MIDI, which allows users to develop their own libraries from data generated by strain analysis.

The automatic sampler allows a maximum of 100 samples to be run without intervention. After injection of the sample, the temperature increases from 170°C to 270°C (increasing 5°C per minute) and is driven by the computer program. The resulting peaks are identified based on their comparison to the retention times of a mixture of known FAs (microbial calibration standard); the calibrator data are used to calculate an equivalent chain length (ECL) for the molecule. The peak area amount of each CFA detected is calculated as a percentage of the total area of CFAs. The multivariate statistical method of PCA is used as the basis of interpreting data and matching an unknown sample with database entries. Numerical analysis of CFA data can also be performed, which results in a computer-generated dendrogram based on Euclidean distance. Dendrograms may be used to show similarities at the genus, species, and subspecies levels (Romesburg 1990).

The usual preparation of samples for CFA analysis consists of hydrolysis of the whole-cell FAs to form sodium salts and then methylation of CFA esters to make them volatile in the GC. A relatively simple four-step process for the preparation of samples has been developed (Sasser 1990a). Culture plates are incubated either 24 or 48 hours (depending on the protocol chosen) and cells (approximately 50 mg wet weight) are harvested. A sodium hydroxide–methanol solution is added and saponification is conducted for 30 minutes at 100°C, thus liberating the FAs. The second step is methylation with HCl in methanol at 80°C for 10 minutes. Third, FAMEs are extracted with a solution of hexane and methyl-tertiary-butyl-ether (10 minutes). The final step consists of washing the extract in aqueous sodium hydroxide for 5 minutes. The organic phase is then transferred to a GLC vial that is then capped and is ready for analysis. Approximately 3 hours are required to process 60 samples and a reagent control.

Samples are then logged into the computer. After initiation of the run, the automatic sampler injects 2 μl of the extract into the sealed injection port (250°C). After the run (approximately 20 minutes) the column is cleaned by heating (310°C for 2 minutes). The FID (300°C) sends the electronic signals produced by the analysis to integrators that amplify and process the signals. These data are passed to the computer, where they are stored and can be compared with the database. Identifications made by the libraries are listed with a confidence measure (similarity index, SI) on a scale of 0 to 1.0. The higher the SI (e.g., 0.900), the closer the sample is to the mean of the library entry. The total run time for each sample is approximately 30 minutes.

The calibration standard used is a mixture of straight-chain saturated FAs from 9 to 20 carbons in length and 5 hydroxy FAs in hexane. Changes in sample injection volume and variables such as carrier gas flow rate, column, and detector temperature will affect the sample retention time. Therefore, the calibration mixture is reanalyzed after every tenth sample to correct for any possible drift of retention time. If the identification of the calibrator does not conform to the programmed standards, the chromatograph is shut down by the computer.

Library Generation

The power of the MIS was expanded by the use of LGS, which allows each operator to create library entries from one's own sample data. This capability expands the flexibility of the system because specialty databases can be created. Researchers can create databases based on their preferred incubation temperature and medium.

There are two steps that are unique for library development—accumulation and generation. Accumulation is the development of composite profiles for samples that are of the same group; this process involves some subjective data assessment. When a large variance in the accumulated data occurs (outliers), decisions must be made as to what to do with the data. If one or two samples cause most of the variance, they can be eliminated since the microorganisms may have been incorrectly classified or contaminated. Alternatively, the presence of taxonomic subgroups may cause the accumulated sample data to be expressed in distinct clusters. Outliers can be resolved either by eliminating the data, correcting the original taxonomic classification, or creating a new classification. After careful evaluation of the accumulated data, the final step is the generation of library entries. In the MIDI system a library entry consists of the FAs present in all the individual samples and the mean values ± standard deviation ($\bar{x} \pm$ s.d.). The coefficient of variation (CV) calculated as the quotient of the s.d. $\div \bar{x}, \times 100$ is also included. Only FAs comprising ≥ 0.3 percent of the total are used in any analysis.

"Tracking" of Strains

The ability to track and differentiate strains is very useful in microbiology. Epidemiological studies often use phage typing, plasmid typing, or antibiograms to characterize isolates. The pharmaceutical and food industries can also use sensitive methods to determine the source of contaminants and to select more active or toxic strains in the case of bacterial insecticides. Multivariate cluster analyses that use FA compositions are very useful (Sasser 1990b). The LGS package of the MIDI MIS contains two cluster analysis packages. The *Dendrogram* and *2-D Plot* programs use data from FA analyses of strains of organisms that have been sampled.

The *Dendrogram* uses cluster analysis techniques to produce unweighted pair matchings based on FA compositions. The MIDI system is based on PCA. In this case PCA is used to take the *n* variables that are the individual FAs and find combinations of these to produce indices that are uncorrelated (Manly 1986; Fisher and van Belle 1993). The Euclidean distance between each strain is then calculated using *n*-dimensional Pythagorean theorem; clusters are created by UPGMA analysis or another hierarchical clustering algorithm. The results are displayed graphically in a tree diagram that depicts the relatedness of pairs of entries.

There are several computer programs (both personal computer and mainframe) that allow the user to enter data on CFA identity and percent composition and then generate Euclidean distance based on dendrograms. In this chapter we have constructed dendrograms from data contained in several of the articles referenced using the JMP® personal computer program, version 3 [Statistical software for the Macintosh (SAS Institute, Cary, NC)]. We found that the JMP method of hierarchical cluster analysis correlated very well with UGMPA of MIS (average option in JMP). If one desired, the mainframe SAS [PROC PRINCOMP with the average option and PROC CLUSTER] could be used instead.

Knowing the Euclidean distances between clusters or individual entries can assist one in making subjective assessments of assigning isolates to genus, species and strains. Based on empirical observations, MIDI states that the boundaries for genus, species, subspecies, biotype, and strain levels are approximately 25, 10, 6, 5, and 2 Euclidean distance units respectively. In our experience species within most genera link within 35 Euclidean distance units. These distances are strongly correlated with the amount of variance of the FAs that comprise each strain.

The dendrograms are constructed using mean, unweighted values. Therefore, for tight clusters to be formed, the strains in an accumulated group need to have low CVs for their principal FA constituents. The peak area CVs are inversely correlated with mean peak areas. Simply stated, minor FAs <3 percent are more variable than major constituents of the cell envelope. This is illustrated by a study of Eerola and Lehtonen (1988) who found that CVs were consistently above 100 when the mean peak areas were <1 percent, were approximately 30 when the mean peak area was >1 percent, and approached values of 5–10 when the mean peak areas were >10 percent. Therefore, when creating entries in the LGS, we strive for low CV values (preferably <7) for the principal FAs (>10 percent). In many cases the CV for the principal FA is ≤3 percent. We believe this helps insure development of entries that allow the greatest discrimination of closely related strains.

We have used our CFA data comprising 11 genera and JMP to produce a dendrogram for purposes of illustration (Figure 9.1). Examination reveals that the most closely related organisms are *Bacillus thuringiensis* serovar *israelensis* and *Bacillus cereus* since they link at 8.5 Euclidean distance units. The species and genus boundaries have been derived empirically to be approximately 10 and 25, respectively, as previously stated. Since *Bacillus thuringiensis* serovar *israelensis*

Figure 9.1. Dendrogram of 11 genera of bacteria, most of which appear to be taxonomically correct, using cellular fatty acid analysis. The *Bacillus thuringiensis* serovar *israelensis* and *Bacillus cereus* link within the boundary of species (8.5 Euclidean distance units), which is consistent with current taxonomy. *Riemerella* and *Weeksella* could belong in the same genus since they link at a Euclidean distance of 18.9. The *Bacillus pasteurii* was grown on medium 1376 for 72 hours instead of the standard medium.

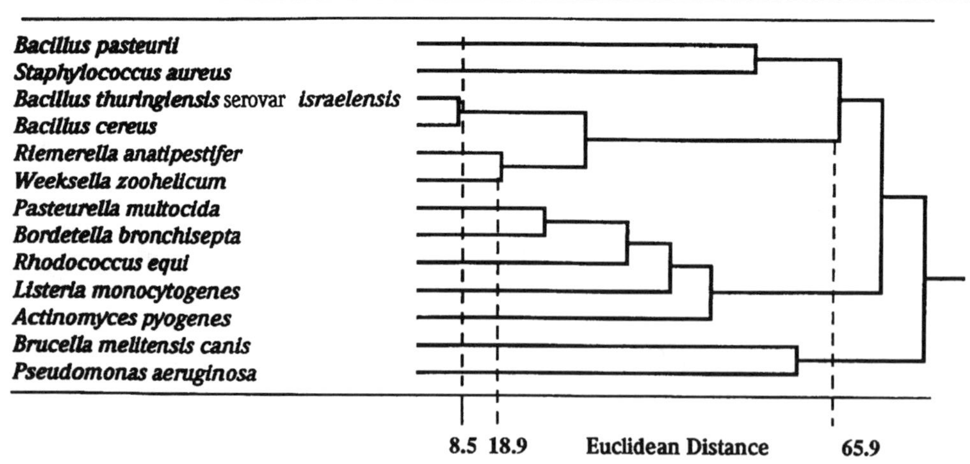

and *Bacillus cereus* link within the boundary of species ($\leq$10), they could be the same species. However, current taxonomical considerations (presence of insecticidal crystal proteins in *Bacillus thuringiensis* serovar *israelensis*) support the validity of maintaining separate species designation for each. Therefore, this is an example of a subjective assessment where FA data is secondary to prevailing taxonomical considerations.

An example of two species which could belong to the same genus are *Riemerella anatipestifer* (*Flavobacterium*-like) and *Weeksella zoohelicum*, which link at a Euclidean distance of 18.9. The fact that both organisms produced nearly identical reactions (*Weeksella* species are similar to *Flavobacterium* species except they are nonoxidizers and usually nonpigmented) led us to conduct FA studies on each organism. The causative agent of new duck disease (septicemia-anserum exsudativa) is *Riemerella anatipestifer*, formerly known as *Moraxella (Pasteurella) anatipestifer* (Segers et al. 1993).

We have a dilemma due to the fact that *Bacillus pasteurii* does not link with the other two *Bacillus* species until 65.9 Euclidean distance units. We feel that *Bacillus pasteurii* should receive a new genus designation on the basis of FA analysis and since a different medium (ATCC media formulation 1376) and incubation time (72 hours) were required for growth. We submit the issue for consideration by the authorities in the genus *Bacillus*.

The taxonomical classification of the other eight genera is supported by CFA data. There is no logical pattern of the organisms chosen for the dendrogram (Figure 9.1) except to include a diverse group of microorganisms (gram positive and gram negative, cocci and rods). If you are skeptical of the validity of the technique, hopefully our review of the literature and dendrograms produced will convince you that CFA analysis has application in microbiology.

In its normal operating mode the *2-D Plot* option in MIS uses PCA of FAME profiles to group entries into a two-dimensional space. We have not used any 2-D Plots in this chapter, preferring the *3-D Spin Plots* generated by JMP. We feel that this option is superior because 3-dimensional relationships between library entries can be visualized.

Reproducibility

The FAs contained in the microbial calibration standard from 84 individual determinations (injections) were analyzed to determine reproducibility of the method. We used the JMP program to construct the dendrogram (Figure 9.2). Eighty-two determinations link at a Euclidean distance of <0.2 and all 84 come together $\leq$0.3. This is probably the basis of why profiles that link at <2.0 are considered to be the same strain by MIDI. We also note that this distance correlates to a very low CV in the calibrator standard, and that the greatest variation was in the FA identified as Sum feature 3[1] (either iso-$C_{16:1}$ or 3-OH-$C_{14:0}$) at 1.8.

We have also made multiple injections of the same clinical sample and have comparable results. Therefore, we feel that it is more productive to concentrate

---

[1]In the MIS, when the FA cannot be resolved with certainty, the FA will be identified as a *Sum feature n*, with the most likely FAs appearing in parentheses. Analysis using mass spectrometry can be used to resolve the identity.

Figure 9.2. Dendrogram constructed from 84 individual determinations of microbial calibrator standard. Different lots of the standard were used and runs made weekly over an 18-month period. All 84 link at a Euclidean distance of ≤0.3.

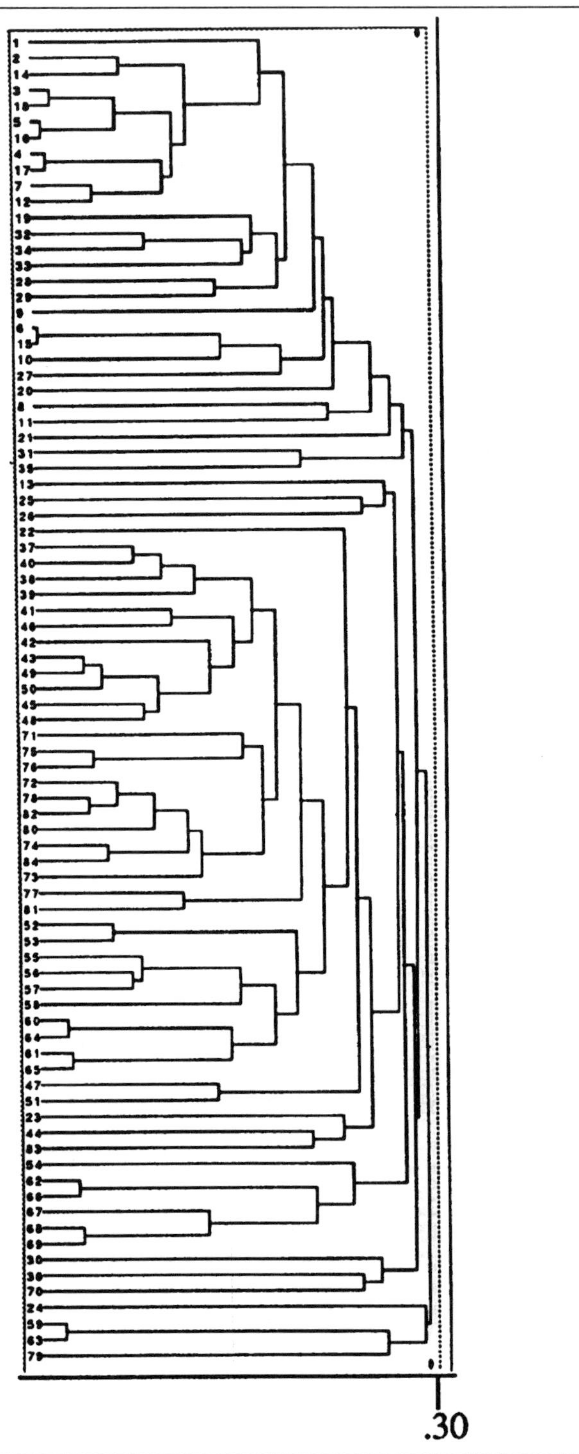

our efforts on extracting additional plates (originating new samples) rather than reanalyzing the same sample (plate) repeatedly (replication versus subsampling).

Evolution of Fatty Acid Data Interpretation

In general, the initial work done with CFA was to qualitatively identify the FAs present and then to visually compare genera on the basis of unique FAs. The next advancement occurred when the peak area of each FA was determined and quantitated against the total peak area. A crude attempt was then made to differentiate genera and species by calculating ratios of either the principal or unique FAs. The disadvantage of this approach is that the majority of FAs are ignored when a researcher only concentrates on two; quantitative data are lost when a ratio is used. For example, take two species X and Y, each with FAs A and B. For species X percentage composition of A and B is 10 percent and 5 percent, respectively. For species Y the respective FA concentrations are 30 percent and 15 percent. Yet, the ratios for both are 2.0. Likewise, discovering and grouping strains on the basis of minor FAs can be dangerous, because the majority of FAs are ignored. We believe that the likelihood of error is greatest when the principal FA is ignored in the grouping scheme. Instead of concentrating on ratios, some researchers have selected one or two FAs of interest and evaluated differences in percent composition by means of the student unpaired t-test. This approach still shares the same drawback of the ratio approach because the majority of FAs are ignored.

A more sophisticated approach, using correlation analysis was taken from numerical taxonomy. All FAs in each strain profile are used, and the presence of all FAs identified is scored as 1 or 0. Similarity indices are then calculated and the strains clustered by hierarchical methods. The disadvantages of this method are twofold. First, the quantitative data for the FAs are ignored, as each FA is scored as a 1 or 0. Second, the technique is particularly vulnerable to inappropriate profile creation. For example, if 10 replicates are made for strain X and 9 do not contain FA "A," some researchers would still score FA "A" as a 1. This character in strain X receives the same weight as FA "A" in strain Y present in all samples.

A survey of the recent literature indicates that some researchers are still unduly influenced by the "ratio mentality" and that not all authors place the same emphasis on the most abundant FAs. It is probably just common sense that the greatest weight should be given to the FAs present in the largest amounts. We believe that some of the difficulty in data interpretation is due to the numerous FAs (in some cases >20) that can be identified in the cell envelope of some species. It is difficult to process this information without computers; therefore, some researchers focus on two FAs, because it keeps their data set manageable. They are willing to sacrifice the bulk of the information present in their samples. The discriminatory ability of the technique improved greatly when PCA was introduced; this technique uses all the FAs as well as the quantitative data for each FA. Optimal distinction between strains and species was then possible.

Philosophical Considerations

How does one obtain low CVs? Only work with genera and species that are homogeneous! No, seriously, one must be aware that isolates and cultures from homogeneous public collections, when analyzed using techniques such as serotyping, may be heterogeneous when analyzed by FAs. The question that every researcher who contemplates using this techniqe must ask is "Do I want to be a *Lumper* or a *Splitter?*" The answer to this question will decide the number of CFA strains created from a given isolate. Our bias makes us *Splitters.* We feel that it is much easier to combine closely related clusters at a later date than it is to subdivide a heterogeneous library entry.

In our work with *Bacillus thuringiensis* serovars, it has been necessary to select single, isolated colonies, because accessions obtained from public collections were heterogeneous. It may be prudent to regard all cultures as mixed. We suggest that one should streak plates from single, isolated colonies for CFA whenever practical and, especially, whenever individual strains or isolates give variable CFA results. We realize that microorganisms that must be grown in liquid medium pose a greater challenge.

There is also the additional issue of how many strains to analyze in order to create a species profile. This is not an easy issue to resolve. The more heterogeneous the species, the more strains that will have to be analyzed and, possibly, the more GC subgroups that will need to be created in order to characterize it. Our general guideline is that 12 isolates (each obtained from a single colony) will need to be run to create a valid library entry for a species. We also place emphasis on inclusion of the type strain and other well-validated isolates for the creation of a species profile. Conventional identification procedures should be performed and the identity verified before being included as a strain used to create a library entry.

LITERATURE REVIEW

Fatty Acid Profiles of Gram-Positive Bacteria

Gram-positive bacteria contain three main groups of CFAs: straight-chain, branched-chain, and complex FA types (O'Leary and Wilkinson 1988). The straight-chain types contain saturated and monounsaturated FAs with a straight-acyl chain— $C_{16:0}$ and $C_{18:0}$ predominate. The branched-chain type possess large percentages of iso- and anteiso-branched acids, but rarely unsaturated FAs. The hydroxy (OH) FAs are examples of complex FA types. Cyclopropanes ($C_{19:0\ cyclo11,12}$) are considered to be derivatives of straight-chain types.

Kaneda (1977) proposed three types of lipid systems occuring in bacteria, each composed of a denovo synthetase with or without a modifying enzyme. Type 1, "branched-acid type," as seen in the genus *Bacillus*, do not require unsaturated FAs. However, such FAs do occur in some of these organisms. Nevertheless, lipid systems of type 2, "anaerobic type," must include unsaturated FAs for proper function in supporting growth. Type 2 organisms include *Clostridium* spp. and *Escherichia coli* (gram-negative). Type 3, "aerobic type," is very similar to type 2, but unsaturated FAs are synthesized by an aerobic pathway. Type 2 is more evolutionarily advanced than type 1, but not as advanced as type 3. Kaneda speculated that type 3 organisms may have led to the emergence of oxygen-producing photosynthetic organisms, introducing oxygen into the atmosphere and allowing the origin of eukaryotic cells.

The FA composition of gram-positive bacteria is especially affected by growth temperature (Suzuki et al. 1993). Cells grown at higher temperatures show a reduction in the ratio of unsaturated to saturated FAs. The branched-chain type of CFAs of gram-positive bacteria contain limited amounts of unsaturated FAs. The ratio of anteiso acids in some *Bacillus* strains decrease in response to higher incubation temperature. This phenomenon counters changes in membrane fluidity caused by increasing temperature.

In a later paper Kaneda (1991) refined his earlier conclusion that bacteria can be divided into three distinct groups on the basis of their membrane lipids. Group I consists of bacteria possessing cell membranes composed of straight-chain acyl esters. Most bacteria are members of this group. Group II has cell membranes composed of branched-chain and alicyclic esters. This includes about 10 percent of the bacterial species. *Staphylococci* spp. and *Arthrobacter* spp. (in addition to *Bacillus*) are gram-positive genera that are homogenous in that all species in the genera possess branched-chain FAs. The Staphylococci (refer to Table 9.1) consist mainly of anteiso-$C_{15:0}$, anteiso-$C_{17:0}$, $C_{18:0}$, iso-$C_{15:0}$, iso-$C_{17:0}$, and $C_{20:0}$ (generally in decreasing amounts of FA present). The anteiso-branched FAs (particularly anteiso-$C_{15:0}$) usually are the predominant FAs in the other two genera as well.

The genera *Micrococcus*, *Clostridium*, and *Corynebacterium* are heterogeneous (i.e., they include species that possess straight-chain FAs alone and those that also possess branched-chain FAs). Most *Clostridium* species possess only straight-chain FAs. *Corynebacterium* species have equal mixtures of the two types. Group III, the *Archaebacteria*, has cell membranes composed of isopenoid ethers.

The 2-OH and 3-OH FAs are examples of the complex type. Interestingly, 2-OH FAs are found in some of the gram-postive organisms. The 3-OH straight-chain FAs usually occur in certain gram-negative bacteria, such as *Pseudomonas* and *Serratia* spp. Usually bacteria with the branched-chain lipid system do not have hydroxy FAs. However, there are exceptions. Thus, their occurrence is rather specific and useful in their systematics.

The Genus *Bacillus*

A characteristic observed in all species of *Bacillus* studied is the predominance of terminally branched iso and anteiso FAs having 12–17 carbons (Kaneda 1977). He grouped the genus *Bacillus* into 6 groups (A–F) based on FA patterns. Groups A–D all had unsaturated FAs of <3 percent. Group E, which contained *Bacillus anthracis*, *Bacillus cereus*, and *Bacillus thuringensis* all had unsaturated FAs of 7–12 percent. Group F, consisting of all psychrophiles, had a large proportion of unsaturated FAs (17–28 percent). The normal saturated FAs, such as $C_{14:0}$ and $C_{16:0}$, the most common FAs in the majority of microorganisms, were generally minor constituents in the genus *Bacillus*. Considering the branched-FAs, iso-$C_{15:0}$ predominated in Groups C and E and anteiso-$C_{15:0}$ predominated in most of the other groups (Groups A, B, and F).

In an extension of Kaneda's (1977) examination of 22 strains, Kämpfer (1994) subjected 313 *Bacillus* strains to FA analysis. Essentially all of the strains were grown aerobically on Tryptone-Soya-Agar (Oxoid) at 30°C for 24 hours. Temperature was modified (20°C for *Bacillus globisporus*, 55°C for *Bacillus stearothermophilus*), and medium was modified (sheep blood agar) for *Bacillus pasteurii*. All strains were characterized by the predominance of branched FAs (ranging from 40 percent to more than 90 percent of total FAs. The strains were grouped into seven clusters (I–VII) using FA analysis by UPGMA clustering based on the correlation coefficient (S_C). Kämpfer's grouping was mainly based on the ratio iso-$C_{15:0}$/anteiso-$C_{15:0}$. The majority of the *Bacillus sphaericus* strains grouped in cluster IV; *Bacillus cereus* and *Bacillus thuringiensis* grouped in cluster VI. Clusters III–VI were characterized by a ratio of iso-$C_{15:0}$/anteiso-$C_{15:0}$ >2. Three strains (*Bacillus brevis*[b], *insolitus*, and *macquariensis*[b]) remained unclustered. Significant quantitative variations were observed for particular FAs in different strains belonging to the species *Bacillus firmus*, *Bacillus firmus-lentus* intermediates, *Bacillus macquariensis*, *Bacillus megaterium*, *Bacillus subtilis*, *Bacillus circulans*, *Bacillus coagulans*, *Bacillus macerans*, and *Bacillus cereus*. The species could be clustered into subgroups at a similarity level of 97.5 percent. Kämpfer could not assign several of the *Bacillus* species unambiguously to one of Kaneda's groups due to intraspecific variability. Kämpfer's conclusion was that with FA analysis, *Bacillus* strains can be assigned to species groups; however, the enormous heterogeneity of fatty acid profiles within several species reflects the unsatisfactory taxonomic situation within the genus, which can be improved by further DNA–DNA hybridization studies.

Relationship of Bacillus anthracis, cereus, and thuringiensis

Kaneda (1977) was the first to show that the *Bacillus* species *anthracis*, *cereus*, and *thuringiensis* cluster together. Three strains of *Bacillus anthracis* and seven strains of *Bacillus cereus* were grown on complex medium (RCM) and on synthetic medium (RM) by Lawrence et al. (1991). Could CFA analysis be used to differentiate these two species? Yes, CFA patterns of strains grown on RM showed a high content of branched-chain FAs and significant differences between the two species were found. The iso/anteiso ratio for the odd numbered carbons was approximately equal in the *Bacillus anthracis* strains, but the ratio was ≥2 in the *Bacillus cereus* strains. However, the CFA patterns of both species were nearly identical when grown on RCM. Lawrence and his coworkers succeeded in using CFA analysis to differentiate the two species in their diagnostic laboratory.

Bacillus thuringiensis serovar israelensis

The technique of CFA was used by Siegel et al. (1993) to determine if commercially produced *Bacillus thuringiensis* serovar *israelensis* mosquito larvacides could be distinguished. We examined five commercial brands (Acrobe®, Bactimos®, Skeetal®, Teknar®, and Vectobac®) as well as the current International Standard (IPS82). Multiple samples were run during a one-year period to maximize heterogeneity. The number of samples representing commercial brands run varied from 10 (Acrobe) to 23 (Teknar). The profile for IPS82 was generated from 32 individual plates. Each plate was inoculated with material derived from a single colony. The CFA profile for IPS82 is shown Table 9.2. The predominant FA was iso-$C_{15:0}$, composing 30.4 ± 2.2 percent ($\bar{x}$ ± s.d., CV 7) of the total. Other

---

[b] Designation by Kämpfer (1994).

principal FAs were Summed feature 4 (either 2OH-iso-$C_{15:0}$ or $C_{16:1 \ trans7}$) 11.8 ± 0.6 percent and iso-$C_{13:0}$ 11.9 ± 1.1 percent (CVs 5 and 9, respectively). The CVs of iso-$C_{13:0}$ were even higher for Bactimos® and Vectobac®, 13 and 16, respectively. Values this high may simply reflect that this FA was quite variable or that a composite profile of several strains was created. We concluded that all but Skeetal® were the same strain (Figure 9.3). We note that Teknar® A was contaminated with a different insecticide (*Bacillus thuringiensis* serovar *kurstaki*) as shown by its

Table 9.2. Mean percent compositon, standard deviation, and coefficient of variation of cellular fatty acids comprising >0.3 percent of the cell envelope of the current international standard for *Bacillus thuringiensis* serovar *israelensis* (IPS82). The profile was generated from the fatty acids extracted from 32 individual plates.

Simplified Name	Mean %	s.d.	CV
iso-$C_{12:0}$	0.9	0.4	49
iso-$C_{13:0}$	11.9	1.1	9
anteiso-$C_{13:0}$	1.4	0.3	24
iso-$C_{14:0}$	5.6	0.9	16
iso-$C_{15:0}$	30.4	2.2	7
anteiso-$C_{15:0}$	4.4	0.2	4
$C_{16:1 \ cis7}$ alcohol	0.7	0.3	50
iso-$C_{16:0}$	5.2	0.4	8
2-OH-$C_{15:0}$	0.6	0.3	49
iso-$C_{17:1 \ cis10}$	2.8	0.6	21
iso-$C_{17:1 \ cis5}$	5.5	1.0	19
anteiso-$C_{17:1 \ A}$	0.8	0.4	44
iso-$C_{17:0}$	6.2	1.0	17
anteiso-$C_{17:0}$	0.7	0.3	48
Summed feature 3[a]	3.8	0.5	14
Summed feature 4[b]	11.8	0.6	5

[a] Either 3-OH-$C_{14:0}$ or iso-$C_{16:1}$

[b] Either iso-$C_{15:0}$ or $C_{16:1 \ trans7}$

linkage in the dendrogram. Our results indicated that CFA could be used to distinguish the strains of *Bacillus thuringiensis* serovar *israelensis* produced by various manufacturers.

In a subsequent paper we demonstrated that our previous profiles were composites, based on CFA of the ONR-60A type strain of *Bacillus thuringiensis* serovar *israelensis* (Siegel et al. 1995). Isolates of ONR-60A were obtained from three culture collections. These isolates were compared to one another, and the culture collections differed in the strains they contained (Figure 9.4 and Table 9.3). Eleven separate strains of ONR-60A were identified using the criterion that profiles that linked at a distance of ≤2 Euclidean units were considered as belonging to the same strain. After IPS82 was reanalyzed, it consisted of three strains using the same definition. Library entries were created for each strain found in ONR-60A and IPS82, and validated against profiles of commercially produced *Bacillus thuringiensis* serovar *israelensis* from two time periods, 1983–1984 and 1990–1994. The same commercial insecticides with the exception of Skeetal® were used as in Siegel et al. (1993). Approximately 84 percent of the commercially produced *Bacillus thuringiensis* serovar *israelensis* (134 out of 160 plates) was identified as belonging to the strains found in ONR-60A and IPS82. The most prevalent strain (L) found in IPS82 was also the most prevalent strain in the recently produced commercial material. However, commercially produced *Bacillus thuringiensis*

Figure 9.3. Dendrogram of five commercially produced isolates of *Bacillus thuringiensis* serovar *israelensis* and the current international standard, IPS82. All but Skeetal® were the same strain. Source: Siegel et al. (1993)

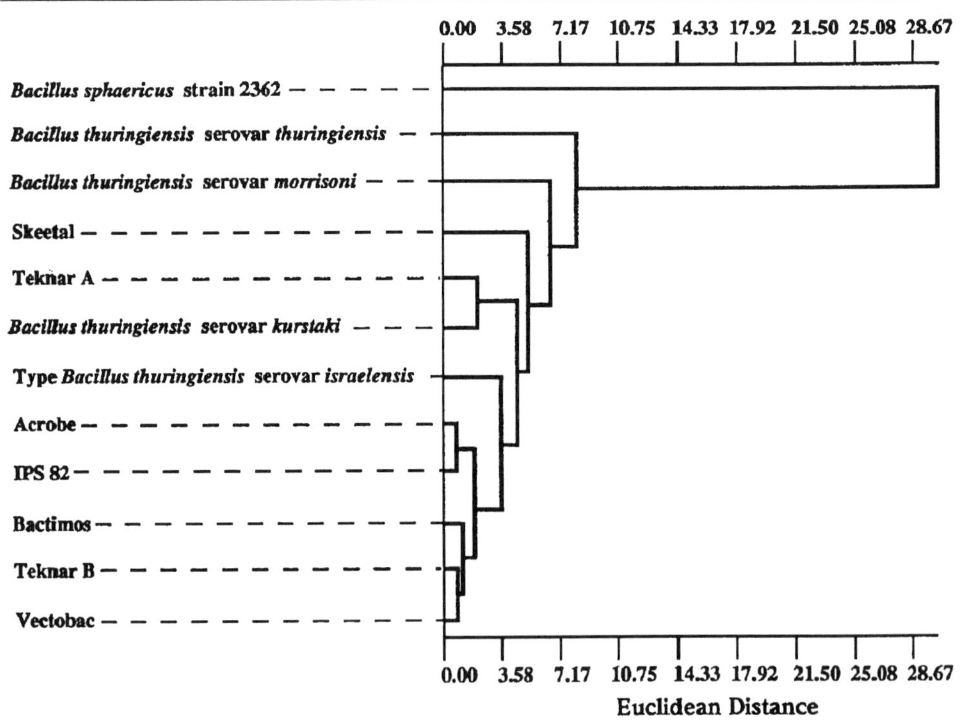

Figure 9.4. Dendrogram of the relationship between isolates of *Bacillus thuringiensis* serovar *israelensis*, IPS82, and isolates of the ONR-60A type strain. The IPS82 profiles are marked by arrows, and identical profiles, which link before two Euclidean distance units, are marked by the same letter. Source: Siegel et al. (1995)

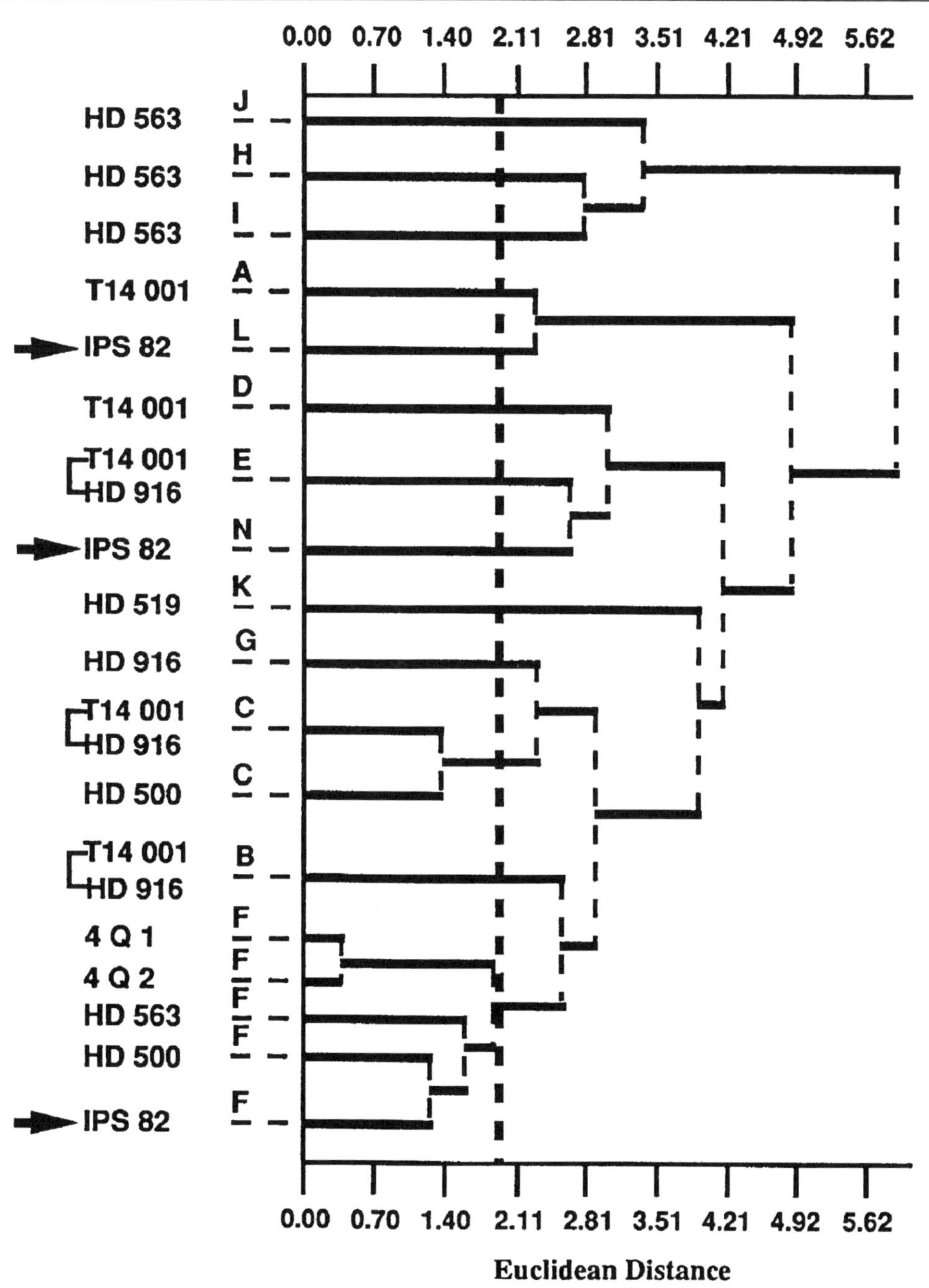

Table 9.3. Mean percent composition and coefficient of variation of cellular fatty acids comprising ≤5 percent of the cell envelope of ONR-60A, WHO 1884, and IPS82. (Source: Siegel et al. 1995)

Strain	13:0 iso	14:0 iso	15:0 iso	16:0 iso	iso 17:1ω5c	17:0 iso	Sum Feature 4[1]	N
HD 500-1	9.2/4	5.1/3	33.8/2	5.4/5	5.5/3	7.8/5	11.5/2	10
HD 500-2	9.5/5	5.6/1	31.4/1	5.6/4	5.7/3	7.8/3	11.8/1	5
HD 519	9.9/7	—	30.9/3	—	8.0/3	8.2/7	11.1/2	8
HD 563-1	8.9/6	5.0/8	31.7/2	5.4/9	6.7/9	7.5/6	12.2/4	7
HD 563-2	9.2/2	—	34.5/2	—	6.7/3	7.4/2	11.6/3	4
HD 563-4	10.4/3	—	37.8/3	—	6.6/13	7.1/7	10.9/2	2
HD 916-4	8.9/1	5.2/4	32.0/2	6.0/3	6.0/4	9.2/4	12.3/4	6
HD 917-1	11.1/4	6.5/7	27.7/3	5.8/5	5.5/8	6.2/10	11.9/2	7
HD 917-2	12.6/1	6.2/3	29.3/2	5.1/3	5.7/2	5.9/2	12.1/1	5
4Q1	10.2/3	—	32.3/2	—	6.7/3	7.3/4	12.1/2	15
4Q2	10.3/4	5.0/4	32.0/2	—	6.7/3	7.3/5	12.2/2	16
T14 001-1	12.9/4	5.8/7	31.3/5	5.2/7	—	6.4/9	11.4/4	7

Continued on next page.

Continued from previous page.

Strain	13:0 iso	14:0 iso	15:0 iso	16:0 iso	iso 17:1ω5c	17:0 iso	Sum Feature 4[1]	N
T14 001-2 + HD 916-1	10.5/3	6.2/11	30.6/8	6.1/8	5.0/15	6.8/10	12.1/5	7
T14 001-3 + HD 916-2	9.0/6	5.4/5	33.1/2	6.1/9	5.5/6	8.4/4	11.2/6	6
T14 001-4 + HD 916-3	9.0/8	5.1/3	31.6/1	6.7/10	5.3/1	9.6/1	10.5/1	2
T14 001-5	8.4/4	5.9/5	29.7/2	7.2/1	—	8.6/6	11.4/6	4
IPS 82-1	12.1/6	6.4/6	29.7/3	5.2/6	5.2/6	5.7/10	12.0/5	48
IPS 82-2	10.0/6	5.7/8	31.0/3	5.4/6	6.0/5	7.0/8	12.0/2	28
IPS 82-3	8.3/6	—	29.4/1	6.3/1	5.9/7	9.1/3	11.4/4	2

[1]Either 15:0 iso 2-OH or 16:1ω7t.

serovar *israelensis* also retained (strains A, B, D, F, and H) a strong ONR-60A ancestry (42 of 120 or 35 percent of the plates analyzed). We attempted to explain the heterogeneity of ONR-60A as possibly due to one or more of the following:

- Contaminants were introduced.

- The original isolate of ONR-60A contained multiple strains of *Bacillus thuringiensis* serovar *israelensis*.

- The original isolate originated from a single colony, but subsequently spontaneous mutants have been selected and propagated.

In view of these findings, it is important to have isolated colonies and to transfer single colonies (one colony per plate) to the proper GC medium. Our paper was written after 355 plates were analyzed. We caution researchers that there is no single strain of ONR-60A and that there is tremendous variation in what is broadly designated as ONR-60A.

Bacillus sphaericus

Bacillus sphaericus can be divided into six homology groups based on DNA hybridization. Only one DNA homology group (IIA) produces toxins with insecticidal activity (Krych et al. 1980). Frachon et al. (1991) used GLC to obtain FAMEs and carried out numerical analysis with 114 strains of *Bacillus sphaericus* in an attempt to link toxin activity with FAME profiles. All of the strains came from the International Entomopathogenic *Bacillus* Center of the Pasteur Institute (IEBCPI). The strains were chosen (according to geographical origin) to include all DNA–DNA homology groups and H serotypes. Thirty-eight strains toxic for mosquitoes were also selected. Numerical analysis revealed two main clusters [cluster A (79 strains) and cluster B (29 strains)]. All toxic strains detected were in cluster A. Cluster A was reevaluated by PCA. There were six groups of strains (I–VI) that converged into one cluster at a Euclidean distance of 12. The 38 toxic strains were found in group III (21 of 29) and group V (17 of 19). Therefore, 80 percent of the strains of groups III and V exhibited toxicity to mosquito larvae. However, there was no explanation for the association between mosquitocidal activity and CFA. Frachon and his coworkers stated that prescreening using CFA could be useful to select strains of groups III and V for bioassay of *Culex pipiens* larvae. The identification and characterization of *Bacillus sphaericus* strains within the same serotype was also another advantage of this technique.

In a later paper (Schenkel et al. 1992) CFA was used to characterize four strains (S1, S2, S5, and L2) of *Bacillus sphaericus* isolated from Brazilian soil that were highly toxic to mosquito larvae. Selected strains from IEBCPI were used for comparison of toxicity (strain 2362) and for CFA (strains 1593, 2362, and 2297). The four Brazilian soil strains and 2362, belonged to serotype H5 and shared proteins (toxins) of 56 and 43 kDa molecular weight. The Euclidean distance between isolates and strains reported was one log too small and this mistake was confirmed by the authors (scale of 0.1–0.7 instead of 1–7).

Purported Bacillus popilliae Actually Bacillus polymyxa

Stahly and Klein (1992) investigated the failure of a commercial spore larvicide to control the Japanese beetle, *Popillia japonica*. This product was supposedly formulated from *Bacillus popilliae* produced by a novel method; however, CFA analysis and conventional analysis indicated that *Bacillus popilliae* was absent from the product. Instead, most batches of the spore powder and fermentation products they examined contained *Bacillus polymyxa* and *Bacillus amyloticus*.

Fatty acid analysis and conventional bacteriology was conducted on type strains and four cultures submitted to the American Type Culture Collection (ATCC) at time of patent application as representative of *Bacillus popilliae*. Three of the four ATCC cultures were identified as *Bacillus amyloticus*, and the fourth (referred to as the preferred strain) was identified as *Bacillus polymyxa*. The conclusions derived from FA analysis were consistent with those derived from morphological, cultural, and physiological characteristics. They concluded that the patented procedure should be reexamined since it was not effective in producing infective *Bacillus popilliae* spores. This example illustrates the utility of FA testing as an in vitro procedure that can help verify or refute identity of species of viable bacteria present in pharmaceutical formulations.

Bacillus brevis

To illustrate the value of CFA in taxonomy, we prepared a dendrogram (Figure 9.5) from data presented by Shida et al. (1995) from six putative *Bacillus brevis* strains (2 each that were formerly assigned to groups 5, 6, and 7) and 5 type strains (27, 28, 29, 31, and 33). Our dendrogram of their data, based only on CFA, reveals 7 species instead of the 8 reported by Shida et al. (using the Euclidean

Figure 9.5. Dendrogram of six putative *Bacillus brevis* strains. Seven clusters were well separated at 10 Euclidean distance units. Source of the data was Shida et al. (1995).

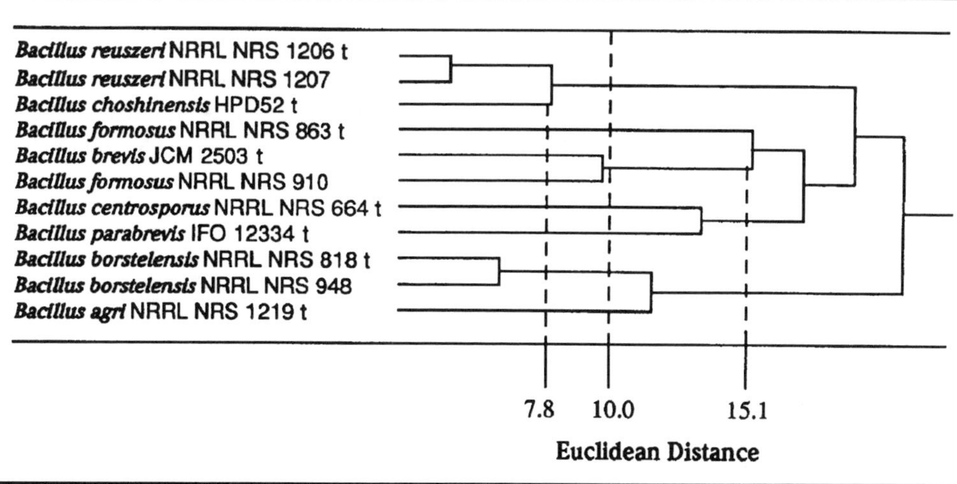

distance of 10 for the species boundary). At a distance of about 35.5, all the *Bacillus brevis*–like strains were grouped together as a single cluster. At a distance of ≤10, the 7 different clusters were well separated, indicating the ability of the technique to discriminate between subspecies.

Asporogenous, Gram-Positive Rods

The gram-positive coryneforms are difficult to identify by classical identification methods. Composition of the CFA of 561 strains (characterized clinical isolates and reference) grown on Columbia agar with 5 percent sheep blood at 35°C in CO_2 were reported by Bernard et al. (1991). Adequate growth was obtained for a majority of the strains after 24 hours of incubation. However, multiple plates and/or 48–72 hour incubation were required for more fastidious strains. They used the LGS to create and update a library for the coryneforms studied.

The coryneforms could be divided into two groups: Group 1 (the branched-chain type) and Group 2 (straight-chain or unsaturated type). Group 1 included coryneform CDC groups A-3, A-4, A-5; some strains of B-1,and B-3; *Corynebacterium aquaticum*; *Brevibacterium* spp.; *Listeria* spp.; and *Rothia dentacariosa*. Group 2 included *Actinomyces pyogenes*, *Arcanobacterium haemolyticum*, *Erysipelothrix rhusiopathiae*, *Corynebacterium* spp., and the remaining coryneform CDC groups. They concluded that CFA composition was sufficiently discriminatory to assist in classification but could not be used as the sole means of identification. We were unable to fully evaluate their data (calculate CVs since they did not include standard deviations).

Wüst et al. (1993) used CFA to determine that five strains of organisms resembling *Actinomyces pyogenes*, which were isolated from human beings with mixed wound infections, were distinct. They evidently grew these organisms and *Actinomyces pyogenes* anaerobically by using the Virginia Polytechnic Institute (VPI) broth medium (PRAS PYG-T). Unfortunately, *Actinomyces pyogenes* was not included in the MIS database library for anaerobes, so their data could not be directly interpreted.

The genus *Eubacterium* may require extensive taxonomic revision if data reported by Itoh et al. (1995) are representative. The genus *Eubacterium* is defined as asporogenic gram-positive rods, which are obligately anaerobic and appear to play a potent pathogenic role in oral infectious diseases. They performed CFA on 10 species, and concluded that they divided into 5 groups by PCA. When we constructed a dendrogram from their data (Figure 9.6), their isolates grouped into 6 clusters (at a Euclidean distance of 32). However, these isolates might encompass at least 6 genera (using the Euclidean distance of 25–30 as the boundary for genus designation).

Gram-Positive Cocci, the Staphylococci

The MIS was used to identify 470 isolates of *Staphylococcus* species (Stoakes et al. 1994). The accuracy of the MIS, which relied only on the clinical aerobic library, was compared to the accuracy of conventional methods. The strains were grown on trypticase soy agar (TSA) supplemented with 5 percent sheep blood and were incubated at 35°C for 24 hours in ambient air. The MIS and conventional methods

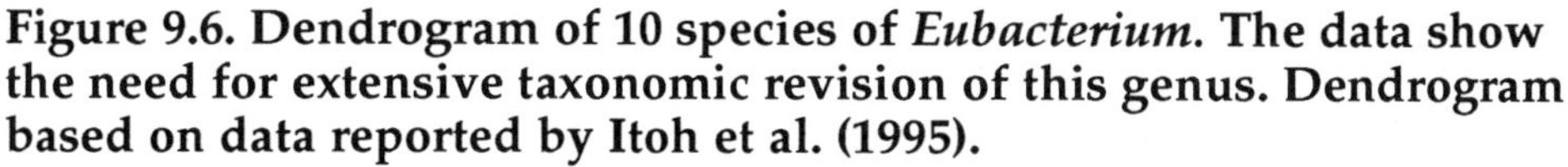

Figure 9.6. Dendrogram of 10 species of *Eubacterium*. The data show the need for extensive taxonomic revision of this genus. Dendrogram based on data reported by Itoh et al. (1995).

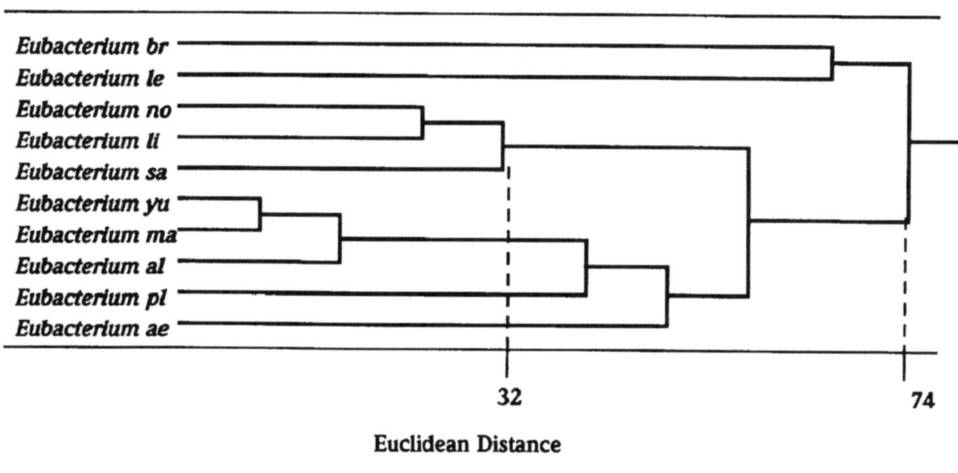

agreed in the identification of 87.8 percent (413/470) of strains tested. Two species of coagulase-negative staphylococci (*Staphylococcus hominis* and *Staphylococcus saprophyticus*) accounted for 52.6 percent of the incorrectly identified strains. The MIS library needed improvement, especially in regard to *Staphylococcus aureus* (where 69/76 strains were correctly identified, but the remaining 7 strains were not identified as staphylococci). However, for the one other coagulase-positive species tested, *Staphylococcus intermedius*, all 12 strains were identified correctly. The authors stated that the performance of the system may be enhanced by creation of user-specific libraries.

Because of the ubiquitous nature of staphylococci in the environment, effective methods are needed for identification at the species level. The close relationship of the coagulase-negative staphylococci was shown based on FA analysis by Kotilainen et al. (1991). We analyzed their Table 1 data and produced a dendrogram (Figure 9.7). All five species cluster at a Euclidean distance of 10.3. They also determined identity, as well as differences between multiple isolates of *Staphylococcus* spp. obtained from blood of hospitalized patients. They speculated that discrimination may be possible between related and unrelated strains isolated during hospital outbreaks.

Fatty Acid Profiles of Gram-Negative Bacteria

The gram-negative bacteria contain primarily straight- and branched-chain FA types. The most common FAs are $C_{16:0}$, $C_{16:1}$, and $C_{18:1}$. Cyclopropane acids are often found ($C_{17:0\ cyclo}$ or $C_{19:0\ cyclo}$). All of the above FAs are usually found in the cytoplasmic membrane. In addition, the gram-negative organisms usually contain 3-OH FAs located as components of their lipid A. The 3-OH FAs are

Figure 9.7. Dendrogram of five strains of coagulase-negative *Staphylococcus* spp. These are closely related species since they form one cluster at 10.3 Euclidean distance units. Dendrogram based on data reported by Kotilainen et al. (1991).

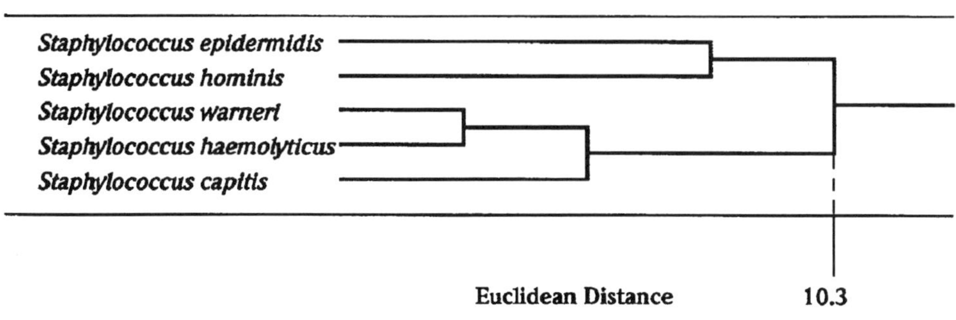

important in the Pseudomonads and *Serratia* spp. (refer to Table 9.1). In contrast to the gram-positive organisms, branched-chain FAs are relatively uncommon. Gram negatives with branched-chain FAs may have iso-$C_{15:0}$ and iso-$C_{17:1}$; they also characteristically produce iso-branched 3-OH FAs (Goodfellow and O'Donnell 1993).

The Pseudomonadaceae

The genus *Pseudomonas* is a large heterogeneous group, which has been difficult to assign to species using conventional bacterial procedures. Palleroni (1984) established five subgroups arranged according to rRNA and DNA homologies. Stead (1992) developed CFA profiles for 340 strains comprising the genus *Pseudomonas*. The 2- and 3-OH FAs were found to be useful in grouping strains into six major groups (1–6); he further differentiated several into subgroups (1a, 1b, 1c, 1d, 1e, 1f, 2a, 2b, 3a, and 3b). His FA results supported the conclusions of other researchers using various techniques—that these pseudomonads should be placed in at least six genera. Some taxonomic changes that have been made were moving Stead's group 3 to the genus *Commonas*, moving Stead's group 4 to the genus *Sphingomonas*, and moving Stead's group 6 to the genus *Xanthomonas*.

Plant pathogenic and related saprophytic bacteria (773 strains representing 25 taxa) were analyzed by FA profiling and compared with two MIDI libraries and one self-generated library based primarily on cultures from the National Collection of Plant Pathogenic Bacteria (NCPPB) (Stead et al. 1992). The most accurate identification was provided by the NCPPB library, which contained only plant pathogenic species. Fourteen of the 25 taxa levels had >95 percent accurate identification with this library. The *Pseudomaonas syringae* pathovars had the lowest percentage of accurate identification (7.2–86.4 percent for the four pathovars). The authors stressed the necessity of standardized cultural conditions for reproducibility, and stated that the accuracy of identification will be increased by improved library development and standardization of cultural and analytical techniques.

Pseudomonas cepacia (presently classified as *Burkholderia cepacia*) belonged in Stead's (1992) Subgroup 2a. He analyzed seven strains using CFA. The high CV values for the principal FAs (23 for $C_{18:1\ cis11}$, 20 for $C_{16:0}$, and 40 for $C_{16:1\ cis9}$) indicated that this species was very heterogeneous (a homogeneous entry would have values <8. Therefore, it was interesting that Mukawa and Welch (1989) had analyzed 42 strains of *Pseudomonas cepacia* from five cystic fibrosis centers in North America, enabling us to compare the findings of these two research groups. Using cluster (numerical) analysis, Mukwaya and Welch separated the strains into five subgroups. With minor exceptions the predominant subgroup identified in each center was different from that identified in the other centers. This suggested that these were nosocomial infections acquired in the hospital. Mukwaya and Welch stated that CFA may provide useful epidemiologic information or provide a basis for further analysis, such as DNA probe analysis. A direct comparison of Stead's data with Mukwaya and Welch's data was not possible since *Pseudomonas cepacia* were grown on a different medium. However, given the heterogeneity of Stead's profile, we believe that the grouping of Mukwaya and Welch is more appropriate.

The MIS using the TSBA library (v. 3.0) was compared to a conventional system for the identification of 573 strains of gram-negative nonfermentative bacteria (Osterhout et al. 1991). The MIS correctly identified 478 of 532 (90 percent) strains contained in the database. However, only 55 percent (314/573) of the strains had a SI of ≥0.5. A SI of 0.5 means that the identified isolate is approximately 3 s.d. from the entry mean. They concluded that the MIS was accurate, efficient, and relatively rapid for the identification of gram-negative nonfermentative bacteria. However, the authors stated that development of a CFA library using media and incubation conditions routinely used in hospital and clinical laboratories was needed.

The Genus Xanthomonas

Xanthomonas, which contains many branched FAs, originated as a reclassification of some misfits from the genus *Pseudomonas*. A total of 975 strains of *Xanthomonas* species [representatives of all 7 species (*albilimeans, axonopodis, campestris, fragariae, maltophila, oryzae,* and *populi*)], 134 *Xanthomonas campestris* pathovars, and two *Xanthomonas oryzae* pathovars were subjected to GLC and their FAs analyzed (Yang et al. 1993). Using UPGMA cluster analysis of the FAME profiles, 31 clusters were delineated (those strains grouping <9 Euclidean distance were considered to be homogeneous). Five of the 7 species (all except *Xanthomonas campestris* and *Xanthomonas oryzae*) each constituted a separate cluster. Two clusters were formed within *Xanthomonas oryzae*, corresponding to pathovars *oryzae* and *oryzicola*. The species *Xanthomonas campestris* were determined to be heterogeneous and comprised 24 clusters. Use of their database allowed Yang and his coworkers to identify unknown xanthomonads rapidly at the genus, species, and often pathovar level. They reported that the analysis provided information at a similar taxonomic level as protein profiles and DNA–DNA hybridization studies within the genus *Xanthomonas*.

Campylobacter

Cellular FA analysis was used to differentiate 368 strains of *Campylobacter* and *Campylobacter*-like organisms (Lambert et al. 1987). Most of the strains (339) were placed in three groups on the basis of FA differences. *Campylobacter jejuni* and most *Campylobacter coli*, were in group A (characterized by $C_{19:0\ cyclo}$ and 3-OH-$C_{14:0}$). Group B was similar to group A, except that $C_{19:0\ cyclo}$ was missing. Group C contained *Campylobacter fetus* and was characterized by 3-OH-$C_{16:0}$ and 3-OH-$C_{14:0}$. The remaining 29 strains were placed in Groups D–G. We have prepared a dendrogram from data presented in their Table 1 (Figure 9.8). The dendrogram verifies the validity of the two profiles for *Campylobacter coli*. Also, the dendrogram suggests that *Campylobacter pyloridis* was wrongly classified as belonging to the genus *Campylobacter* since it links at a Euclidean distance greater than 30 (53.3). This organism is presently classified as *Helicobacter pylori*. We believe the data of Lambert and his coworkers validate the role of CFA in taxonomy.

Aeromonas

The genus *Aeromonas* contains at least 11 phenospecies (biochemically distinct species) and at least 14 DNA hybridization groups (HGs) or genospecies (Janda 1991). Ninety genotypically characterized strains (all 14 HGs were represented) were studied by the CFA technique (Huys et al. 1994). The four FAs present in highest concentration in the genus *Aeromonas* were $C_{16:1\ cis7}$, $C_{16:0}$, Summed feature 7 (one or more of the $C_{18:1}$ isomers), and Summed feature 3 (either $C_{16:1\ iso}$ and/or 3-OH-$C_{14:0}$). We used Huys' data to produce the dendrogram shown in Figure 9.9 (which is similar to their Figure 1). Cluster analysis of the mean FA values revealed five clusters (A–E) at 7.1 Euclidean distance. Cluster A contained 27 strains comprising HGs 8 (*Aeromonas veronii* biogroup *sobria*), 13 (*Aeromonas trota*), 9 (*Aeromonas jandaei*), and 12 (*Aeromonas schubertii*), and was separated from the other four clusters at a Euclidean distance of 12.8. Cluster B (consisted of 10 *Aeromonas caviae* HG4 strains) and linked to the remaining clusters at a Euclidean distance of 10.7. Cluster C comprised HGs 5A (*Aeromonas*

Figure 9.8. Dendrogram of seven clusters of *Campylobacter* and *Campylobacter*-like organisms. Reclassification of *Campylobacter pyloridis* to *Helicobacter pylori* has occurred. Dendrogram based on data of Lambert et al. (1987).

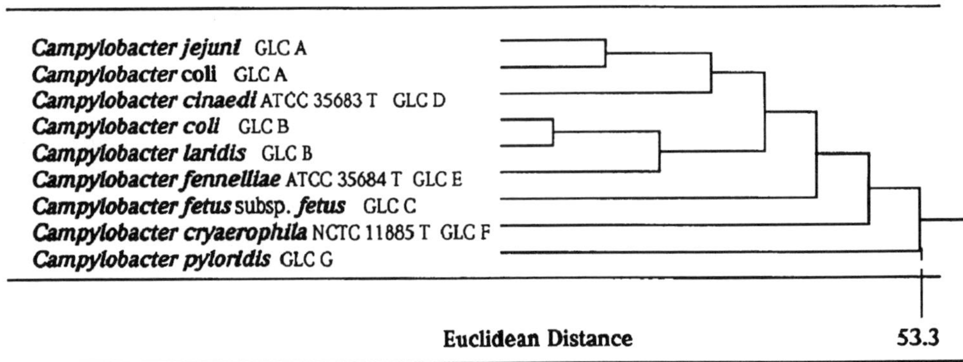

Figure 9.9. Dendrogram of 14 DNA hybridization groups of the genus *Aeromonas*. Five cluster groups form at 7.0 Euclidean distance units. Cluster E contained 7 *Aeromonas sobria* (HG7) strains and 10 *Aeromonas hydrophila* (HGs 1, 2, and 3) strains. Dendrogram based on data of Huys et al. (1994).

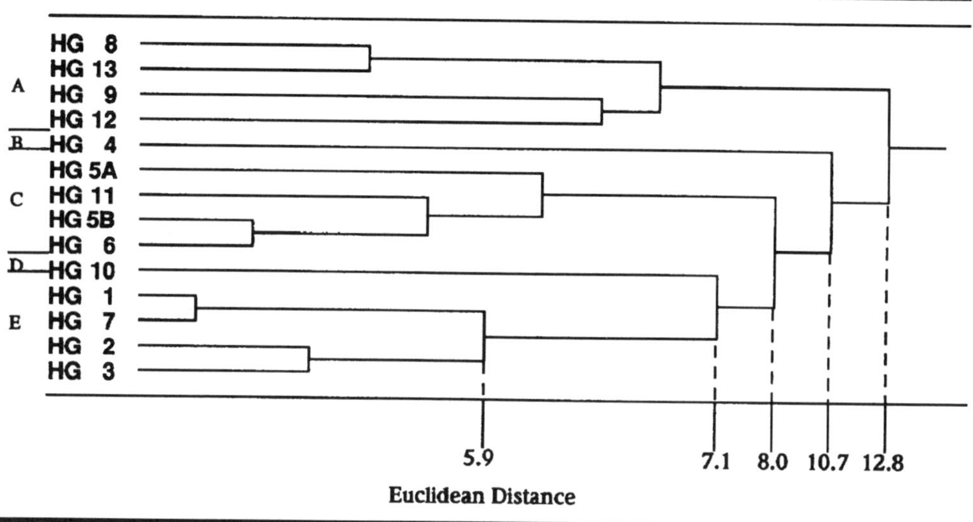

caviae), 5B (*Aeromonas caviae* and *Aeromonas media*), 11 (*Aeromonas veronii*), and 6 (*Aeromonas eucrenophila*), its members linked at a Euclidean distance of 8.0. Cluster D consisted of 4 *Aeromonas veronii* biogroup *veronii* (HG10) strains and joined at a Euclidean distance of 7.1. Cluster E consisted of 7 *Aeromonas sobria* strains (HG7) and 10 *Aeromonas hydrophila* strains (HGs 1, 2, and 3), linking at a Euclidean distance of 5.9. Hybridization groups that constituted the *Aeromonas hydrophila* complex, the *Aeromonas caviae* complex, and the *Aeromonas sobria* complex were basically grouped into distinct FAME clusters. The proper taxonomic position of HGs 7 and 11 in these clusters remained unclear. However, they demonstrated that FA analysis can be used to differentiate the majority of the phenospecies and/or HGs in the genus *Aeromonas*.

In contrast, the conclusions reached by Kämpfer et al. (1994), who also attempted to characterize *Aeromonas* using FA analysis, were quite different. They stated that although some differences in fatty acid profiles between the genomic species could be observed, a clearcut differentiation of all species was not possible.

Gram-Negative Anaerobic Bacilli

The genera *Bacteroides*, *Wolinella*, and *Campylobacter* contain several similar species that require taxonomic revision (Brondz and Olsen 1991). Asaccharolytic gram-negative rods have been difficult to classify since these organisms do not catabolize carbohydrates and are inert biochemically in most tests. The genus *Bacteroides* has been recently been reclassified. Only members of the "*Bacillus fragilis* group" have been retained in the genus *Bacteroides*. The genus *Porphyromonas* consists of pigmented, asaccharolytic species (previously classifed as *Bacteroides*). The genus *Prevotella* was created for moderately saccharolytic species like *Bacteroides melaninogenicus*, *Bacteroides oralis*, and related species. Brondz and Olsen used FA analysis to assess taxonomic relationships and increase the number of reliable characteristics for classification and identification of the above five genera. Their data were subjected to PCA. The dendrogram (Figure 9.10) and plot (Figure 9.11) were prepared from the FA compositions they reported. The dendrogram and plot validate the reclassification of the genus *Bacteroides* and support their conclusions.

Bacteroides fragilis was distinct from the other genera. They reported that *Bacteroides gracilis*, *Campylobacter fetus* subspecies *venerealis*, and three species of *Wolinella* were close to each other (Figure 9.11; Figure 9.10 establishes the Euclidean distance as 25.3, which validates them as being in the same genus). However, Brondz and Olsen distinguished the six species of *Prevotella*, *Porphyromonas endodontalis*, and *Bacteroides ureolyticus*. Thus, their FA data supported the current belief that *Bacteroides fragilis* should be in a distinct genus and that *Bacteroides gracilis* and *Bacteroides ureolyticus* are not "true" *Bacteroides*.

A recent article (Vandamme et al. 1995), based in part on CFA, has reclassified *Bacteroides gracilis* as *Campylobacter gracilis*. However, *Bacteroides ureolyticus* (considered a *Campylobacter* on a genotypic basis) was excluded from the *Campylobacter* taxon on results obtained by its FA components and proteolytic metabolism. They stated that before proper classification of *Bacteroides ureolyticus*, more strains

---

Figure 9.10. Dendrogram of strains comprising the former genus *Bacteroides*, which confirms the validity of the reclassification. Dendrogram based on data of Brondz and Olsen (1991).

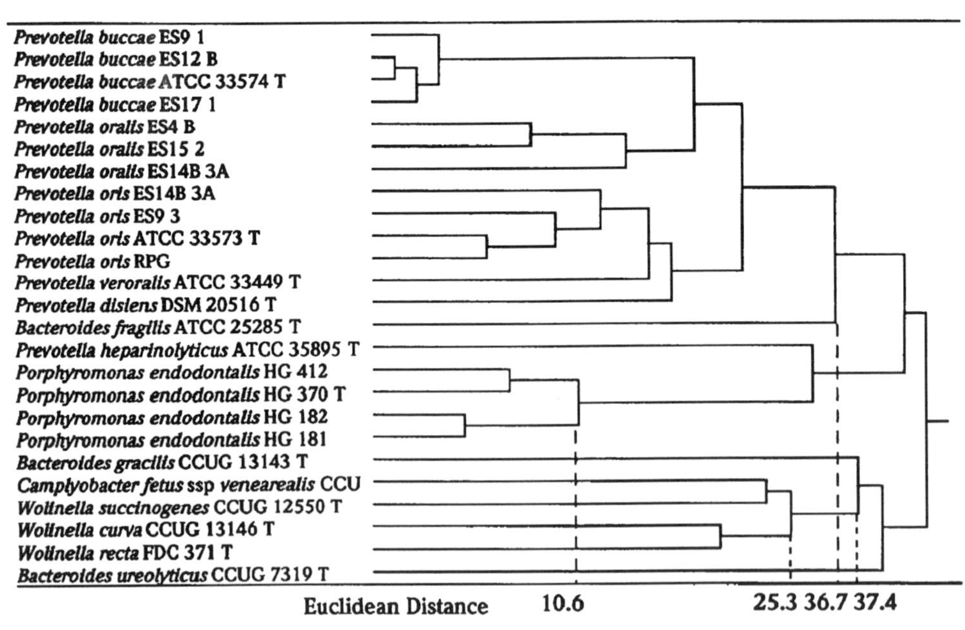

Figure 9.11. Plot of the first three principal components of strains comprising the former genus *Bacteroides*. The principal components are represented by the x, y, and z axes. Plot taken from data of Brondz and Olsen (1991).

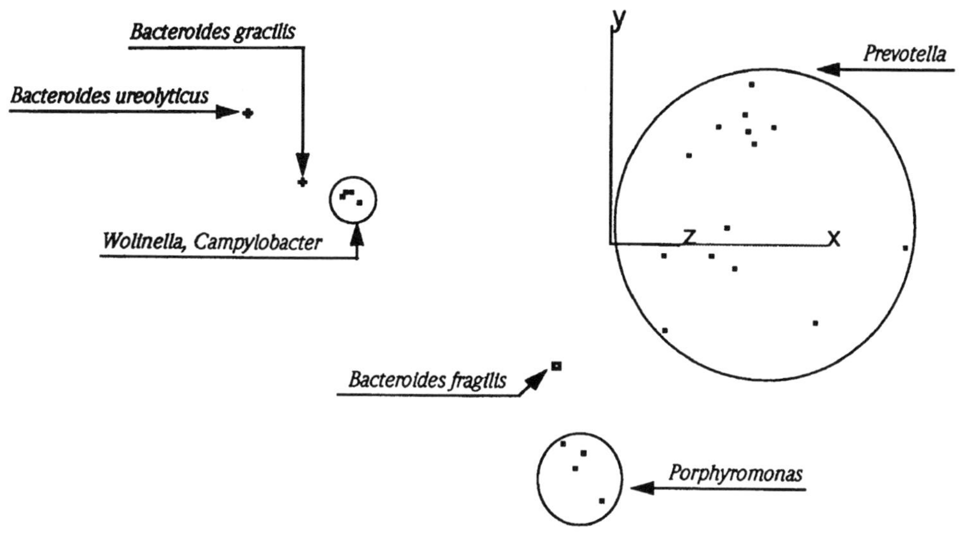

would need to be isolated and characterized. Figure 9.12 is a dendrogram from their CFA data (Table 4, Vandamme et al. 1995).

An evaluation of gram-negative anaerobic bacilli using MIS and the VPI broth-based anaerobe library (v. 3.2) was conducted by Stoakes et al. (1991). The genus *Bacteroides* was identified correctly at the species level 76.4 percent (71/93) of the time. However, none of the 12 *Bacteroides thetaomicron* were correctly identified at the species level (10 were misidentified as the closely related *Bacteroides caccae*). The library performance was better for the genus *Fusobacterium*, where 90 percent (19/21) of the identifications were species correct. Overall, identification was 100 percent correct at the genus level for both *Bacteroides* and *Fusobacterium* genera. The genera *Porphyromonas* and *Prevotella* were also analyzed. Only 69.2 percent of the *Porphyromonas* strains (9/13) were correctly identified at the species level (with 2 strains being incorrectly identified at the genus level). The system did worse within the genus *Prevotella*, where its accuracy at the species level was only 65.3 percent (64/98, with 13 strains incorrectly identified at the genus level). Considering the 225 strains tested, 163 were correctly identified at the species level (72.4 percent) and 210 strains were correctly identified at the genus level (93.0 percent). Stoakes and his coworkers concluded that the system needed refinement and improvement of its database before it could be recommended for general use.

In 1994 species from 33 genera of gram-negative anaerobic bacteria were analyzed by CFA to determine profiles. The method was found beneficial as a

Figure 9.12. Dendrogram supporting reclassification of *Bacteroides gracilis* to *Campylobacter gracilis* based on data of Vandamme et al. (1995).

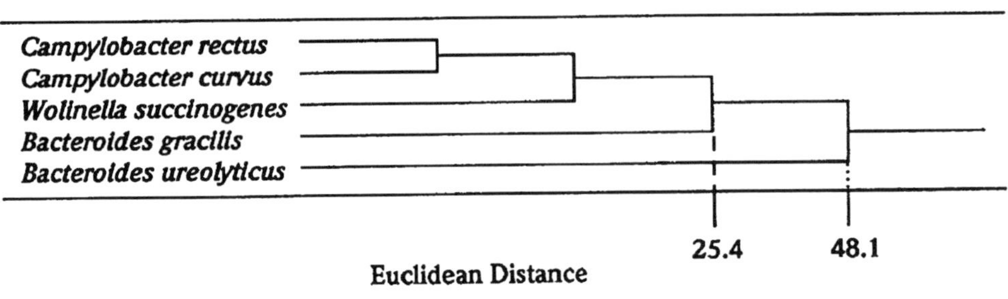

taxonomic tool (Moore et al. 1994). Most of the genera could be differentiated by visual examination of the FA profiles. We used their data (Table 2, Moore et al. 1994), and constructed the dendrogram (Figure 9.13). The genus *Johnsonella* is the most distantly related to the other genera (separated from the other 32 genera at a Euclidean distance of 57.1). However, they stated that the species were differentiated by the anaerobic library and stressed the necessity of using the same cultural (medium) and analytical conditions for valid comparisons with the library entries.

The Spirochetes

Some recent FA analyses have been performed with the spirochetes. The FA profiles of *Borrelia*, *Serpulina*, and *Leptospira* were compared (Livesley et al. 1993a). Our analysis using PCA of their data verified the validity of separate genera. The linkage occurred at 37 for *Leptospira icterohaemorrhagiae* and *Borrelia burgdorferi*, and at 38.4 for *Serpulina hyodysenteriae*. A follow-up study was done where FA analysis was determined for 12 *Borrelia burgdorferi* and other *Borrelia* species (Livesley et al. 1993b). The dendrogram reveals that all the isolates belong in the genus *Borrelia* and clustered in five groups (Figure 9.14). All strains of *Borrelia burgdorferi* cluster in group 1 (at a Euclidean distance of <9) in Figure 1 of Livesley et al. (1993b), which supports the delineation as *Borrelia burgdorferi* sensu stricto. The borreliae associated with tick-borne relapsing fevers (*Borrelia parkeri*, *Borrelia turicatae*, and *Borrelia hermsii*) are found in clusters 2 and 4. The first two are in cluster 2 and thus combined, but *Borrelia hermsii* is kept distinct since it links at Euclidean distance 11.4. The VS461 (cluster 3) links with cluster 2 (all are relapsing fever spirochetes). They concluded that the diversity of Lyme disease spirochetes has implications for the protean clinical manifestations of the disease.

The data of Moore et al. (1994) support the reclassification of *Treponema hyodysenteriae* to *Serpulina hyodysenteriae* (Figure 9.13). The genus *Treponema* links with *Serpulina* at a Euclidean distance of 37.7. An early attempt was made by Matthews and Kinyon (1984) to differentiate *Treponema hyodysenteriae* and *Treponema innocens* (both have been reclassified as *Serpulina*) using cellular lipid

Figure 9.13. Dendrogram showing relationship of 33 genera of gram-negative anaerobic bacteria. Data is from Moore et al. (1995).

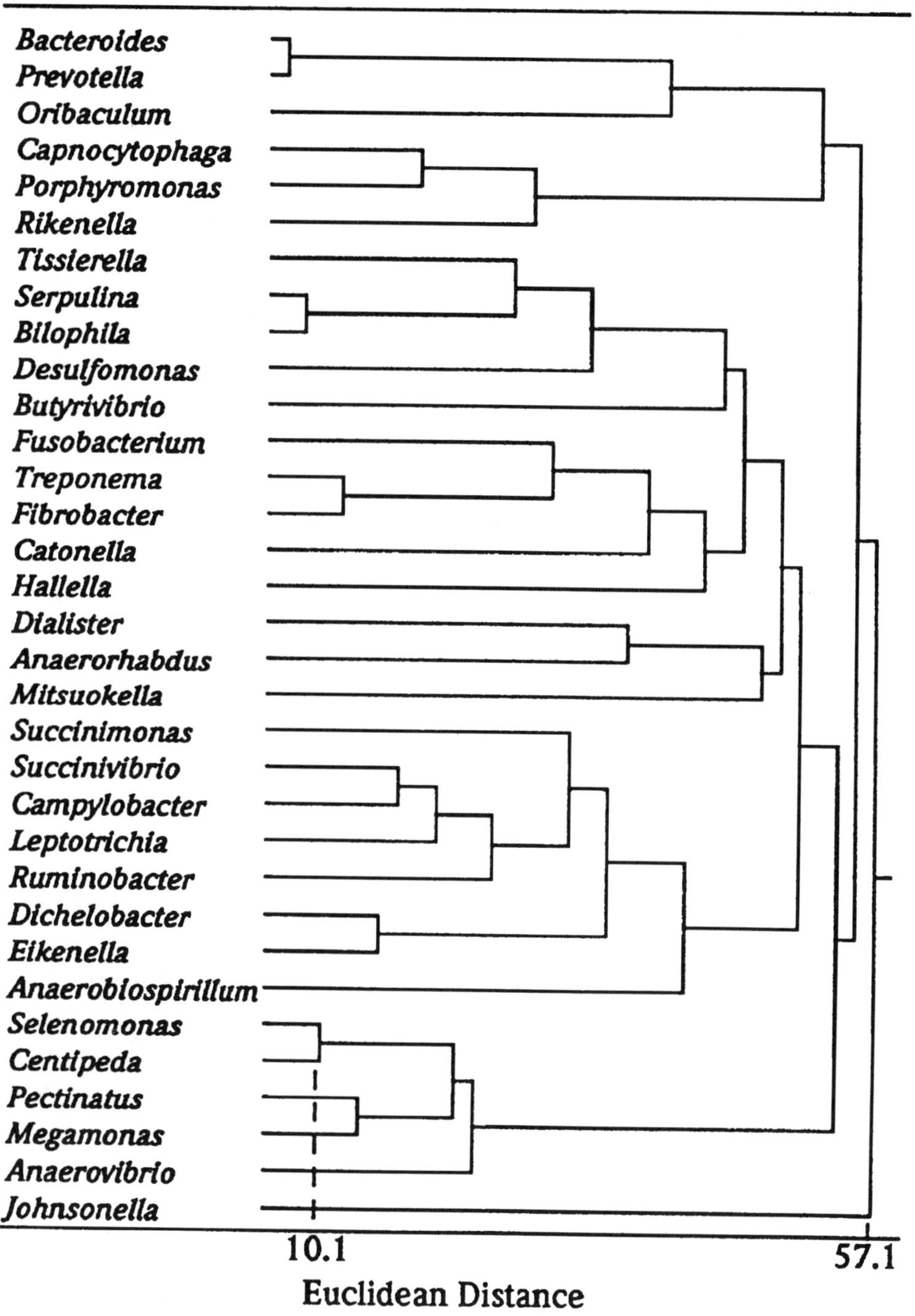

Figure 9.14. Dendrogram of various *Borrelia* spp. from data of Livesley et al. (1993).

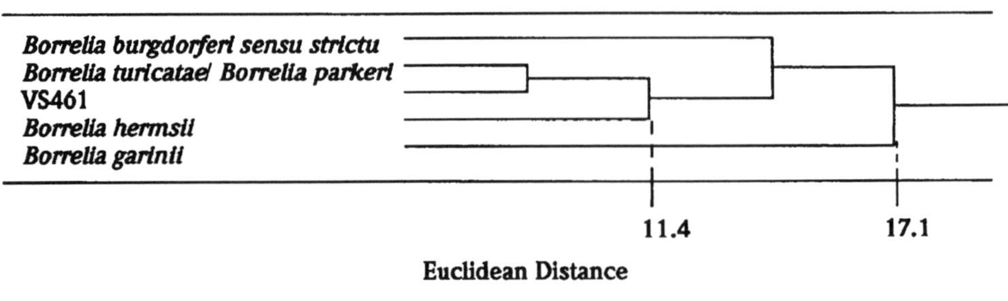

comparisons. Livesley and his coworkers concluded that cellular lipids provided useful information for differentiation of the two organisms. The significance of this is that the two are remarkably similar morphologically and biochemically. The organisms are usually distinguished by the degree of beta hemolysis produced on blood agar (*Serpulina hyodysenteriae* produces more complete hemolysis). The etiological agent of swine dysentery is *Serpulina hyodysenteriae*, while *Serpulina innocens*, is generally considered a nonpathogenic spirochete. Thus, FA analysis may be used successfully to distinguish similar microorganisms objectively.

DISCUSSION

In this chapter we have reviewed the recent literature to acquaint the reader with the diversity of organisms that have been characterized by FA analysis and to show a range of applications of the technique. The system is being used in a myriad of applications and can be used to identify microorganisms ranging from obligate aerobes to obligate anaerobes, from organisms with simple growth requirements to fastidious organisms, from organisms grown on solid medium to those grown in liquid medium, and from prokaryotes to eukaryotes. The applications appear limited only by the imagination of the researcher.

The number of recent publications indicate that FA analysis is a promising and useful method for the identification of bacteria. Most of the journal articles report good correlation between FA results and DNA or rRNA homology groupings.

Discussion Relevant to Publications Cited

The recent reports were more statistically sophisticated than the reports of the 1980s. For example, early publications contained figures that showed peak areas of the various FAs within analyzed strains. The tables listed the major FAs detected and showed ratios of the most abundant FAs. Attempts to establish differences were primarily based on visual observations. In contrast, recent reports used PCA and hierarchical clustering based on UPGMA algorithms.

The Genus Bacillus

The genus *Bacillus* was broken into six groups (Kaneda 1977) or into seven cluster groups, with groups I and II most closely related (Kämpfer 1994). Kämpfer also stated that for exact phenotypic identification at the species level (preferably genomically defined species) the use of physiological tests and numerical identification is recommended. We generated a dendrogram (Figure 9.15) from his data (Table 2, Kämpfer 1994) that approximates his results by only utilizing values for the iso-$C_{15:0}$ and anteiso-$C_{15:0}$ FAs. Some members of his Cluster I clearly belong in other clusters. We then created an alternative dendrogram, using all the FAs >0.3 percent (Figure 9.16). Kämpfer's groups I and II certainly do not cluster together, and individual strains were clearly grouped inappropriately by him. The "unclusterable" *Bacillus brevis*[b] and *Bacillus insolitus* now cluster as well.

Kämpfer's paper illustrates that one can make mistakes using CFA. He undertook the vast task of analyzing 313 strains of *Bacillus*. However, it is doubtful that all of these were valid strains since the CVs for the principal FAs in a number of clusters approximated 50. This variation is indicative of inappropriate combining of strains within clusters. The main value of this paper was in verifying the considerable heterogeneity at the species level within the genus, and that the individual species will have to be resolved before the entire genus can be reorganized.

Bacillus anthracis and Bacillus cereus

We have taken the data of Lawrence et al. (1991) reported in their Table 2, and produced the dendrogram based on PCA (Figure 9.17). *Bacillus anthracis* (cluster A) could be clearly distinguished from *Bacillus cereus* (clusters B, C, D) when RM was used. In fact, the virulent *Bacillus anthracis* isolates (18–74 and Zambia) link at a Euclidean distance of 6.2. The avirulent vaccine strain (Sterne) links with the virulent strains at a distance of 7.5. The *Bacillus cereus* grown on RM cluster in three groups (using a Euclidean distance of 10 as a point of reference). The diversity of *Bacillus cereus* is shown by the fact that 18.5 Euclidean distance units were required for the three groups to link. The *Bacillus anthracis* and *Bacillus cereus* species link at a distance of 21.6.

Their conclusion that the two species could not be differentiated on RCM was valid (Figure 9.18). It has been well documented that the medium and the incubation temperature influence the amounts of FAs present. The influence of the medium on CFA present in *Bacillus anthracis* and *Bacillus cereus* is clearly shown.

Bacillus sphaericus

Our interpretation of the *Bacillus sphaericus* data (Schenkel et al. 1992) is that two of the soil isolates, S1 and S5, were the same strain; strains 2297 and 2362 were identical (since they link at a distance <2). The other two soil isolates (S2, L2), and

---

[b] Designation by Kämpfer (1994).

Figure 9.15. Dendrogram of *Bacillus* species (or subgroups combined by Kämpfer), which used values of only two principal fatty acids—iso-$C_{15:0}$ and anteiso-$C_{15:0}$. Results approximate Kämpfer's conclusions. Dendrogram taken from data of Kämpfer (1994).

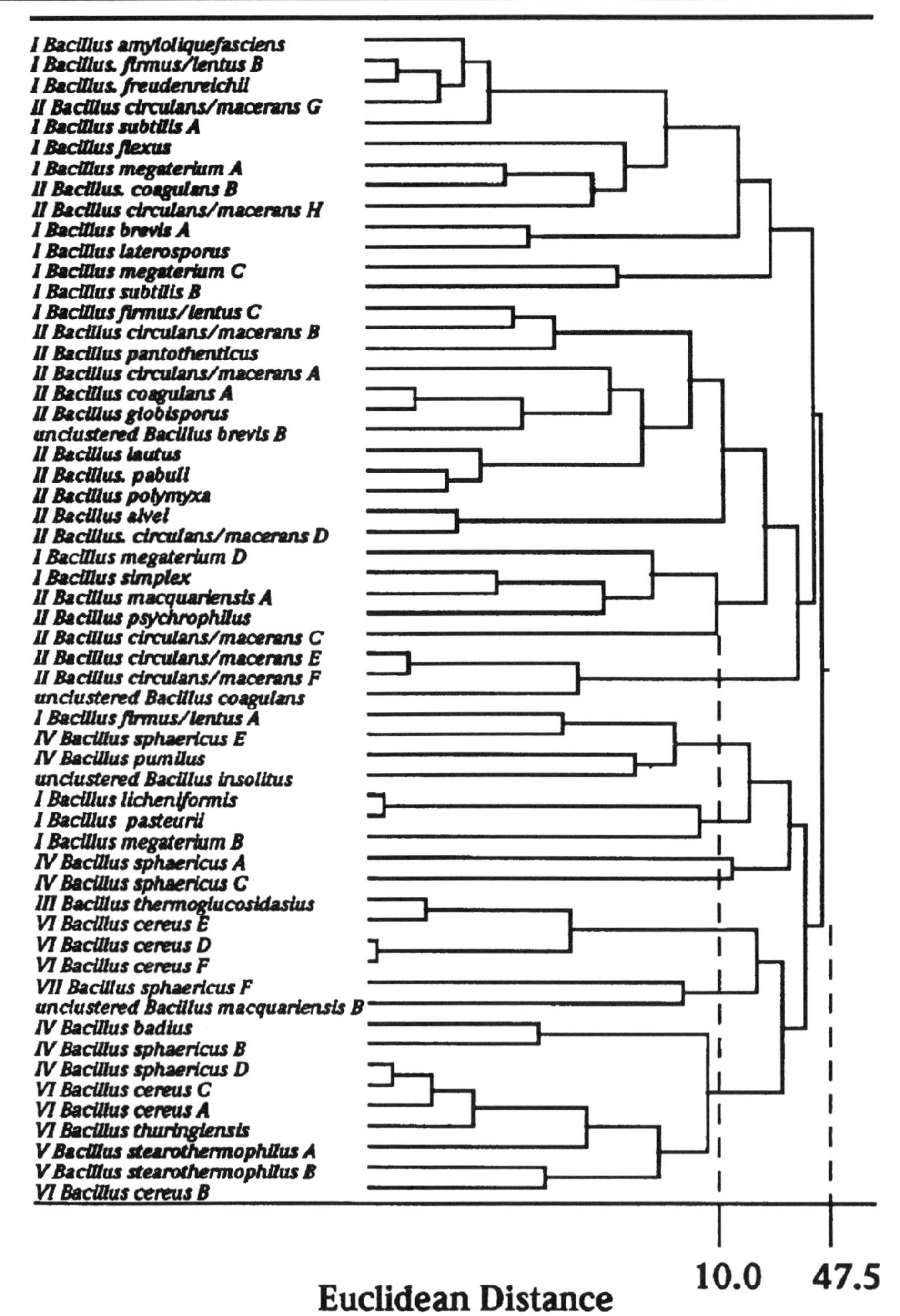

Figure 9.16. Alternative dendrogram of *Bacillus* species or subgroups using all fatty acids (>0.3 percent) and prinipal component analysis. Dendrogram based on Kämpfer's data (1994).

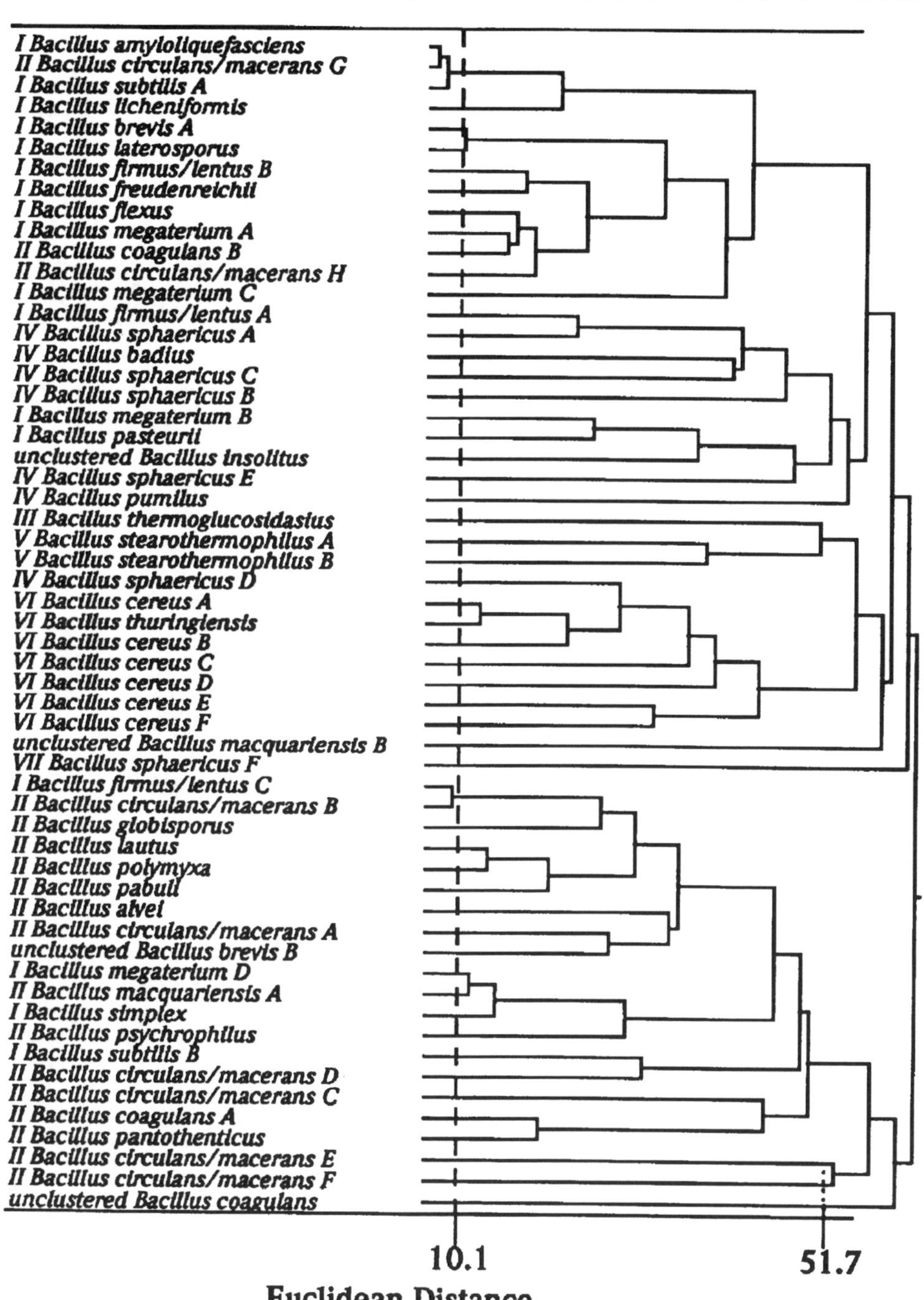

Figure 9.17. Dendrogram of *Bacillus anthracis* and *Bacillus cereus* grown on synthetic medium from the data of Lawrence et al. (1991).

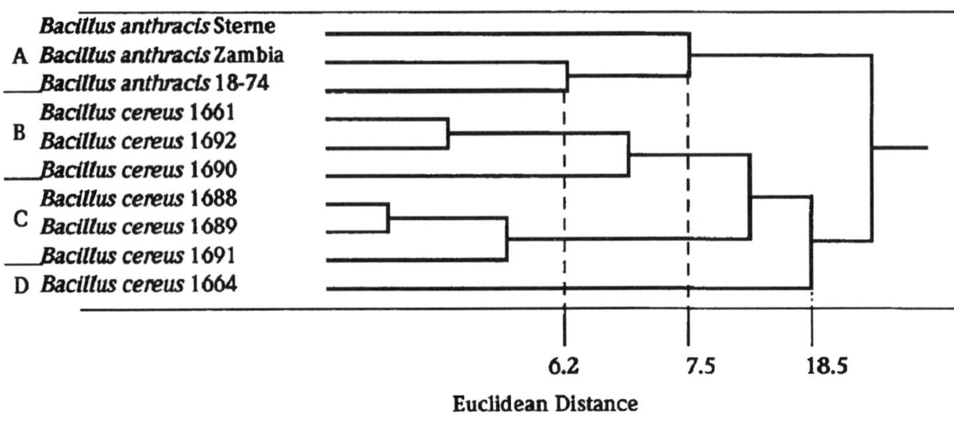

Figure 9.18. Plot of *Bacillus anthracis* and *Bacillus cereus* grown on synthetic medium and complex medium. Effect of medium is demonstrated. Data of Lawrence et al. (1991) was used.

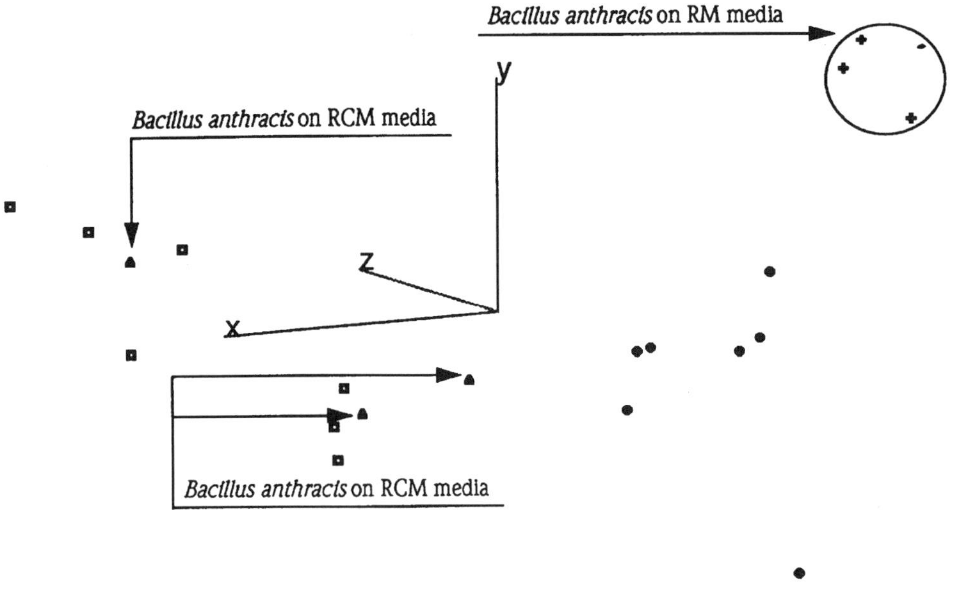

1593 were distinct strains (but all belonged to the same biotype since they link at a distance of about 5). S1–S5 might be a subspecies of the biotype group. The logical strain for further research would be S2, on the basis of CFA.

Bacillus brevis

The CFA data (our dendrogram, Figure 9.5) suggest two problems with the *Bacillus brevis* results (Shida et al. 1995). The first is that the type strain of *Bacillus choshiensis* links with *Bacillus reuszeri* at a distance of 7.8, indicative of a single species, and the second is that the two strains of *Bacillus formosus* do not cluster together until about 15.1. Additionally *Bacillus formosus* NRS-910 links with the type strain of *Bacillus brevis* at a distance of about 9.6, suggesting that it is a true *Bacillus brevis*. What could be done to make the CFA data more useful? Data shown in the dendrogram were obtained after a single determination. The reproducibility should be assessed by analysis of each strain at least three times under the same standardized conditions: Grow each strain on three plates of media, harvest each plate individually, and process and analyze each vial (replication), rather than reanalyze a single vial three times (subsampling). Increasing the number of valid strains evaluated could also allow one to obtain groupings with acceptable CVs.

The Pseudomonaceae

We subjected Stead's *Pseudomonas* data (1992) to cluster analysis based on PCA and obtained a dendrogram (Figure 9.19) consisting of 21 clusters (using a Euclidean distance of 10). Generally, the results support Stead's conclusions. However, some of Stead's species assignments to subgroups may be invalid due to the fact that his grouping was based on -OH patterns, instead of PCA. For example, *Pseudomonas rubrisubalbicans* (Stead's subgroup 1c) clusters in cluster 1 (a subgroup of 1a). Cluster 2 is all subgroup 3a and cluster 3 is mainly a subgroup 1a, but *Pseudomonas corrugata* (Stead's subgroup 1b) and *Commonas acidovorans* (Stead's subgroup 3a) and *Commonas testeroni* (Stead's subgroup 3b) cluster together. Stead stated that only the absence of 2-OH-$C_{12:0}$ and 3-OH-$C_{12:0}$ differentiated the group 3 strains from most group 1 strains. The clustering of the two *Commonas* strains supports Willems et al. (1991) suggestion that although the plant-pathogenic strains belonged to the family Commanadaceae, they were not members of the genus *Commonas*. Cluster 4 is all subgroup 1a, including *Pseudomonas putida* biovar B. However, *Pseudomonas putida* biovar A is the single entry in cluster 5, linking with *Pseudomonas cattleyae* (cluster 6, Euclidean distance 11.4). All of Stead's subgroup 2c are in cluster 7. There is some question concerning the validity of Stead's *Pseudomonas flectens* (group 5), which is our cluster 8. Our single entry cluster 9 is *Pseudomonas aeruginosa* (Stead's subgroup 1a). We question the validity of the 1a grouping for *Pseudomonas aeruginosa* since its predominant FA was $C_{18:1\ cis11}$, while for the other fluorescent strains this was the third most common FA. The Euclidean distance required for *Pseudomonas aeruginosa* to link with other subgroup 1a's is 32.7 units (Figure 9.19). The plot of Stead's subgroup 1a (Figure 9.20) illustrates the isolated position of *Pseudomonas aeruginosa*. However, *Pseudomonas aeruginosa* links more closely (11.8 Euclidean distance units) with *Pseudomonas caryophyll*. Cluster 10 consists of *Pseudomonas caryophyll* and *Pseudomonas glumae* GC subgroup A (both in Stead's 2c subgroup) and

Figure 9.19. Dendrogram of 21 clusters of Pseudomonads (based on 10 Euclidean distance units). Note the position of *Pseudomonas aeruginosa*. Data of Stead (1992) was used.

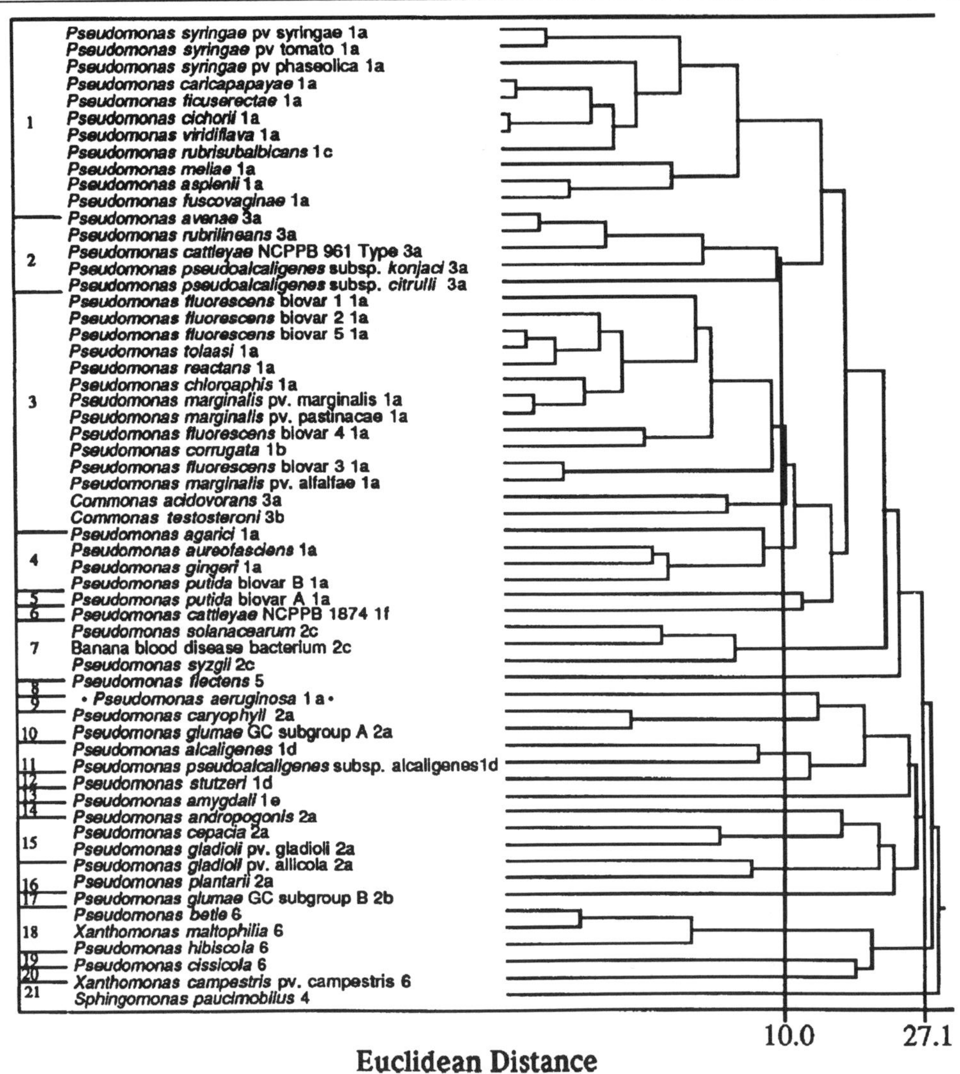

cluster 11 has *Pseudomonas alcaligenes* and *Pseudomonas pseudoalcaligenes* subspecies *alcaligenes* (Stead's 1d subgroup). Another subgroup 1d, *Pseudomonas stutzeri*, was our single entry cluster 12. Clusters 13 and 14 were single entry clusters, *Pseudomonas amygdali* (subgroup 1e) and *Pseudomonas andropogonis* (subgroup 2a), respectively. *Pseudomonas cepacia* and the closely related *Pseudomonas gladioli* pathovar *gladioli* (both subgroup 2a) were in cluster 15. Cluster 16 also had two subgroup 2a's, *Pseudomonas gladioli* pathovar *allicola* and *Pseudomonas*

Figure 9.20. Plot of the first three principal components of Stead's entire subgroup 1a of *Pseudomonas* is shown. Data of Stead (1992) was used to construct the plot.

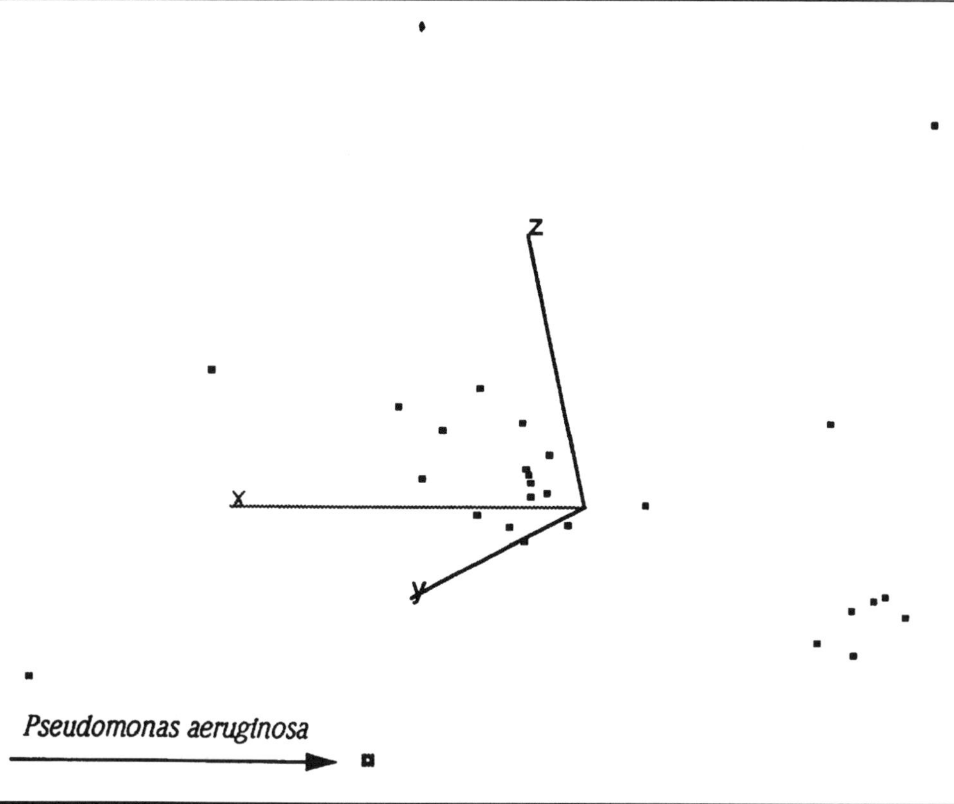

plantarii. The remaining subgroup 2b, *Pseudomonas glumae* GC subgroup B was the single entry in cluster 17. Clusters 18, 19 and 20 (all group 6) are probably all *Xanthomonas* species. Cluster 21, consists of the group 4, *Sphingomonas paucimobilis.*

Thus, our analysis of Stead's FA data confirms his conclusion that the Pseudomonaceae should be split into at least three genera. In fact, reclassifications have being made. For example, *Pseudomonas cepacia* (our cluster 15) is now classified as *Burkholderia cepacia* (Bevivino et al. 1994) and *Xanthomonas maltophilia* (our cluster 18) is now known as *Stenotrophomonas maltophilia* (Palleroni and Bradbury 1993). We are confident that more genera will arise due to reclassifications of the Pseudomonaceae.

The Genus Aeromonas

The genus *Aeromonas* was not completely resolved to the satisfaction of Huys et al. (1994), especially HG groups 7 and 11 (refer to Figure 9.9). Only two strains each were used to represent HGs 7 and 11. Also, the FA profiles for the two strains in HG 7 were very heterogeneous. According to Popoff (1984), HGs 7, 8, and 9 were phenotypically highly related. On the basis of other phenotypic data, HG 11 does not properly belong in cluster C. Inclusion of additional strains of HG 7 and 11 should help resolve their proper position.

How does one resolve the discrepancies between Kämpfer et al. (1994) and the results of Huys et al. (1994)? We have used Table 2 data (Kämpfer et al. 1994) to construct the dendrogram shown in (Figure 9.21). There are five clusters present (Huys et al. data also revealed five clusters) but the *Aeromonas hydrophila* are the most unrelated. A Euclidean distance of 12.2 units is required for *Aeromonas hydrophila*-A to link with *Aeromonas hydrophila*-C. However, according to Huys et al. data, (1994) the *Aeromonas hydrophila* complex (cluster E) link at a Euclidean distance of 5.9 (Figure 9.9). The cells were grown on a different medium (Kämpfer cultured on R2A agar, Huys cultured on TSA) as well as at different temperature regimes (30°C and 28°C, respectively), and there were equipment differences as well. However, we speculate that the extremely large CV values of Kämpfer and his coworkers indicate inappropriate grouping of strains and species. In some of their groupings the CV for the major FAs exceeded 100. In contrast Huys et al. (1994), again working with *Aeromonas*, created clusters that had 90 percent less variation. We conclude that the results of Huys et al. (1994) are valid.

Gram-Negative Anaerobic Bacilli

We predict that there will be some future taxonomic changes for the genera *Porphyromonas* and *Prevotella* based on the publication by Brondz and Olsen (1991). Our prediction using the dendrogram shown in Figure 9.9 (based on FA analysis only) is that additional new genera and species will be created. For example, one of the four strains of *Porphyromonas endodontalis* linking at 10.6 probably is a new

---

Figure 9.21. Dendrogram of strains of *Aeromonas*. Note that the *Aeromonas hydrophila* strains do not cluster together. Taken from data reported by Kämpfer et al (1994).

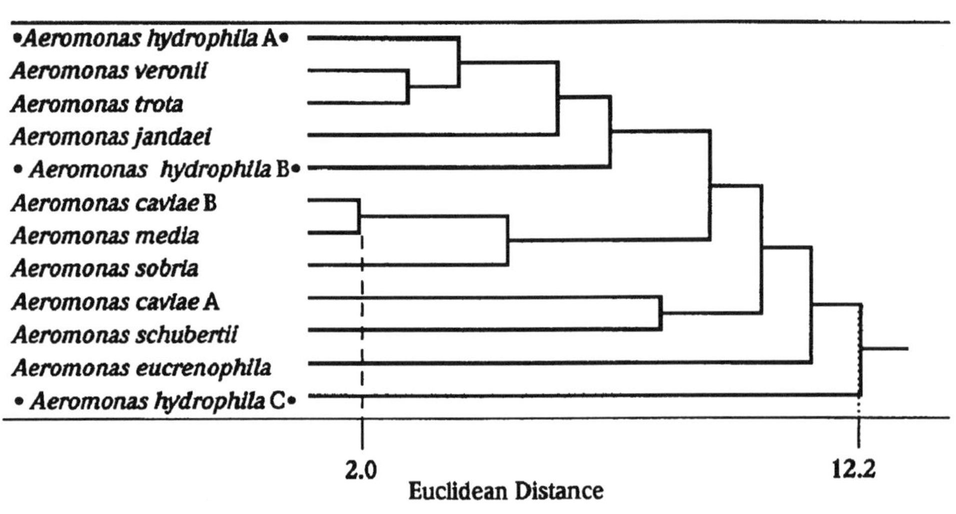

species. The *Prevotella* strains are shown, in Figure 9.11, to be very diverse. Data for *Prevotella* indicate possible subdivision into a new species for *Prevotella oris* (linkage at 11.1). However, including more strains may tighten the clusters and decrease the distance below the general distance rule of 10 units for species designation. There is a problem with the strain of *Prevotella heparinolyticus* used (it linked with the other *Prevotella* at a distance of 67.8, thus suggesting it belongs in a different genus).

Again, using FA data only, with the reclassification of *Bacteroides gracilis* to *Campylobacter gracilis*, why not reclassify *Wolinella succinogenes* as a *Campylobacter* also (Vandamme et al. 1995)? They all link at a Euclidean distance of 25.4 (see Figure 9.12). At the present time, all the data support the concept that a new genus be created for *Bacteroides ureolyticus*. The Euclidean distance is reported at 48 by both Vandamme et al. (1995) and Brondz and Olsen (1991).

The dendrogram (Figure 9.13) confirms why *Bacteroides* and *Prevotella*, *Pectinatus* and *Megamonas*, and *Serpulina* and *Bilophila* were difficult to distinguish visually (Moore et al. 1994). These genera link at a distance of 10, which we regard as the species limit. However, properly constructed library entries should discriminate each genus of closely linked pairs without difficulty.

Present Assessment

The potential of the method increased dramatically when computers were used for the comparisons of the FAs (both qualitatively and quantitatively). One of the strengths of the MIS is that principal components and cluster analysis are utilized. The fact that the data are processed automatically and objectively reported expands the application for use in clinical laboratories.

In general, every researcher has found the MIS libraries inadequate when compared to their own specialized laboratory-generated libraries. However, MIDI updates its libraries annually and strives to improve them. In many cases an increase in the number and diversity of strains used would help. Changes in medium and cultural conditions to mimic those used in clinical laboratories are needed in some cases. To illustrate the potential value of a better clinical library, Osterhout et al. (1991) used 15 clinical isolates of *Pseudomonas aeruginosa* grown at 35°C for 22–24 hours on TSA supplemented with 5 percent sheep blood (BA) to establish a library entry. *Pseudomonas aeruginosa* ATCC 27853 was grown on TSA at 28°C and the MIS library entry tested (10 analyses). The mean SI was 0.71 with a CV of 23. The same strain was also grown on BA at 35°C and their created library entry tested (also 10 analyses). In this case, the mean SI was 0.88 with a CV of 4. Both libraries identified the strain as *Pseudomonas aeruginosa*. On the basis of this result, Osterhout et al. speculated that a marked improvement in the SI values could be achieved if standardized growth temperature was changed from 28° to 35°C, which would allow all organisms to attain the same physiological age (i.e., stationary growth phase).

We also feel that reporting certain minimal statistical data should be a requirement for publication, mean values with s.d.'s for each FA reported. This allows one to calculate CVs, which in turn allows one to determine if isolates and/or strains were inappropriately combined. Values of FAs should be reported to the nearest 0.1 percent instead of rounding to the closest whole number. When dendrograms or data are published, the number of strains constituting each subgroup in a cluster should be stated. Clearly stating the number of replications for each strain analyzed is also important; many authors use replications and repeated injections of the same sample (subsamplings) interchangeably.

Lastly, the literature is skewed relative to genera and species subjected to CFA analysis. This can be explained by the axiom, "Necessity is the mother of invention." For example, the literature is very scant on Enterobacteriaceae and FA methodology. However, existing conventional methodology is adequate for their identification. When present methods of identification are inadequate, CFA is more likely to be investigated. If the microorganism cannot be identified solely by CFA, a few supplemental biochemical tests may be required to resolve its identity. Sasser and Wichman (1991) stated that it may be unreasonable to expect CFA to differentiate *Escherichia coli* and *Shigella flexneri*, due to their close relatedness based on DNA homology.

We feel that this is an emerging technique in the clinical laboratory. We anticipate that the method will be expanded, found useful, and practical to characterize other microbial species (i.e., *Rickettsia*, *Chlamydia*, *Mycoplasma*, and Fungi). Also, sensitivity may be increased allowing smaller samples to be analyzed. We eagerly anticipate the future uses of this methodology.

REFERENCES

Abel, K., H. deSchmettzing, and J. I. Peterson. 1963. Classification of microorganisms by analysis of chemical composition. I. Feasibility of utilizing gas chromatography. *J. Bacteriol.* 85:1039–1044.

Bernard, K. A., M. Bellefeuille, and E. P. Ewan. 1991. Cellular fatty acid composition as an adjunct to the identification of asporogenous, aerobic gram-positive rods. *J. Clin. Microbiol.* 29:83–89.

Bevivino, A., S. Tabacchioni, L. Chiarni, M. V. Carusi, M. del Gallo, and P. Visca. 1994. Phenotypic comparison between rhizophere and clinical isolates of *Burkholderia cepacia*. *Microbiology* 140:1069–1077.

Brondz, I., and I. Olsen. 1991. Multivariate analyses of cellular fatty acids in *Bacteroides*, *Prevotella*, *Porphyromonas*, *Wolinella*, and *Campylobacter* spp. *J. Clin. Microbiol.* 29:183–189.

Eerola, E., and O.-P. Lehtonen. 1988. Optimal data processing procedure for automatic bacterial identification by gas-liquid chromatography of cellular fatty acids. *J. Clin. Microbiol.* 26:1745–1753.

Fisher, L. D., and G. D. van Belle. 1993. Biostatistics: A methodology for the health sciences. New York: John Wiley.

Frachon, E., S. Hamon, L. Nicolas, and H. de Barjac. 1991. Cellular fatty acid analysis as a potential tool for predicting mosquitocidal activity of *Bacillus sphaericus* strains. *Appl. Environ. Microbiol.* 57:3394–3398.

Goodfellow, M., and A. G. O'Donnell, eds. 1993. *Handbook of new bacterial systematics.* San Diego: Academic Press.

Huys, G., M. Vancanneyt, R. Coopman, R. Janssen, E. Falsen, M. Altwegg, and K. Kersters. 1994. Cellular fatty acid composition as a chemotaxonomic marker for the differentiation of phenospecies and hybridization groups in the genus *Aeromonas. Int. J. Syst. Bacteriol.* 44:651–658.

Itoh, U., M. Sato, H. Tsuchiya, and I. Namikawa. 1995. Cellular fatty acids and aldehydes of oral *Eubacterium. FEMS Microbiology Letters.* 126:69–74.

Janda, M. J. 1991. Recent advances in the study of the taxonomy, pathogenicity, and infectious syndromes associated with the genus *Aeromonas. Clin. Microbiol. Rev.* 4:397–410.

Kämpfer, P. 1994. Limits and possibilities of total fatty acid analysis for classification and identification of *Bacillus* species. *Syst. Appl. Microbiol.* 17:86–98.

Kämpfer, P., K. Blasczyk, and G. Auling. 1994. Characterization of *Aeromonas* genomic species by using quinone, polyamine, and fatty acid patterns. *Can. J. Microbiol.* 40:844–850.

Kaneda, T. 1977. Fatty acids of the genus *Bacillus*: An example of branched-chain preference. *Bact. Rev.* 41:391–418.

Kaneda, T. 1991. Iso- and anteiso-fatty acids in bacteria: Biosynthesis, function, and taxonomic significance. *Microbiol. Rev.* 55:288–302.

Kotilainen, P., P. Huovinen, and E. Eerola. 1991. Application of gas-liquid chromatographic analysis of cellular fatty acids for species identificaion and typing of coagulase-negative staphylococci. *J. Clin. Microbiol.* 29:315–322.

Krych, V. K., J. L. Johnson, and A. A. Yousten. 1980. Deoxyribonucleic acid homologies among strains of *Bacillus sphaericus. Int. J. Syst. Bacteriol.* 30:476–484.

Lambert, M. A., C. M. Patton, T. J. Barrett, and C. W. Moss. 1987. Differentiation of *Campylobacter* and *Campylobacter*-like organisms by cellular fatty acid composition. *J. Clin. Microbiol.* 25:706–713.

Lawrence, D., S. Heitefuss, and H. J. Seifert. 1991. Differentiation of *Bacillus anthracis* from *Bacillus cereus* by gas chromatographic whole-cell fatty acid analysis. *J. Clin. Microbiol.* 29:1508–1512.

Livesley, M. A., I. P. Thompson, M. J. Bailey, and P. A. Nuttal. 1993a. Comparison of the fatty acid profiles of *Borrelia, Serpulina,* and *Leptospira* species. *J. Gen. Microbiol.* 139:889–895.

Livesley, M. A., I. P. Thompson, L. Gern, and P. A. Nuttall. 1993b. Analysis of intra-specific variation in the fatty acid profiles of *Borrelia burgdorferi. J. Gen. Microbiol.* 139:2197–2201.

Manly, B. F. J. 1986. *Multivariate statistical methods: A primer.* New York: Chapman and Hall.

Matthews, H. M., and J. M. Kinyon. Cellular lipid comparisons between strains of *Treponema hyodysenteriae* and *Treponema innocens*. *Int. J. Syst. Microbiol.* 34:160–165.

Miller, L., and T. Berger. 1985. Bacteria identification by gas chromatography of whole cell fatty acids. Hewlett-Packard application note 228–41. Avondale, PA: Hewlett-Packard Co.

Moore, L. V. H., D. M. Bourne, and W. E. C. Moore. 1994. Comparative distribution and taxonomic value of cellular fatty acids in thirty-three genera of anaerobic gram-negative bacilli. *Int. J. Syst. Bacteriol.* 44:338–347.

Mukwaya, G. M., and D. F. Welch. 1989. Subgrouping of *Pseudomonas cepacia* by cellular fatty acid composition. *J. Clin. Microbiol.* 27:2640–2646.

O'Leary, W. M., and S. G. Wilkinson. 1988. Gram-positive bacteria. In *Microbial lipids*, vol. 1, edited by C. Ratledge, and S. G. Wilkinson. New York: Academic Press.

Osterhout, G. J., V. H. Shull, and J. D. Dick. 1991. Identification of clinical isolates of gram-negative nonfermentative bacteria by an automated cellular fatty acid identification system. *J. Clin. Microbiol.* 29:1822–1830.

Palleroni, N. J. 1984. Genus 1 *Pseudomonas*. In *Bergey's manual of systemic bacteriology*, vol. 1, edited by N. R. Krieg, and J. G. Holt. Baltimore: Williams & Wilkins.

Palleroni, N. J., and J. F. Bradbury. 1993. *Stenotrophomonas*, a new bacterial genus for *Xanthomonas maltophilia* (Hugh 1980) Swings et al. 1983. *Int. J. Syst. Bacteriol.* 43:606–609.

Popoff, M. 1984. Genus III. *Aeromonas* Klluyver and van Niel 1936. In *Bergey's manual of systematic bacteriology*, vol. 1, edited by N. R. Krieg, and J. G. Holt. Baltimore: Williams & Wilkins.

Romesburg, H. C. 1990. *Cluster analysis for researchers*. Malbar, FL: Krieger Publishing Co.

Sasser, M. 1990a. Identification of bacteria through fatty acid analysis. In *Methods in phytobacteriology*, edited by Z. Klement, K. Rudolph, and E. Sands. Kiado, Budapest: Akademia.

Sasser, M. 1990b. "Tracking" a strain using Microbial Identification System. Technical note #102. Newark, DE: MIDI.

Sasser, M., and M. D. Wichman. 1991. Identification of microrganisms through use of gas chromatography and high performance liquid chromatography. In *Manual of clinical microbiology*, 5th ed., edited by A. Balows et al. Washington, DC: American Society of Microbiology.

Schenkel, R. G. M., L. Nicolas, E. Frachon, and S. Hamon. 1992. Characterization and toxicity to mosquito larvae of four *Bacillus sphaericus* strains isolated from Brazilian soils. *J. Invert. Path.* 60:10–14.

Segers, P., W. Mannheim, M. Vancanneyt, K. Debrandt, K. H. Hinz, K. Kersters, and P. Vandamme. 1993. *Riemerella anatipestifer* gen. nov., comb. nov., the causative agent of septicemia-anserum exsudativa, and its phylogenetic affiliation within the *Flavobacterium-Cytophaga* r-RNA homology group. *Int. J. Syst. Bacteriol.* 43:768–776.

Shida, O., H. Takagi, K. Kadowaki, S. Udaka, L. K. Nakamura, and K. Komagata. 1995. Proposal of *Bacillus reuszeri* sp. nov. *Bacillus formosus* sp. nov. nom. rev. and *Bacillus borstelensis* nov. nom. rev. *Int. J. Syst. Bacteriol.* 45:93–100.

Siegel, J. P., A. R. Smith, J. V. Maddox, and R. J. Novak. 1993. Use of cellular fatty acid analysis to characterize commercial brands of *Bacillus thuringiensis* var. *israelensis. J. Am. Mosquito Control Assoc.* 9:330–334.

Siegel, J. P., A. R. Smith, and R. J. Novak. 1995. Cellular fatty acid analysis of *Bacillus thuringiensis* var. *israelensis* (ONR-60A). *J. Am. Mosquito Control Assoc.* 11:176–185.

Stager, C. E., and J. R. Davis. 1992. Automated systems for identification of microorganisms. *Clin. Microbiol. Rev.* 5:302–327.

Stahly, D. P., and M. Klein. 1992. Problems with *in vitro* production of spores of *Bacillus popilliae* for use in biological control of the Japanese Beetle. *J. Invert. Path.* 60:283–291.

Stead, D. E. 1992. Grouping of plant-pathogenic and some other *Pseudomonas* spp. by using cellular fatty acid profiles. *Int. J. Syst. Bacteriol.* 42:281–295.

Stead, D. E., J. E. Sellwood, J. Wilson, and I. Viney. 1992. Evaluation of a commercial microbial identification system based on fatty acid profiles for rapid, accurate identification of plant pathogenic bacteria. *J. Appl. Bacteriol.* 72:315–321.

Stoakes, L., T. Kelly, B. Schieven, D. Harley, M. Ramos, M. Lannigan, D. Groves, and Z. Hussain. 1991. Gas-liquid chromatographic analysis of cellular fatty acids for identification of Gram-negative anaerobic bacilli. *J. Clin. Microbiol.* 29:2636–2638.

Stoakes, L., M. A. John, R. Lannigan, B. C. Schieven, M. Ramos, D. Harley, and Z. Hussain. 1994. Gas-liquid chromatography of cellular fatty acids for identification of staphylococci. *J. Clin. Microbiol.* 32:1908–1910.

Suzuki, K., M. Goodfellow, and A. G. O'Donnell. 1993. Cell envelopes and classification. In *Handbook of new bacterial systematics*, edited by M. Goodfellow, and A. G. O'Donnell. San Diego: Academic Press.

Tornabene, T. G. 1985. Lipid analysis and the relationship to chemotaxonomy. *Microbiol Methods.* 18:209–234.

Vandamme, P., M. I. Daneshuar, F. E. Dewhirst, B. J. Paster, K. Kersters, H. Goossens, and C. W. Moss. 1995. Chemotaxonomic analyses of *Bacteroides gracilis* and *Bacteroides ureolyticus* and reclassification of *Bacteroides gracilis* as *Campylobacter gracilis* comb. nov. *Int. J. Syst. Bacteriol.* 45:145–152.

Welch, D. F. 1991. Applications of cellular fatty acid analysis. *Clin. Microbiol. Rev.* 4:422–438.

White, M. A., M. D. Simmons, A. Bishop, and H. A. Chandler. 1988. Microbial identification by gas chromatography. *J. R. Nav. Med. Serv.* 74:141–146.

Willems, A., J. De Ley, M. Gillis, and K. Kersters. 1991. Commonadaceae, a new family encompassing the acidovorans rRNA complex, including *Variovarax paradoxus* gen. nov., comb. nov., for *Alcaligenes paradoxus* (Davis 1969). *Int. J. Syst. Bacteriol.* 41:445–450.

Wüst, J., G. M. Lucchini, J. Lüthy-Hottenstein, F. Brun, and M. Altwegg. 1993. Isolation of gram-positive rods that resemble but are clearly distinct from *Actinomyces pyogenes* from mixed wound infections. *J. Clin. Microbiol.* 31:1127–1135.

Yang, P., L. Vauterin, M. Vancanneyt, J. Swings, and K. Kersters. 1993. Application of fatty acid methyl esters for the taxonomic analysis of the genus *Xanthomonas. Syst. Appl. Microbiol.* 16:47–71.

10

Enumeration and Characterization of Microbial Populations Using Optical and Electrical Zone-Sensing Particle Counters

Thomas A. Barber

Baxter Healthcare Corporation

Classical methods of enumerating microbial populations have involved the use of either dilution culture or microscopy. The latter mode of analysis may be used to perform counts of organisms in fixed smears on filters or in liquid suspensions that are illuminated in dark field or bright field or by more technically specialized microscopic methods. These historical methods remain extremely valuable; however, they are time-consuming and yield little information regarding the size distribution of the organisms that are being observed or grown. Epifluorescence microscopy has proven useful in detecting microcolonies and very low numbers of organisms in dilute environments, but it requires a high level of technology and sophisticated instrumentation.

More recently, accurate, real-time measurements of microbial cell concentration, size distribution, and growth phase have been shown to be possible using automated particle counters of various types; while dilution of some suspensions remains necessary, contemporary electronic or optical counters can count as high as 5.0×10^4 cells per ml of sample, and can sample at rates of 20–100 ml per minute. The application of instrumental particle counting and sizing technology to populations of microorganisms of various types shows promise as an ancillary technique for use by the microbiologist.

This chapter comprises a summary discussion of (1) particle counting technology applicable in counting and sizing bacteria and other isolated cells; and (2) applications of this methodology to microbial populations in the air or in aqueous media. A summary literature review on the subject is provided; it is not intended to be comprehensive, but will provide the reader with an interest in this area with a starting point for investigating the application of the technology.

INSTRUMENTATION AND METHODS

Instruments useful in the characterization of microbial populations generally fall into two categories: (1) single particle counters and (2) population analyzers or instruments that measure dynamic collective properties of particle populations (sometimes called "ensemble" instruments). Nephelometers and turbidimeters have long been applied in measuring microbial growth based on the static light scattering or absorptive properties of microbial suspensions; since these instruments are not particle counting or sizing instruments, they have been omitted from this discussion.

A. Single Particle Counters

1. Electrical zone-sensing (Coulter®-type) counters

2. Optical particle counter (OPC) systems

   — Light extinction (liquid)

   — Light scattering (air, liquid)

   — Flow cytometry devices

B. Population Analyzers

1. Dynamic light scattering instruments (photon correlation spectrophotometers)

2. Laser diffraction analyzers

Light extinction counters, Coulter® counters, and flow cytometers deal with particles as discrete entities, and size and count individual particles. In these three types of counters, each individual particle passing the sensor is represented by a single electronic pulse. Photon correlation spectrometers and laser diffraction instruments measure collective or interactive particle properties and provide average values and generalized population distribution histograms, but no discrete count data. Commonly used microscopic and instrumental methods of analyzing microbial cells are summarized by measurement range and data output in Table 10.1.

Table 10.1. Particle Population Analysis Methods

Method	Nominal Range	Size Determined
Microscopic		
Optical	0.5–500 μm	Martin's, Feret's, or equivalent
Electron	0.002–15 μm	circle diameter
Electrical zone sensing	0.1–500 μm	Volume-weighted diameter
Light extinction	1–500 μm	Equivalent circle diameter
Light scattering	0.02–5.0 μm	Volume-weighted diameter
Photon correlation spectroscopy	0.003–3.0 μm	Volume-weighted mean diameter
Laser diffraction	0.5–600 μm	Volume-weighted mean diameter
Nephelometry	> 0.1 μm	Total light scattering (size independent)

The methods listed in Table 10.1 include both direct and indirect methods of measuring cell particle size. Direct methods are those in which the response of a sensor of some type reflects a phenomenon related to an individual particle or particles. This classification includes single particle-counting instruments. Those that produce "ensemble" measurements, such as laser diffraction instruments, measure particle size indirectly and produce data in some average form, such as average size or population distribution.

The following descriptors define the operational characteristics of instruments used in population analysis:

- **Accuracy:** Defines how close an experimental value is to the "true" value.

- **Precision:** Describes the variation in repeated runs.

- **Resolution:** Measures the minimum detectable differences between features in a size distribution.

- **Reproducibility:** Assesses the extent of variation between assays or between different instruments, sample preparations, operators, and so on.

- **Dynamic Range:** Measures the range (typically μm) over which instrument response is proportional to actual particle size.

Dynamic range is a critical parameter in selecting an analytical method for use in the analysis of microbial populations. Single particle counters employ aperture tubes and sensors that have a finite particle size range for which the produced pulse response will be proportional to the particle size. Some dynamic range limitation also holds for dynamic light-scattering instruments and those that generate data based on diffraction of light. A general rule of thumb is that any single particle counter will have a dynamic range of 50–70, which means that the largest organism counted can be no greater than 70 times the size of the smallest organism. The skewed nature of many mixed microbial distributions can cause problems, even if the entire width of the distribution falls within the dynamic range of the instrument. The large numbers of smaller cells present may necessitate dilution of the sample to stay within specified concentration limits; large particles may at the same time be diluted out of the sample and, thus, not be represented in correct proportions in the data collected. Issues regarding dynamic range are of decreased importance in microbial analysis when the range of cell sizes of interest is narrow. Such would be the case if bacteria of a single species were studied. In some applications, however, where both bacteria and protozoans must be counted (e.g., studies of marine plankton), dynamic range will be a real concern.

THE COULTER® PRINCIPLE

The Coulter® technique allows determination of the size and number of particles suspended in an electrolyte solution based on their passage through a small orifice in a glass tube that separates two electrodes of opposite potential. The principle is illustrated in Figure 10.1. The changes in resistance as particles pass through the orifice generate voltage pulses of amplitudes proportional to the volume of the particles. The pulses are amplified, sorted, and counted; from the resulting data the number and size distribution of the particulate matter may be determined (Lines 1967a, 1967b; Lines 1981). Coulter® instruments were the first automatic instruments to be widely applied for counting particles in parenteral injectable solutions (Kinsman 1969; Amaker and Boymund 1967). A similar type of instrument, the Elzone™ counter (Particle Technology Labs, Ltd., Downers Grove, IL), also operates on the electrical resistance principle and has applications comparable to the Coulter® instruments.

On some Coulter® counter models the threshold levels and pulse amplitudes may be visualized on an oscilloscope on the instrument as the analysis is conducted. In the total mode of pulse collection, all pulses above a given threshold level are counted, and this count represents the number of particles larger than some volume proportional to the appropriate size threshold setting. The Coulter® instruments currently marketed, such as the Multi-Sizer™, allow counts to be separated into a large number of channels so that a relative frequency distribution or population analysis may be performed. This is accomplished by simply collecting the counts between each of a number of sequentially increasing threshold settings.

In Coulter® counters the ratio of particle diameter to orifice diameter should be 2–60 percent for the optimum proportionality of resistance change and particle volume. A 15 μm or 20 μm aperture tube is thus generally acceptable for counting bacteria; a 30 μm aperture is appropriate for counting many larger organisms. Apertures suitable for counting over a range from 0. 3 μm to 1200 μm

Figure 10.1. Operational principle of a Coulter® counter.

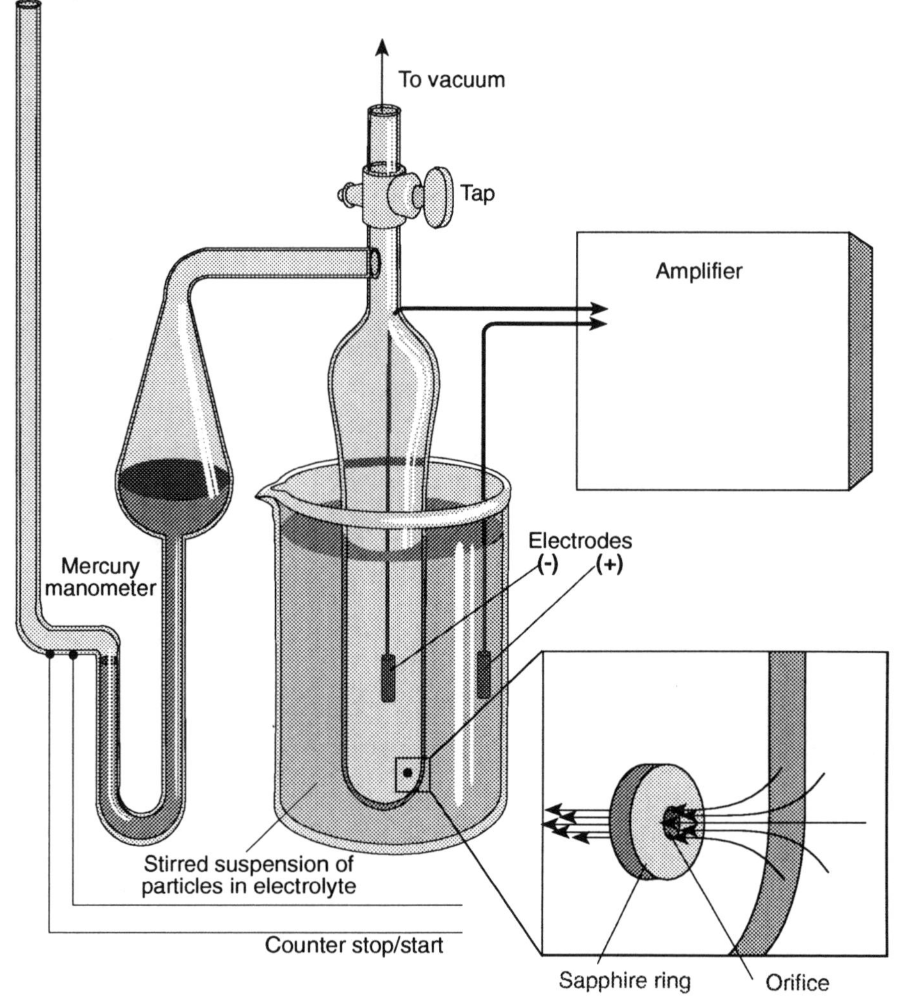

are available. As noted, the Coulter® principle is based on the fact that change in resistance between the instrument electrodes is directly proportional to the volume of the particle. The Coulter® counter thus sizes a particle as a three-dimensional object, in contrast to light extinction counters that see a particle in two dimensions and size particles based on area of light extinction. Therefore, the Coulter® principle of measurement is insensitive to the shape of the microbial cell counted.

One unique property of the Coulter® method is the requirement for an ionic (conducting) diluent if the solution to be tested is not itself ionic in nature. Sodium chloride (0.45–0.9 percent concentration) is commonly used as a diluent. The diluent solution itself (or the dilution process) may also add a significant number of particles to the sample, and thorough filtration of the diluent solution is desirable. Blank controls must also be utilized, and the particle contribution caused by the added electrolytic solution may be subtracted if desired. A more acceptable procedure is to filter the diluent at the 0.45 μm size using a recirculating filtration device offered by the vendor. For apertures <30 μm in size, a 0.45 percent sodium chloride diluent is satisfactory, but for larger apertures a 1 percent electrolyte concentration should be used.

The Coulter® instruments employ a mercury-filled manometer to sample from 500 μl to 2 ml of the suspension being tested. Due to the positioning of the aperture and the low flow rate of solution through it, large cells may be drawn into the orifice in disproportionately low numbers, resulting in undercounting (Harfield and Wood 1990, 293–298; Blanchard et al. 1977). As with any instrumental analysis, air bubbles adversely affect accurate counting. Air bubbles, a significant potential error source, are eliminated either by stirring during sampling, or by application of a vacuum before measurement or sonication. Electrical background noise, vibration, and electrical fields also contribute to errors in counting. The Coulter® analysis is also more time-consuming than other types of instrumental analysis. Despite these drawbacks, cell matter in the submicrometer size range can be accurately counted using Coulter® counters, if the user takes the time necessary to understand and master the technology before beginning to count for record (Barnett 1987, 222–233; DiGrado 1970). One advantage of the Coulter® counter in the analysis of elongate cells, such as bacilli, is the precision of the count result. The cells will be sized based on the conversion of their volume to an equivalent spherical volume. This is a decided improvement over the light extinction counter that "sees" rodlike cells as being different sizes depending on the aspect presented to the sensor.

The Coulter® principle for particle counting was first described in the late 1940s. The available instruments have been updated extensively since that time, and the current microprocessor-based instruments provide for a high level of versatility in use. The Multi-Sizer II™ model allows data collected to be displayed on the instrument monitor in a wide range of formats. Optimum resolution for the sample tested may be obtained by selecting the number of size classes (channels) into which the size distributions can be subdivided. The instrument contains a multichannel analyzer board, and from 16 to 256 channels of data can be obtained over a 30:1 diameter size range on the full range analysis. Use of *Narrow* and *Window* modes enables the theoretical equivalent of 25,600 channel size resolution to be obtained.

LIGHT EXTINCTION COUNTING

The original reference for this means of counting was presented in a 1969 paper by Carver. The data obtained from light extinction counters result from complex interactions between a small particle moving at high velocity and an intense light beam in the narrow confines of the particle counter sensor (Figure 10.2). In the sensor the sample fluid travels through a narrow rectangular passage 0.025–0.5 μm^2 in cross-section in front of an illuminated window. The light from an

Figure 10.2. Light extinction particle sensor.

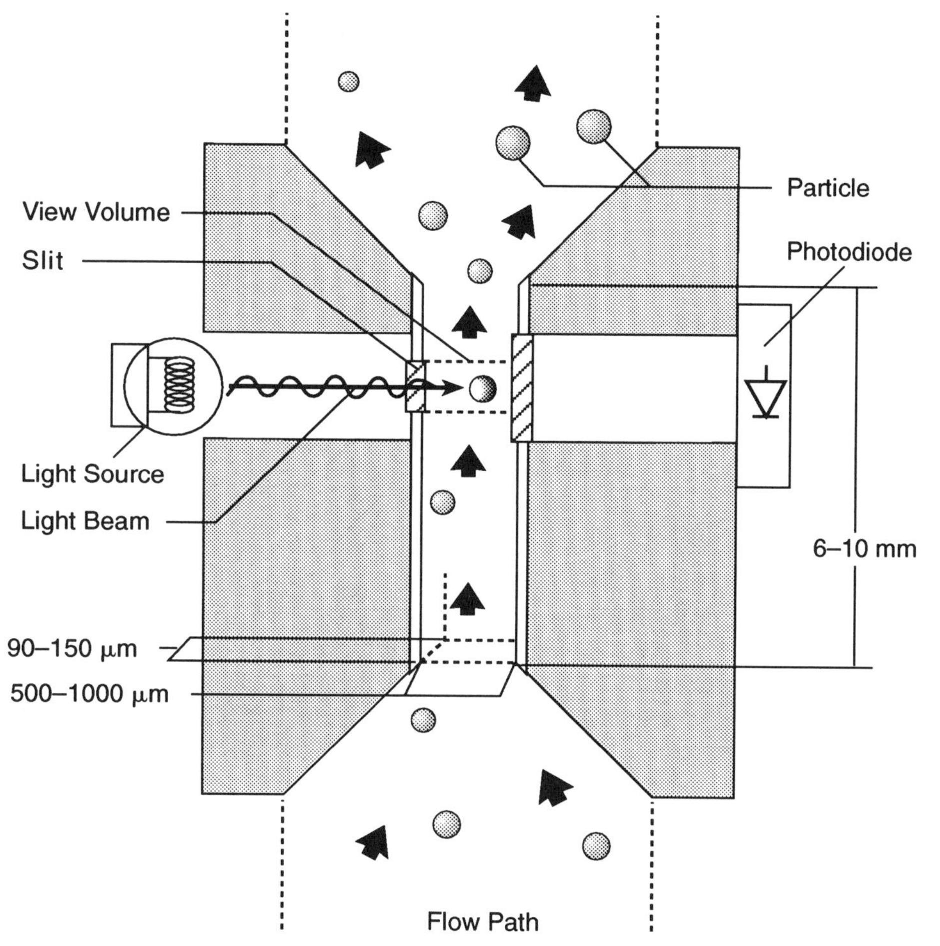

incandescent lamp or laser is formed into a collimated beam at the window, and directed through the liquid sample stream onto a photodiode. As long as the number of particles in suspension does not exceed a specified concentration, the particles will pass through the sensor view volume individually. Whenever a particle (cell) traverses the light beam, the intensity of light reaching the photodiode is reduced and an amplified voltage pulse is produced by the sensor. In theory, the amplitude of the pulse is proportional to the projected area of the particle in a plane normal to the light beam, and the particle size is registered as the diameter of the circle having an equivalent projected area. A nonspherical cell passing through the sensor may, therefore, appear to be a number of different sizes, depending on its orientation. The sample solution may be passed through the sensor cell by applying either pressure or vacuum.

The analog signal from the sensor is screened in the counter, and converted into a digital form that can be displayed or read and stored in memory by a dedicated microprocessor. As with the Coulter® counter, pulses are sorted and counted within channels determined by preset voltage amplitude thresholds in order to measure particle size distributions. In the differential mode of counting, the pulse produced by the particle is recorded as a count only in that channel with a range in mV that includes the pulse produced by the particle; in the total or cumulative mode, the pulse from the particle triggers a count in all channels with a threshold setting in mV below that of the pulse produced by the particle. Contemporary light extinction sensors can count up to 40,000 cells/ml at a flow rate of 5 ml/minute.

The stated theoretical relationship between the size of the particle and the amplitude of the voltage pulse produced in light extinction counting is given as

$$E_0 = \frac{\lambda}{A} E_b$$

where E_0 = pulse amplitude from photodetector, λ = maximum projected area of the particle, A = area of the window, and E_b = 10 V, the base voltage from the photodetector.

The interaction of a particle with light in the view volume of a light extinction sensor is not as simple as this relationship suggests (Figure 10.3). This equation is generally appropriate, but it does not take into consideration refractive index, light absorption factors, and light scattering effects—all of which can impact the result obtained in light extinction counting. The "extinction" effect is due to the removal of light from the beam by both scattering and absorption. While particles

Figure 10.3. Interaction of a particle with incident light.

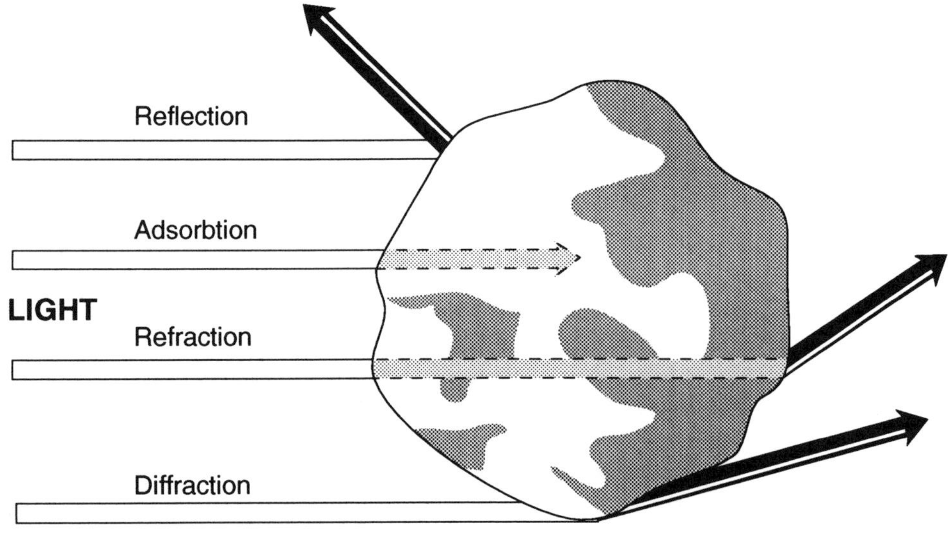

that are transparent can be expected to be sized somewhat smaller than opaque particles, the refractive index effect (refraction) predominates for particles over 5 μm in size, so that transparency is not a significant consideration for particles greater than this size.

Resolution is a key factor in the performance of light extinction sensors. This term describes the range of pulse voltages that will be produced by the sensor for particles of a single size. Poor resolution (i.e., broadening of the pulse response distribution) will result in decreased accuracy and precision of count data. Nonuniform illumination in the sensor is the primary cause of variances in sensor response for particles of the same size. This variance is typically several percent of the mean diameter of particles in the 5–20 μm range, and can be much greater for cells in the range of 1–2 μm, the minimum size that can be counted with this type of instrument. Under normal sampling conditions, however, the particles are randomly distributed throughout the cell, and a statistical averaging process applies to the pulses generated.

Most currently used light extinction counters have white light sensors with size measurement ratios of 1:60. Vendors commonly specify dynamic measurement range in the sensor technical description or model designation. A 1:60 specified sensor can measure particles from 1 μm to 60 μm, while a 2.5:150 sensor can measure particles ranging from 2.5 μm to 150 μm. Over the past several years light extinction counter systems have progressed significantly with regard to technology, user friendliness, and automation. HIAC/Royco (Silver Spring, MD), Climet Instruments (Redlands, CA), Particle Measuring Systems (Boulder, CO), Rion Co., Ltd (Tokyo, Japan), and Met-One (Grants Pass, OR) all currently offer systems operating on this principle.

As with the Coulter® method, there are potential pitfalls in using this method. Results of the instrumental assay cannot be expected to agree in any predictable fashion with those of a microscopic test (Rebagay et al. 1977; DeLuca 1977; DeLuca et al. 1987, 376–380; Schroeder and DeLuca 1980; and Hopkins and Young 1974). The various potential sources of error are outlined in detail in Barber (1987, 317–375) and Barber and Williams (1990, 502–537). The user should also be aware that although light extinction sensors with a similar dynamic range have a generally similar response to particles of the same type, such as bacterial cells, there may be specific sensitivity differences related to particle refractive index or type of sensor illumination used (white light, diode laser, gas laser) that should be investigated before an instrument is purchased for a specific application.

LIGHT–SCATTERING COUNTERS

Light-scattering counters, which collect scattered light rather than measuring a decrease in transmitted light as a means for counting particles, are applicable to both solution-borne and airborne counting. Their principal application lies in the counting of particles that range from a few micrometers down to submicrometer size. These instruments differ from light extinction units based on their measurement of light scattered by a particle in the view volume, rather than measurement of the decrease in intensity of the beam of illumination. Since the detection of scattered light can be made with a higher level of sensitivity than the measurement of a slight decrease in total sensor illumination intensity, these particle counters can detect and size particles smaller than those that can be analyzed using the extinction instruments. On this basis they may be useful in studies of organisms such as bacteria. These instruments are most highly developed for use in counting particles in air or gases, and the following discussion is based on the type of counter applied with gases.

At this time light-scattering instruments are capable of counting and sizing 0.05 μm particles in air at a flow rate of 2.8 ℓ/minute (0.1 cfm). A variety of particle-counting instruments are available for measurement of particles larger than 0.1 μm at sample flow rates of 28.3 ℓ/minute (1.0 cfm). The generalized function of a light-scattering counter is shown in Figure 10.4. Particles pass through the light beam (sensing volume) one at a time. Intensity of scattered light is measured by a photodetector. Photodetector pulses are analyzed by a pulse height analyzer (PHA). Calibration permits conversion from pulse height to particle size. Relating the pulse height data to the particle size information yields a response curve for the instrument and allows particles counted at given size thresholds to be quantitated.

The basic, most critical differences between single particle, light-scattering counters marketed today involve source illumination (laser or white light) and the geometry of scattered light collection (forward, near forward, or wide angle) (Liu et al. 1985; Montague and Sommer 1990). A white light sensor with wide angle collection geometry is shown in Figure 10.5. Quartz-halogen bulbs are typically used as a light source for white light counters. Laser counters may use He-Ne hard-seal lasers or a solid state laser diode. Other factors being equal, sensitivity (i.e., the smallest particle that may be sensed and counted) will be critically dependent on the intensity of the illuminating radiation and its wavelength

---

Figure 10.4. Operational principles of a light-scattering optical particle counter.

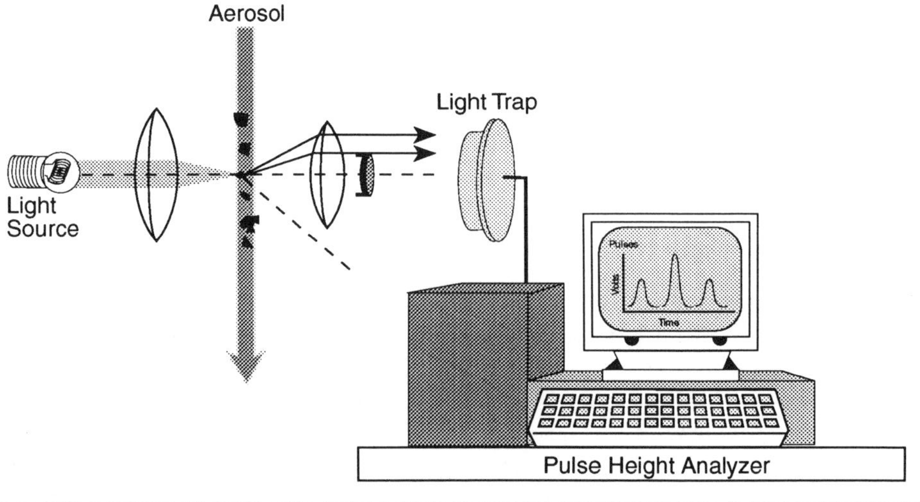

Figure 10.5. White light–based light-scattering counter with ellipsoidal mirror collection optics.

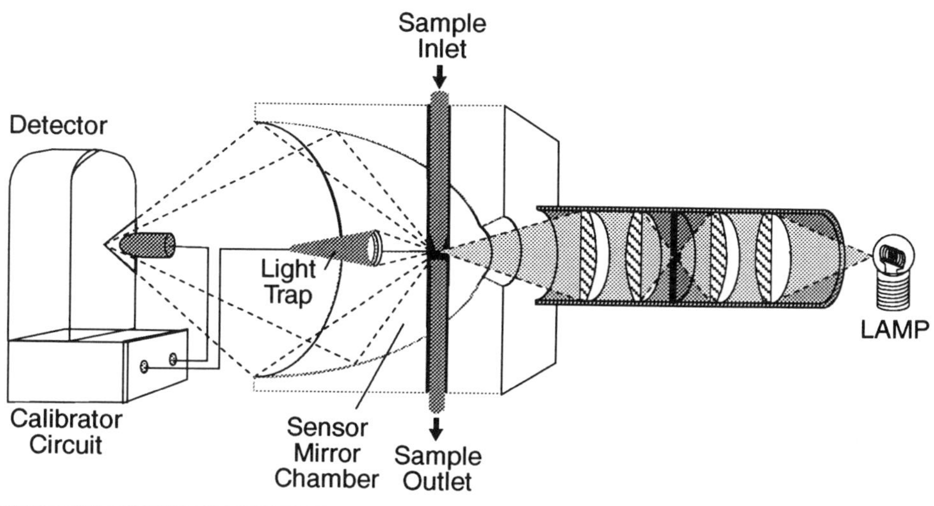

range. A hard-seal laser used in the active cavity mode gives an essentially monochromatic light with an extremely high photon density and results in the best sensitivity available (0.05 μm). Generally, more intense illuminating radiation results in a greater amount of scattered light for particles of a given size.

The response of a light-scattering counter to aerosol particles is a complex relationship between functional properties of the instrument and physical and optical properties of the particles in question (Horvath et al. 1990; Hovenac 1990, 108–117). The critical factors of the instrument in this regard are sensor design, type of illumination, electronic amplification, and noise. Particle size, shape, and refractive index also have a significant effect on how a particle is sized.

Important in this consideration is that light-scattering counters calibrated by different methods or of different design must be expected to yield different counts for the same particle populations (Bemer et al. 1990; Buettner 1990). Particle population distribution information from different manufacturers' counters, particularly, may be expected to show wide variation.

The response curve of light-scattering counters is typically nonmonotonic for particles that have different sizes and refractive indices; the shape of the response curve may also differ depending upon whether the collection optics used are of near-forward, wide angle, or off-axis design (Figure 10.6). At sizes <1 μm, particles of different sizes with different shapes and refractive indices can, in many instances, give the same pulse height, thus appearing to be the same size. While counter pulse height response generally increases with particle diameter, resonances in the response curve may result in very different responses for particles of closely similar sizes (Caldow and Blesner 1989). This may be an important

Figure 10.6. Generalized response curves for counters of different optical design.

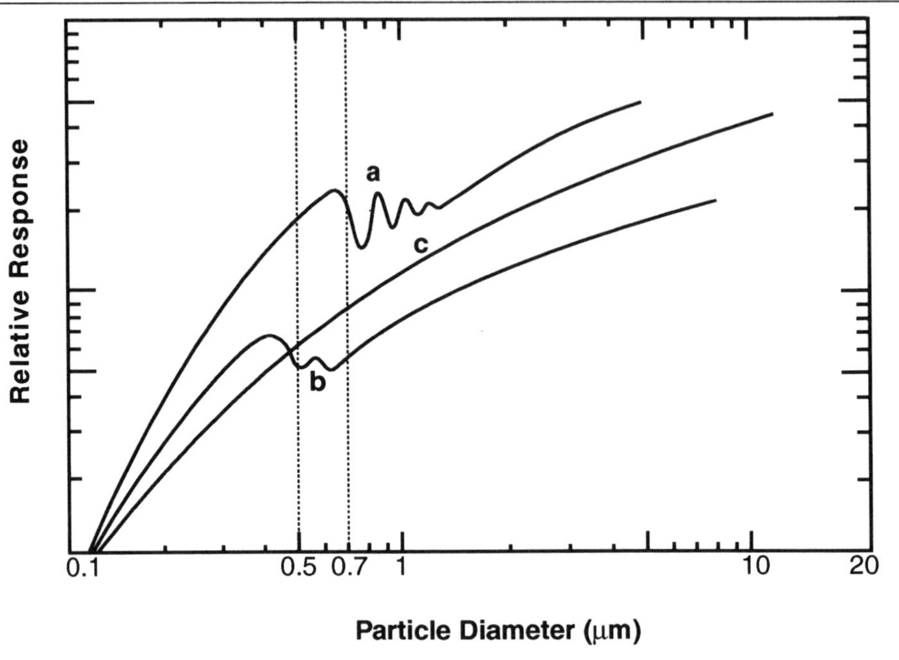

consideration for microorganisms of sizes that fall into the resonance area of the calibration curve.

Shape and orientation of nonspherical particles also affect the counter response (Cooper and Grotzinger 1989). If particles have dimensions less than the wavelength of illumination, the measured size may be considered equal to the volume equivalent diameter. For particles with dimensions above the wavelength of illumination, the measured particle size approximates the projected area equivalent diameter for instruments with wide-angle collecting optics. Refractive index differences between particles of the same size may cause light-scattering counters to see them differently (Knollenberg 1989). The refractive index of a particle is described by a complex number. The real (scattering) and imaginary (absorption) components of this number are both functions of the wavelength. Both components of the refractive index affect the response of a light-scattering counter. Generally, for vegetative bacterial cells, one may expect the response to be similar for cells of a wide range of species. Spores or highly hydrated cells, such as protozoans, may elicit a different response, requiring care in calibration.

FLOW CYTOMETRY

Flow cytometry measures physical and, in some cases, chemical characteristics of cells that are suspended in a liquid and pass singly by one or more optical sensors. This means of counting and sizing single cells is well covered in the text by Coons and Weinstein (1991). In the almost three decades during which flow cytometers have been used for blood cell counting, ongoing development of instruments, reagents, and analytical methods has greatly expanded the usefulness of the technique. Flow cytometry is now in common use for classifying normal and tumor cells, blood cells, and cells from the reticuloendothelial system; it has been applied by researchers in a wide range of other fields, including bacteriology, protozoology, microbial ecology, and pharmacology. Flow cytometric techniques have become increasingly important in diagnostic procedures, such as those performed on cerebrospinal fluid (Kleine et al. 1990).

Modern flow cytometers permit multiple measurements to be made on the particles studied. Fluorescent reagents or markers may be used to perform quantitative or qualitative analyses of specific cellular properties, such as surface antigens. It is this capability for the use of fluorescent markers that sets flow cytometers apart and makes them of somewhat greater usefulness for some specific types of studies than simple light-scattering, light-extinction, or electrical zone sensing counters. Cellular parameters evaluated by fluorescence flow cytometry can also be evaluated by other means of analytical cytology, but flow cytometry is particularly useful in that it often gives a more rapid analysis than other modes. Flow cytometric instruments have become progressively more available and simpler to use, and are presently applied in research as well as clinical roles. A typical instrument schematic is shown in Figure 10.7.

The most versatile flow cytometry instruments can measure multiple parameters of the cells analyzed. Several detectors and filters may be used to measure fluorescence at a number of wavelengths, together with low-angle forward and/or 90° light scatter. Fluorescence detectors are placed at angles of 45–90° to simplify the requirement for barrier filters and/or beam splitters. Multiple detectors permit the use of multiple fluorochromes for the measurement of several cellular parameters in a single analysis. In some instruments multiple sources may

Figure 10.7. Flow cytometer of current design.

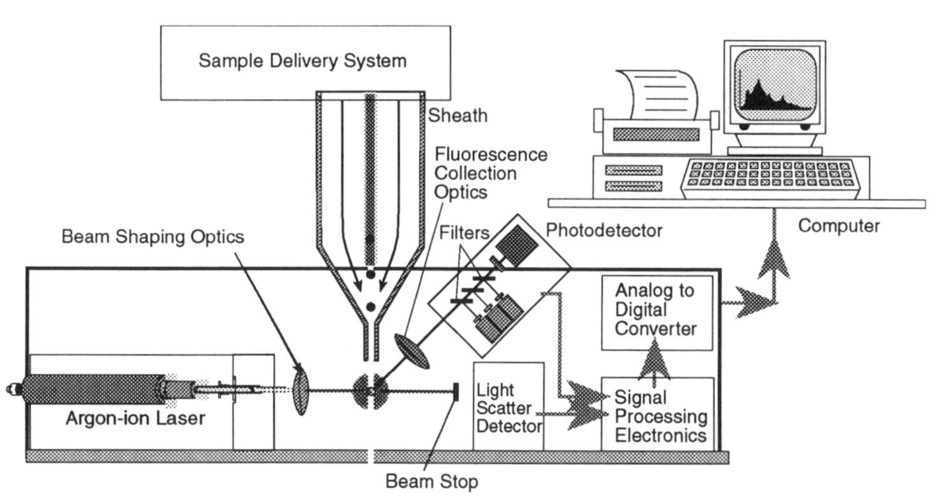

be used to excite fluorescence in cells at two different wavelengths. Apertures in the flow stream may also be used in order to gather electrical conductivity measurements on the cells analyzed.

Controlled transport of cells in suspension through the view volume of the instrument is essential to the function of flow cytometers. Cells in suspension are placed in a sample chamber from which they are delivered by pressure to the flow chamber. Because many cell suspensions of interest include cells of different sizes, continuous sample mixing must occur in the sample chamber to assure the delivery of a representative cell suspension to the view volume. Typically, this is accomplished through mechanical agitation or stirring of the sample. Prefiltration of some samples may be required to remove cell aggregates or debris that might occlude the small flow passages in the sensing zone.

In the flow chamber the sample stream containing cells is injected into the center of a cell-free stream of sheath fluid. The sample and the coaxial sheath liquids then enter the view volume, where their diameters are further reduced and the flow velocity is commensurately increased. The end result is a thin stream of sample fluid with the cells flowing in a column at the center surrounded by the sheath fluid. Sheath and sample pressures are adjusted to collimate the sample stream to a size of 10–30 μm. The sheath stream diameter is generally 70–300 μm, and flow velocity ranges from 10 m/sec to 20 m/sec. The sheath flow collimates and aligns the flow of cells through the measuring volume so that interaction of cells with the illuminating radiation is precise and reproducible. The optics used in flow cytometric systems have a limited depth of field, and the slight movements of the sample stream within the sheath stream may cause changes in the measured values. As with other optical particle counters, the beam intensity distribution in the view volume is Gaussian, with the center of the beam having a higher photon density than the periphery; this inherent unevenness of illumination makes precise intersection of the flow stream and the beam all the more important.

In the most common current application, fluorescence from a fluorochrome marker is excited as the cell passes through the beam in the view volume. Multiple detectors are then used to measure the secondary fluorescence of the cell at single or multiple wavelengths. The range of measurements made on cells in flow cytometry includes (1) the total light scattered by a cell, which is related to the reflective, refractive, and diffractive properties of the cell; (2) the volume of a cell, which is determined by measuring the electrical resistance change as a cell passes through a small Coulter®-type aperture; (3) light extinction loss, which is the total reduction in light intensity caused by a cell passing through the light beam in the view volume; and (4) morphology, as derived from the pulse shape or duration of the various pulse outputs. Laser sources of various types are the most widely used illumination in flow cytometers. They offer high beam density stability, are monochromatic, and can be focused down to areas approximating the dimensions of the cell; condensing lenses are used to focus the laser beams down to a diameter of 20–100 μm in the view volume of typical instruments. This provides a collimated beam that is of requisite intensity across the region of intersection with the cell.

LASER DIFFRACTION

Laser diffraction is sometimes called Fraunhofer diffraction, but instruments operating on the principle of laser diffraction may not be based on Fraunhofer theory (Elias 1972, 397–457). The Malvern MasterSizer® instrument, probably the most widely used of this type, is in fact based on Mie rather than Fraunhofer theory. In this instrument, as shown in Figure 10.8, the light from a 1 mW He-Ne laser is passed through a dispersion of particles in some suspending medium. The diffracted light is focused onto a multiple-element annular ring detector. For a particle of given size in the range of 0.1–600 μm, the intensity maxima in the diffraction pattern will be size determined, so that the light intensity at different radii on the detector serves as a measure of particle size. The measurement thus obtained for the components of a population are volume dependent to a first approximation. The data from this instrument are highly reproducible and rapidly obtained (Dodge 1984); calibration is also rapid and reproducible (Hirleman 1987). The instrument is widely applied in the measurement of particle-size distributions of powders dispersed in air or water (e.g., drug powders or emulsions). In applications for counting bacterial cells, relatively high cell concentrations may be counted, but debris present in significant quantities may destroy resolution and lead to erroneous results. The investigator should be forewarned that instruments of this type will produce only a generalized description of biological materials, such as a microbial cell suspension. If precision of particle sizing or the analysis of mixed populations is to be performed, a single particle counter should be the method of choice.

Figure 10.8. Laser diffraction analysis.

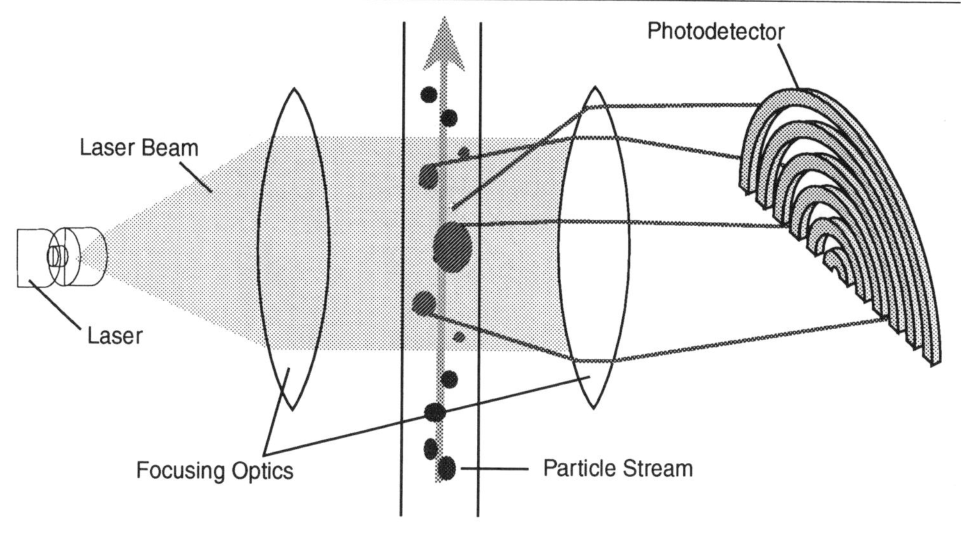

PHOTON CORRELATION SPECTROSCOPY

Conventional (static) nephelometry provides no measurement of the sizes of particles involved in total scattering interactions (Hirleman 1990, 159–168). Dynamic light scattering or photon correlation spectroscopy (PCS) is, in essence, a dynamic nephelometry whereby the size-dependent Brownian motion of particles (3 nm– 3 μm) is used to perform a population distribution analysis. These instruments (Malvern, Nicomp, Coulter®, Brookhaven) all operate according to the same general physical principle shown in Figure 10.9 (Swithenbank et al. 1977). A laser beam is focused into a cell containing a solution of suspended particles. A small fraction of the incident light is scattered by the particles and collected at some angle *j* (usually 35–90°) by a sensitive photodiode or photomultiplier tube (PMT) detector. Each particle illuminated by the laser beam produces a scattered light wave whose phase at the detector depends on the position of the particle in solution. For a population of particles, such as a suspension of bacteria, the total scattered intensity at the PMT is the result of the superposition of all the individual scattered waves. The Brownian motion of the particle causes the relative phases of the light scattered from different particles to vary, which in turn causes the intensity at the detector to fluctuate in time.

The stationary detector records fluctuations in the scattered light intensity as particles undergo Brownian motion. These intensity changes occur within a millisecond or less, depending on particle size. Small particles produce fast fluctuations, while large particles produce slower fluctuations. Although individual fluctuations occur randomly, there is a well-defined time for their buildup and decay, roughly equal to the average time required for a pair of particles to change

Figure 10.9. Photon correlation spectroscopy.

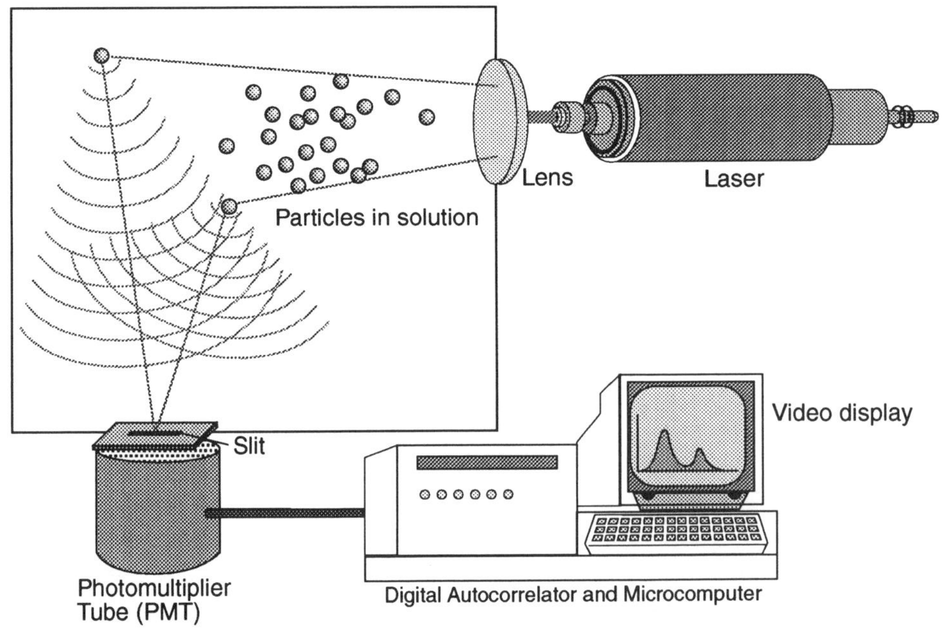

their separation by one-half the laser wavelength l. To perform the analysis, one computes electronically the autocorrelation function $C(t)$ of the scattered intensity $I(t)$,

$$C(t) = <I(t)' \times I(t'- t)>$$

This function is constructed by computing the product of the scattered intensity at an arbitrary time t' with the value at the earlier time $t'_n - t$. The symbol $<>$ indicates a sum of such products taken at different times t'. A sufficient number of pairs of intensities are sampled (e.g., 10^5–10^7) to yield a reliable statistical average of $C(t)$. These data are used for the calculation of the population distribution of the scattering particles (Hirleman 1990, 159–168). A digital autocorrector is used to define the function $C(t)$ for many values of t, resulting in a smooth quasi-continuous function (curve or histogram).

While the physical construction of these instruments is quite similar, the programs by which they compute the autocorrelation function and distribution analysis are not (Weiner 1979; Elias 1972, 397–457). A simple test for how well the autocorrelation function works is to evaluate the instrument in which one is interested, using monosized and mixed latex sphere standards. The instrument that performs the analysis the quickest with the highest accuracy and whose resolution provides the best estimate of the known distribution is probably the most suitable instrument for purchase.

The principle advantage of the PCS technique for particle sizing is that it yields an absolute measurement; that is, a PCS–based instrument is inherently self-calibrating, requiring no particle size standard per se. The calibration is determined by the wavelength of a laser, the period of a crystal-controlled clock, the scattering angle, and physical parameters of the suspending medium (viscosity and index of refraction), all of which can be determined. A PCS measurement of particle size is immune to normal analog-type drifts, such as changes in laser power or detector sensitivity. Furthermore, the measured diffusion coefficient is unaffected by either the composition of the particles or their concentration, provided the suspension is sufficiently dilute so that interparticle interactions are negligible. Additional advantages of the PCS technique include speed (a typical measurement requires only a few minutes) and the fact that it does not in most cases disturb the system under investigation. This latter feature is especially important for emulsions and suspensions whose particle properties are a sensitive function of concentration and/or solvent composition (e.g., micelles and microemulsions). Most PCS instrumentation is simple mechanically, possessing no moving parts or sophisticated optics that require careful alignment. This method has in common with laser diffractometry the ability to work with higher concentrations of cells than single particle counters; it also has a sensitivity to debris that will tend both to broaden the apparent size distribution of a sample, such as bacterial cells, and to affect the apparent average size of the cells.

CHOICE OF INSTRUMENTAL METHODS

In assessing the possible usefulness of instrumental particle counting in evaluation of populations of bacteria or other single-celled organisms, a number of considerations must be addressed. The first and most important is the nature of the data that must be acquired to answer the questions at hand. The second is a careful consideration of whether or not some instrumental particle counting or particle sizing assay will yield the required type of data. The initial study design process is an excellent point at which the microbiologist can enlist the services of a particle analyst; getting help and advice may prevent the wasting of valuable time and resources.

One most important factor is the lowest size threshold and dynamic range of analyses available. If the microbe to be counted is on the order of 0.2 μm average size (e.g., *Pseudomonas diminuta*), the analysis is immediately relegated to light-scattering, single particle counters or Coulter®-type instruments, since light extinction counters have a lower end sensitivity limit of 1.0 μm at best. If population distribution data is needed, laser diffraction instruments or PCS may yield the required data based on average size and shape of the distribution, and incorporate significant time advantages. These instruments do not provide adequate data from samples containing cells of widely disparate sizes, such as bacteria and protozoans. Single particle counters will have a 50- to 70-fold dynamic range; this factor will dictate the width of the distribution that can be critically measured and the upper limit of the count data.

Data collected from dispersed cells by particle analyzers and single particle counters will be powerfully modulated by the presence of other particles in the test suspensions. Debris or microbial populations other than the one being tested may mask the specific results the researcher needs. Airborne sampling must be carried out so that particles of one size or another are not selected against by the means of sample transport or collection. Most debris populations have high numbers of counts at ≥1 μm; pulses from this material may fall atop the counts from bacteria so that the counts from bacterial cells either cannot be discriminated from the background or appear larger than their actual size. Sampling methodology is in itself most critical. Bacterial cells are subject to sedimentation and aggregation, so that steps must be taken to ensure even distribution and isolation of the target organism in suspension. Airborne organisms typically are found in aggregates or in "rafts" of larger particles, such as skin cells.

The environments in which cells to be tested are often found (e.g., media or water samples) may be detrimental to the counter sensor, causing blockage or other artifact. Calibration of counters is critical, and specific calibration may be necessary using the medium in which the cells will be suspended when counted. In addition to dynamic range, resolution and concentration limits are also critical instrumental parameters. Poor resolution is an inherent feature of many older counters at submicrometer sizes and their count sensitivity may also be unacceptable at these small sizes. Poor resolution results in peak broadening and sensitivity may be <50 percent at the lower sensitivity limit of an older instrument, meaning that only half the cells present there are counted. If numbers of cells in excess of concentration limits are counted, coincidence counting or the counting of multiple particles as a single large particle may cause count loss and change distribution shape. Again, expert help may be invaluable in the study design phase. Finally, the nature of the data obtained from a given type of counter must be considered. Light extinction counters will produce a range of pulse amplitudes for a single rod-shaped cell, depending on its orientation in the light beam. Light-scattering pulses or Coulter® pulses from the same cell will relate to volume and are generally a more uniform means of measurement.

MICROBIAL SIZING AND COUNTING APPLICATIONS

Published results regarding use of particle counting instruments to size and count microbial cells generally relate to four areas of study:

1. Use of instruments to determine numbers of bacteria or other cells in air or aqueous environments

2. Tracking phagocytic or immunologic phenomena

3. Cell sizing or general population analyses

4. Tracking cell size changes due to processes such as sporulation or mutation

A discussion of application of particle sizing and counting instruments to microbial samples is grouped below by these four categories. Again, this collection of literature is less than comprehensive, but will serve to provide an idea of the range of applications of the technique.

Counting of Organisms

The relationship of airborne particle counts to microbial content of air is a subject of great interest to pharmaceutical manufacturers. Whyte (1983) carried out experiments in four manufacturing pharmacies to assess the effects of ventilation and airborne contamination on terminally sterilized products. Each pharmacy manufactured containers of 0.9 percent NaCl and carried out experiments in widely differing conditions of airborne cleanliness. No correlation could be demonstrated between the cleanliness of the air with regard to particles and the microbial quality of the product (as established by bacterial and airborne particle counts). The solution containers and closures (especially after autoclaving) were evidently the main source of contamination. Interestingly, complex ventilation and filtration schemes for the production of terminally sterilized product were shown to be unnecessary for the control of airborne bacteria.

Rhame et al. (1986) used light-scattering, airborne particle counters in an assessment of numbers of bacteria in aerosols produced by humidifiers. These workers studied this aerosol generation in a clean room using an airborne particle counter and samplers for airborne bacteria. At gas flow rates of up to 80 ℓ/minute, a bubbling humidifier produced between 450 and 1000 water droplets/ℓ humidified gas with a total water volume aerosolized of approximately 10^{-8} ml per liter of humidified gas. Approximately 75 percent of these particles had diameters between 1 and 5 μm. With the reservoir containing 6.4×10^6 *Pseudomonas aeruginosa* per ml, the humidifier produced between 2 and 9 *Pseudomonas aeruginosa* cells per liter humidified gas. Most of the bacteria were in particles of a size likely to be deposited in the lung (<10 μm). In some cases, this bacterial carryover was greater than 100 times the amount predicted by comparison of bacterial concentration and the water volume aerosolized. An ultrasonic humidifier produced fewer particles <10 μm in size, but when the reservoir contained *Pseudomonas aeruginosa*, it also aerosolized bacteria. Wick-type humidifiers did not produce detectable aerosol or bacterial carryover. Use of the particle counter in this study provided a practical way to relate particle size to rate of bacterial aerosolization.

Suzuki et al. (1984) studied airborne contamination in an operating suite with a slit sampler, settling plates, and a light-scattering particle counter. There was, in general, a significant difference in airborne contamination as determined by particle counting between the empty rooms and rooms in use; the mean total bacterial count by a slit sampler changed from 1.1 colony forming units (CFU)/m^3 in an empty room to 42.5 CFU/m^3 during use (a 40-fold increase). The settling plate count changed from 1.5 to 17.4 CFU/m^2/minute (a 12-fold increase), and the mean total particle count changed from 56.9 to 546.7/ℓ (a 10-fold increase). The increase was found to be due mainly to personnel present in the room. Another difference was found between zones in the operating room; the bacterial count in the cleanest area was doubled in the next cleanest area and doubled again in the least clean area based on slit sampler counts as well as settling plate counts. Particle counts in the cleanest area increased by 14 times in the less clean and least clean areas. Air cleanliness of operating rooms in use dropped to a level between the clean and the less clean area in spite of the high quality of the ventilating system. Bacterial species identified were mostly coagulase-negative staphylococci and micrococci. This study indicated that airborne counts track microbial counts, but without a specifically predictable relationship; it points to the generation of both particles and bacteria by personnel.

Kogure and Koike (1987) investigated the applicability of the Elzone™ (Coulter®-type) particle counter to the determination of marine bacterial biomass. The biomass of bacterially pure cultures and a mixed natural population were followed by using the particle counter, a carbon hydrogen nitrogen (CHN) analyzer, and an adenosine triphosphate (ATP) analyzer. The particle counter showed the precise size distribution of number and volume of submicrometer particles in sea water. For the pure cultured bacterial strains, the conversion factor from volume to carbon was 0.209 mg of C per mm^3, and for natural bacterial cells of <0.6 μm in diameter it was 0.184 mg of C per mm^3. Particle counts obtained tracked biomass measurements, suggesting the means of analysis to be useful in studies of this type.

Perez-Rosas and Hazen (1988) applied electronic counters in their investigations of bacterial survival in marine environments. *Escherichia coli* and *Vibrio cholerae* were inoculated into membrane diffusion chambers and placed adjacent to coral reef islands near Puerto Rico and monitored for five days; chambers were also buried in the sands of reefs. Both *Escherichia coli* and *Vibrio cholerae* densities declined by two orders of magnitude in five days, as measured by direct particle counts with a Coulter® counter. However, neither of the cell types changed dramatically in density when the same samples were analyzed by epifluorescent direct counts. Differences in the two methods were believed to be accounted for by changes in cell morphology that occurred in both bacteria after exposure to sea water and affected the size registered by the particle counter. Morphological changes occurred more rapidly in *Escherichia coli* than in *Vibrio cholerae*. Bacteria in chambers exposed to sediment did not show significant changes in morphology and declined only slightly in density. Physiologic (metabolic) activity declined by more than 40 percent for both bacteria within 24 hours; the decline in activity was less severe in the sediments. Tropical coral reef sands and turtle grass beds were shown to be less stressful environments for *Vibrio cholerae* and *Escherichia coli* than would have been predicted from temperature and microcosm studies. Importantly, *Vibrio cholerae*, a pathogen, can survive the in situ conditions of a tropical coral reef and could become a source of bacterial contamination for fish and shellfish in this environment. The simultaneous monitoring of *Escherichia coli* levels established that this organism is unsuitable as an indicator of *Vibrio cholerae* or other fecal-borne pathogens in coral reef environments because of the comparatively greater stress these environments put on *Escherichia coli*.

Submicrometer particle counting was evaluated by Maerz et al. (1990) as a method to continuously monitor the integrity of reverse osmosis (RO) membranes. For this purpose the method proved to be superior to the conditioning measurements that are widely used. At high membrane retention rates, particle rejection corresponded well to microbial retention. The particle count method was shown to be prone to artifacts from particles that are released from the inner surface of the cartridges, and so it would be better suited for use after sample washout time. Although the counter had a lower detection limit of 0.5 μm, it failed to detect some portion of the bacteria that grew in the RO cartridges. Likewise, water-borne bacteria cultivated in bottles filled with pure water produced particle counts lower than expected from their number, as detected by dilution plate technique. The basis of the undercount could not be determined; total airborne counts have been found in other studies to exceed plate counts. This finding may well have to do with disaggregation of clumps of cells during culturing.

Butera et al. (1991) used pigs housed in pens in two separate rooms in order to assess the effects of ventilation rate, temperature, relative humidity, and source of air on bioaerosol levels and dust with particle sizes <10 μm. A six-stage Andersen sampler and a light-scattering particle counter were used to determine bioaerosols and respirable dust (0.1 to 10 μm) in the two areas. Total bioaerosols were assessed using trypticase soy agar (TSA) and potato media. Dextrose agar was used for fungal aerosols and Baird-Parker Agar for isolation of *Staphylococcus aureus*. Molds amounted to less than 1 percent of total microorganisms; grampositive bacteria made up 72 percent of the bacterial isolates. Respirable dust counts did not correlate well with respirable bioaerosols. Ventilation rate (2, 5, or 8 changes/hr) did not affect bioaerosol levels or respirable dust. Total bioaerosols were significantly reduced ($P < 0.05$) in higher temperatures only, and relative humidity did not influence total bioaerosols. This and the previous study (Maerz et al. 1990) point to the need for a better understanding of the relationship of viable counts versus counts obtained from particle counters.

Masuko et al. (1991) described a novel method for the enumeration of bacteria based on the principle that small, light-emitting particles on a flat surface can be easily and rapidly detected and counted using an ultrahigh sensitivity TV camera. To test this method, TV images of individual cells of a luminous bacterium on a membrane filter were obtained without the use of a microscope. The positions of the luminous points in the TV images were the same as the positions of the bacterial colonies after growth. These results show that single cells can be detected and counted by instrumental methods based on luminosity if they emit light or can be stimulated to emit light. This technique, though not based on a conventional particle counter, illustrates a principle for possible applications with counters, such as the low angle laser scanners used to measure fallout on wafer surfaces in microelectronics manufacture.

A bacterial aerosol challenge filter test bed was developed by Kastelein et al. (1992) for the direct assessment of the effectiveness of bioreactor gas filters as an alternative to routinely applied indirect wet integrity testing (IT). The test apparatus was based on bacterial aerosol challenging with *Pseudomonas diminuta*, dual monitoring by laser particle counting, and Andersen impactor microbial sampling of viable cells. The filter test apparatus was found to reproduce usefully the various conditions encountered in fermentation processes. In experiments with several filters from one class, it was demonstrated that some filters were actually penetrated by up to 3,000 viable cells per test, despite approval of these filters by commercially available IT test equipment that assesses average efficiency. Repetitive filter use, prolonged use, and the autoclaving of filters resulted in an increase in pressure drop over the filter, but improved the performance of leaking/deviant filters due to buildup of a filter cake. The integrity tests used were found inadequate for accurate assessment of filter quality; certification of filter lots by random testing was also found inadequate for accurate assessment of overall filter quality. Counting by particle counter yielded a 40-fold higher count level than with cell counts obtained with the Anderson impactor; the ratio allowed counts originating from the filter surface to be differentiated from those passing through the filter.

One of the most useful applications of straightforward instrumental counting of microorganisms has proven to be in the area of water quality. Semiconductor manufacturers, whose processes require ultrapure water, are concerned about their ability to size and enumerate bacteria that may multiply in areas of piping downstream of filters. Similarly, water purification facilities may obtain worst-case estimates of their microbial water quality by assuming all counts in specific size ranges represent microbes. Electronic particle counters have proven useful in both applications. In the case of bacterial counting, benefits to public health accruing to the use of this easily applied test methodology are significant. Clancy and Hango (1992) determined the response of Particle Measuring Systems' HSLIS 0.05 μm OPC to challenge known bacteria concentrations in ultrapure water. With the test procedures used, some discrepancies between the OPC data and epifluorescence microscopy data were found, but the particle counting methodology was found to be an extremely useful tool, and instrumental counts generally agreed with epifluorescence data.

The same authors in a later article (Hango and Clancy 1993) explored the question of whether or not a laser particle monitor could quantify bacteria in clean water. The monitor response was compared with scanning electron microscopy (SEM) and epifluorescence counts. The laser monitor reported rod-shaped bacterial cells ranging in size from 0.5–1.0 μm to 0.8–5.0 μm as having a scattering cross-section equal to that of latex spheres of 0.18–0.4 μm in size. Using this conversion factor, good agreement was found between the laser monitor data and those from direct count methods.

A study was undertaken by Jacangelo et al. (1989) to evaluate the efficacy of membrane filtration for the removal of particulate material from two untreated water supplies in northern California. The permeate produced by the hollow-fiber ultrafiltration module was assessed by monitoring four water quality parameters: turbidity, suspended solids, direct particle counts, and selected indicator bacteria. The study showed that the ultrafiltration process was very effective in removing particles from both sources. The data lead to the conclusion that a constant final water quality in relation to particulate material can be achieved regardless of the differences in concentration or numbers of particles in the raw waters. Lewis et al. (1993) found that instrumental counting provided detailed information about the effectiveness of water treatment processes, since individual particle size and number are defined. The use of light extinction counters to detect *Giardia*, as well as *Giardia*-sized particles, was shown to be possible by monitoring size channels into which *Giardia* cysts were counted by the counter used.

McCoy and Olson (1987) evaluated turbidity standards for water quality in relation to total particle counts, heterotrophic plate counts, and epifluorescence direct cell counts in the drinking water supplies of three municipalities. Turbidity and particle counts were directly proportional, although there was no predictable relationship between bacterial quality and turbidity or particle counts. Heterotrophic plate counts underestimated epifluorescence direct cell counts significantly, as has been shown to be the case in previous studies. Water quality degradation occurred in municipal drinking water systems because of short-duration events that resulted in high turbidity, particle counts, and heterotrophic plate counts. For all three parameters measured, variability increased with the distance that the water traveled within a pipe, except for ground water. It is interesting that in the ground water and reservoir water samples tested there were approximately 9.4 times more cells than particles. Typical bacterial aggregates from these samples showed that one "particle" might very well consist of a clump of bacterial cells. Thus, it is easy to conceptualize that the ratio of cells to particles is greater than one in certain types of drinking water. As might be expected, the ratio of cells to particles in the Los Angeles (LA), California, drinking water (surface water directly from the LA aqueduct) was less than one. This result was supported by direct examination of the LA samples by epifluorescence microscopy. The authors believed that the heavier, nonbacterial aggregates in ground water and reservoir water may have settled out in the storage reservoir of the well and in the reservoir, respectively, or that, alternatively, lower or no chlorine residuals may have allowed more bacterial growth or longer survival.

Importantly, the predominant "particles" in the ground water and reservoir drinking waters were aggregates of bacterial cells (on an average there were about 9 cells per particle). These bacteria were not necessarily associated with or colonized onto nonbiological particles, nor were they necessarily viable. The existence of significant numbers of bacterial aggregates was believed by the authors to be an important factor that could potentially contribute to a substantial underestimation of total cell concentration by CFU assays and, in some cases, by particle counting.

Immunologic and Phagocytic Processes

As discussed earlier, flow cytometry provides the researcher with a significant advantage in analyzing microbial or other single cells due to the wide range of information that can be obtained. Bassoe and Bjerknes (1985) and Bassoe et al. (1983) described the use of flow cytometry in measurement of the phagocytosis of bacteria by leukocytes, and proposed that such measurements could prove useful in clinical studies for the assessment of cell-mediated immune function of patients suffering from the effects of severe burns or chronic infections. The fluorescence detection mode of these instruments allowed the numbers of ingested particles, such as labeled bacteria, to be determined by fluorescence intensity. Latex beads were also used to determine a nonspecific index of phagocytic efficiency. In a similar study Bjerknes (1984) used a trypan blue dye to differentiate between cells that had been adhered to phagocytes and those that had been ingested. Gorman et al. (1886) in a related application used a Coulter® counter to measure uptake (adherence) of bacteria to mucosal cells.

Flow immunofluorescence techniques were utilized by Philips and Martin (1988) for the specific detection of the bacteria *Escherichia coli*, *Legionella pneumophila*, and *Bacillus anthracis* spores after staining with a fluorescein-conjugated antibacterial antibody. For detection of each bacterial type, a comparison was made of channel selection based on narrow forward angle (NFA) light scatter and on the red fluorescence signal pulse available from staining with the nucleic acid dye propidium iodide. Results were inconclusive based on the fact that the *Bacillus* spores did not take up propidium iodide and only a fraction of the *Legionella* population gave NFA scatter signals. The efficiency of detecting bacteria stained with antibody remained constant with differing cell concentrations, and the estimate of the count for specific bacteria expressed as a fraction of the total cytometer count decreased proportionally to total bacterial concentration. Cytometer background signal noise inherent in the high signal amplification needed for bacteria proved to be a difficulty. The noise was found not to originate in the photomultipliers and was evidently the result either of light scatter from subcountable particles in the sheath fluid or anomalous scatter from optical components. Part of the noise could be removed by channel selection, but there remained a noise component overlapping with the NFA scatter and fluorescence signals from the heterologous bacteria (i.e., those not stained with a specific antibody). At the low bacterial concentration, no meaningful flow cytometric count could be obtained for unstained bacteria due to noise effects.

Cambiaso et al. (1990) devised a combined particle count immunoassay for the detection of mycobacterial antigens in cell lysates and in tissue extracts, which was based on the agglutination of latex particles coated with anti-*Mycobacterium bovis* F(ab')2, followed by counting nonagglutinated particles. These workers were able to detect mycobacterial antigens in cell lysates from bronchoalveolar washings and in spleen and liver lysates obtained from experimentally infected rabbits. Antigens were also detected in 10 out of 11 samples obtained from patients with proven tuberculosis infection. These samples were readily distinguished from 32 negative control samples after pepsin treatment. Periodate treatment of samples to destroy cell surface carbohydrates abolished all reactivity. Following gel filtration chromatography, three peaks were identified with antigenic properties in samples of all types. The uniform detection of mycobacterial carbohydrate antigens by latex agglutination and particle counting was believed to be a potentially useful ancillary technique in the diagnosis of tuberculosis.

Cambiaso and Limet (1989) reported the development of a precise and rapid anti-Brucella IgM antibody (Ab) latex agglutination assay based on particle counting. The interference of IgG Ab was eliminated by the addition of antigamma Fc and free Brucella-lipid polysaccharide (LPS). Anti-mu Fc MAb enhanced the agglutinating activity of IgM antibodies and improved the sensitivity of the test. The possible interference of rheumatoid factor was eliminated by adding human aggregated IgG. The assay was complete in 45 minutes, with an interassay variation of 9 percent. The assay correlated well ($r = 0.94$) with the accurate but time-consuming enzyme-linked immunoadsorbent assay (ELISA) methodology.

A particle-counting immunoassay (PACIA) was compared with the BACTEC® system for detecting mycobacterial growth after short-term culture and was used to identify *Mycobacterium tuberculosis* by Drowart et al. (1993). Latex particles were coated with polyclonal anti-*Bacillus calmette guerin* (BCG) or with specific 2A1-2 monoclonal antibody (MAbs). Culture flasks containing Middlebrook 7H9 liquid medium and BACTEC® 12B vials were inoculated with equal amounts of mycobacteria from four reference strains (*tuberculosis, kansasii, avium,* and *xenopi*). With the use of anti-BCG, PACIA detected mycobacterial antigens 3 to 6 days before the BACTEC® system. *Mycobacterium tuberculosis* was differentiated from other mycobacteria. Seventeen clinical samples were also studied. In the same 10 samples the two techniques detected mycobacteria, PACIA with anti-BCG after 9 days while BACTEC® became effective 1 to 5 days later. The authors concluded that the particle count assay could detect mycobacterial growth earlier than BACTEC® and that *Mycobacterium tuberculosis* could be distinguished from other mycobacteria in PACIA performed with specific MAbs.

To address issues related to accurate enumeration and sizing of bacterial suspensions, Cantinieaux et al. (1993) used a flow cytometric method for determining bacteria concentrations by comparison with a standardized fluorescent latex bead solution. Relative counts of beads and bacteria were established in a system using both fluorescence and light scatter for the two types of particles using the beads or internal standard. The latex bead size (0.98 μm in diameter) permitted counting on traditional hematological counters and, on the other, a flow cytometric detection with the same conditions for bacteria. The reproducibility of the study of bacteria concentration measurements gave a coefficient of variation (CV) of <5 percent.

Flow cytometry was also investigated by Pinder et al. (1990) as a rapid detection and counting method for bacteria in pure cultures. A two-parameter detection scheme was employed: Particle size was measured by forward angle light scatter and nucleic acid content was measured by fluorescence of the selective DNA/RNA dye ethidium bromide. The technique gave results that correlated exceptionally well with conventional plate counting for four species of bacteria, and concentrations in the range 10^2 to 10^7 CFU/ml. Cytometric counts were obtained in a few minutes, in contrast to the 48 hours typically required for plate counts. Under ideal conditions each bacterial species examined exhibited a characteristic population distribution on the cytometer, which could be explained by its known properties and morphology.

Avesani et al. (1994) evaluated the potential of PACIA for the direct detection of *Clostridium difficile* serogroup G-specific antigen. As a label, F(ab')2 fragments from a rabbit anti-serogroup G antiserum were covalently coupled to carboxylated latex beads. This reagent was mixed with acid extracts of fecal specimens and the reaction was assayed with an optical counter, which allowed the differentiation of unagglutinated and agglutinated latex particles. Culture for *Clostridium difficile*, fecal cytotoxin detection, PACIA, and serogrouping of *Clostridium difficile* isolates were performed on a total of over 200 stools. Of 71 culture-negative specimens none gave a positive result in the cytotoxin assay or in PACIA. Fecal cytotoxin was detected in 100 of the 178 culture-positive specimens. The PACIA was positive for 63 of the 71 fecal specimens that yielded serogroup G *Clostridium difficile* on culture and gave negative results for all other culture-positive stools tested, with one exception. PACIA detection of serogroup G antigen in fecal specimens showed a sensitivity of 88.7 percent, a serogroup specificity of 99.7 percent, a predictive efficiency value of a positive culture with a serogroup G strain of 98.4 percent, and a predictive efficiency value for specimens that were culture negative for a serogroup G strain of 95.6 percent. The results indicated that particle-counting techniques with specific antiserum can be a reliable method for detecting serogroup specific antigens of *Clostridium difficile* in fecal specimens.

Microbial Size Distribution Studies

Poole (1982) investigated the Coulter® Nano-Sizer™ for use in sizing bacterial cell populations. The author found that the advantages of this photon correlation spectrophotometer for measurements of sizes of microorganisms included

1. The measurements are rapid (typically 1 to 4 minutes), and average particle diameters are presented directly with no further processing required.

2. Provided that the temperature, refractive index, and viscosity are known, any diluent may be used, including media in which the organisms are growing.

3. There are no special requirements for conductivity properties.

4. Measurements can be made independent of particle shape. For rod-shaped cells, which have slower diffusional motions than a sphere of equal volume, an axial ratio (i.e., major axis:minor axis) of 3:1 was found to lead to an overestimation of size by only 10 percent. When the ratio is 5:1, the overestimation was 20 percent.

5. The particle sizes determined showed a good correlation with independent (infrared microscopic) techniques not only for inert colloidal particles, but also for a wide range of microorganisms within the dynamic size range of the instrument (0.03–3 μm).

Consequently, the Nano-Sizer™ may be superior to the conventional Coulter®️ counter, in which the choice of electrolyte must be made carefully for true measure of cell volume. Even with the appropriate electrolyte, anomalous size data can result from changes in the conductivity of the particles, nonuniform flow through the sensing aperture, and sampling errors. The second item listed is perhaps the greatest asset of this instrument in studies of cell growth and the cell cycle, since it allows one to avoid removal of cells from their environment.

The most serious shortcomings of the Nano-Sizer™ regarding its widespread use in sizing organisms were found to be (1) the upper size limit for particles, which excludes many yeasts, fungi, and protozoa; (2) the fact that only superficial information on the distribution of particle volumes can be obtained; and (3) the sensitivity of the instrument to differences in refractive indices of various microbes. Despite these considerations, it is likely that the techniques of laser light scattering and PCS could enjoy extensive use in studies of microbial growth.

Reconstituted, lyophilized, attenuated *Mycobacterium bovis, Bacillus calmette guerin* vaccine (Tice substrain) was characterized by Zhang and Groves (1988) using a Coulter®️ Multi-Sizer™ and a HIAC/Royco counter. The primary organism had an equivalent spherical diameter approximating 1 µm but the BCG cell suspensions were found to be heavily aggregated. The authors found that the cumulative size distribution of the suspensions fitted a log-probit plot; this information was used to determine the total number of particles per ampoule of suspension. The instrumental count was found to be related in a predictable fashion to the viable count. The state of cell dispersion was unaffected by mild shear (syringe aspiration or ultrasound) and only slightly affected by the addition of cetylpyridinium chloride or sodium tauroglycolate.

Airborne particle counting in eight size ranges (0.5 µm to greater than 20 µm) by computerized single particle-counting equipment was compared by Seal and Clark (1990) with the numbers of bacteria-carrying particles (BCP) assessed by slit sampling in ultraclean and turbulently ventilated operating theaters. In an ultraclean theater the number of 5–7 µm particles correlated well with BCP, while peaks in the numbers of particles less than 3 µm and greater than 15 µm related to human activity. Counts could also be related in the turbulently ventilated theater, but the use of particle counting equipment in that environment was not believed adequate to replace counts of airborne bacteria. The authors believed that electronic particle counting in the 0–20 µm size range could be used to assess the performance of a clean air operating theater distribution system, including efficiency and integrity of the filter/seal systems and the presence or absence of entrainment of bacteria and other particles.

One of the most promising applications of instrumental particle counting would seem to be the study of microbial ecosystems. A natural subject of this type of study is marine and aquatic environments, where the presence of bacteria, phytoplankton, and zooplankton are the key descriptors of the productivity of the system. Tungate and Reynolds (1980, 1–11) conducted the first study of this type in which light extinction particle counting was applied to obtain a comprehensive population profile of microorganisms and other particulate matter in marine waters. These researchers overcame the approximate 50-fold dynamic range limitations of the sensors used by coupling a total of 6 sensors with a combined range of 2–480 µm into a single pulse height analyzer. The sensors were operated sequentially to obtain a continuous size spectrum of particle numbers over the measured range. The sequencing of sensors, data reduction, and sampling was computer-controlled to permit untended operation for long sampling intervals.

The data collected in this study conformed generally to a log-log distribution that has been found to occur with many other types of particulate matter, including contaminant particles in widely diverse materials, such as pharmaceutical solutions and hydraulic oil, as well as in powders produced by comminution (Groves 1969). Although these investigators did not discriminate between viable particles (bacteria, phytoplankton, zooplankton) and detritus in their samples, peaks in the general distribution (2, 4, 22, 38, and 120 μm) corresponded closely with the size of planktonic organisms known to be present in high quantities in the samples.

Recent enumeration and identification of marine particles that are less than 2 μm in diameter, suggested to Stramski and Kiefer (1991) that these organisms may be the major source of light scattering in the open ocean. The authors presented a mathematical consideration of scattering phenomena related to numbers of organisms of various types. The living components of these small particles include viruses and the smallest eukaryotic cells. In order to examine the relative contribution by these various microorganisms to scattering, these workers calculated a balance for both the total scattering and backscattering coefficients (at 550 μm) of suspended particles. This balance was determined by calculating the product of the numerical concentration of particles of a given category and the scattering cross-section of that category. In order to make such a comparison, the authors estimated both the total scattering and backscattering cross-sections of various microbial components in which the authors included viruses, heterotrophic bacteria, prochlorophytes, cyanobacteria, ultrananoplankton (2–8 μm), larger nanoplankton (8–20 μm), and microplankton (>20 μm). In addition, the authors gathered published information on the numerical concentration of living and detrital marine particles in the size range from 0.03 μm to 100 μm. The results of the study were summarized as follows: The size distribution of microorganisms in the ocean roughly obeys an inverse fourth power law over three orders of magnitude in cell diameter, from 0.02 μm to 100 μm. Thus, the size distribution of living organisms is similar to that for total particulate matter in the same environment as determined by electronic or particle counters.

The authors concluded that, for representative values of refractive index, most of the scattering in the sea comes from particles less than 8 μm in diameter, and that most of the backscattering comes from particles less than 1 μm. Among the microorganisms that are found in this size range, free-living heterotrophic bacteria may be the most important. The second most important source of microbial light scattering is cyanobacteria (especially in tropical and temperate waters) and ultrananoplankton. Viruses, which may be very abundant, make little contribution because of their extremely small cellular scattering cross-sections. Larger microorganisms, which include nanoplankton >8 μm and microplankton that efficiently scatter light, contribute little to total scattering because of their low numerical concentrations. While a significant fraction of the total scattering coefficient appeared to come from the combined contributions of viable prokaryotic and eukaryotic cells, only a small fraction of the backscattering would appear to be related to these microbes. Instead, it appears that the major source of particulate matter backscattering is small (<0.6 μm), numerically abundant detrital particles. The conclusions of the authors have important implications in the application of OPCs in studies of marine microbes.

Size Changes and Growth Phenomena

White and Attwell (1993) used a novel approach for the measurement of endospore germination in solution using *Thermoactinomyces vulgaris* as a model organism. The method employed an Elzone™ (Coulter®-type) particle counter to monitor spore development from activation to outgrowth. The advantages of using this technique over microscopic or spectrophotometric techniques included the ease of monitoring germination and the ability to determine the relative frequencies of different germination stages in the bacterial population based on particle size. Histograms showing population size distribution at hourly intervals over a five-hour incubation period corresponded to the various germination stages. The use of the particle counter to measure changes in bacterial size in contexts other than germination was also evaluated. Direct viable counting was applied to *Escherichia coli* and the formation of enlarged cells was monitored by means of the instrument. Changes in the frequency of different cell sizes within the total cell population over a seven-hour incubation period were detected easily. The authors believed that the Elzone™ particle counter and similar instruments offer a potential time saving alternative method for microbiological investigations.

Salmeen and Durisin (1981) used instrumental (Coulter®) counting in concert with conventional methodologies (Ames assay) to investigate numbers of mutant bacterial cells. Ames assays were carried out with the directly acting mutagen 2-nitrofluorene on *Staphylococcus typhimurium* TA98 as a function of initial innoculum over the range 10^4–10^8 bacteria per plate (estimated from 100× micrographs of the background lawn). The initial innoculum was determined by counting background colonies in the micrographs, by dilution plating and by electronic particle counting. The data gathered in the study explained at least some reported variations in quantitation by the Ames assay. The data also showed that the slopes of dose-response curves depended on the numbers of organisms in the original assay and explained the range observed for spontaneous revertants, which suggests some limitations to quantifying the Ames assay and suggests a method for normalizing independently obtained Ames assay data. In this study the particle counter served as a useful ancillary means of gathering count data for comparison to that gathered by conventional methods.

SUMMARY

The studies reviewed show that electronic particle counters are widely applicable to the study of microbial cell suspensions. The implementation of unfamiliar technologies has historically moved slowly in the initial stages. This adaptation of instrumental counters to microbiological work is an evolving technology, and more rapid progress can be expected as the methods of applying particle counters in the study of cell populations become more widely used. A difficulty that will presumably exist for some time relates to the lack of interchange of information between the technologist well versed in particle counting and the microbiologist. Aside from the use of Coulter® counters for counting blood cells, knowledge of the highly specialized instrumentation used in particle counting has been closely restricted to the particle analyst.

Despite known detractions based both on instrumental operating characteristics of the instruments in question and nature of the cell populations that are to be assayed, this methodology shows promise in the study of bacterial cell populations and other microbial suspensions. For successful application a sufficient knowledge base must be available regarding both the nature of the phenomena to be studied and the application of the mode of particle analysis chosen. While the counters and population analyzers discussed can undoubtedly size and count microbial cells as particles, there are a number of specific questions that must be answered before a study is undertaken.

First and most basic is the magnitude of the size or number change that must be quantitated for generation of meaningful data. Size and number parameters must mesh appropriately with the sensitivity, dynamic range, and concentration limit of the instrument to be used. Said another way, the instrument must be able to see the size or number change that will occur and resolve that change from various interfering influences. The latter originate from various sources. The instrument itself contributes "noise" both in the form of extraneous pulses at or near its lower sensitivity limit, and in the form of peak broadening due to imprecision in its electronic circuits or in its sensing mechanism. Instrumental resolution is typically poorest at the smallest measured sizes, and if an instrument cannot resolve (discriminate) between the organism before or after a size change (e.g., growth, sporulation), or between the organism of interest and debris, it will not yield useful data.

"Noise" of another type is contributed by the sample itself. Any noncellular particles or debris in the analytical sample will generate spurious or misleading pulses that detract from the visualization of the phenomena being measured. This is an extremely important consideration when environmental samples such as marine or fresh water ecosystems are being evaluated. In these situations rigorous "cleanup" of the sample may not be possible, and the only recourse may be to rely on a thorough understanding of the sample material to assist in data interpretation.

The data gained from particle counters and population analyzers would seem to be of greatest value in a correlative or ancillary role rather than as a "standalone." An example is provided by the counting of bacterial cells in fairly dilute environments such as the air of an operating theater or in drinking water. In these cases particles counted are of interest not only at the size range of bacteria but at the size of much larger particles (10–20 μm), which represent clumps of bacteria in water or "rafts" of skin cells in air on which bacteria ride. This clumping phenomena may be responsible for gross discrepancies between cultivable numbers of organisms and counts in the size range of individual cells.

If attention to these caveats and cautions and having a comprehensive understanding of the mode of analysis before starting a study, particle-counting technology appears a very useful ancillary technique for the microbiologist or cytologist. It is a promising consideration for the investigator who wishes to quantify cell size or number changes, or physical changes such as morphotype alteration or sporulation in bacteria; it also would appear to be a method of choice for the time-effective collection of data regarding cell-cell interactions, such as cell fusion or budding, adherence, or phagocytosis. The investigator should also keep in mind that single particle counters are eminently useful in the collection of negative data (i.e., the determination of cleanliness or the absence of cells), as in filtration testing or in enumeration of low numbers of cells present as contaminant particles in high purity liquids.

REFERENCES

Amaker, P., and P. Boymund. 1967. The control of the presence of foreign particles in an injectable solution with the aid of the Coulter counter. *Pharm. Act. Helv.* 42:340–357.

Avesani, A., C. Cambiaso, J. L. Guarin, A. Lick, and M. Delmee. 1994. A particle counting immunoassay for the direct detection of *Clostridium difficile* serogroup specific antigen in fecal specimens. *J. Med. Microbiol.* 40 (4): 270–274.

Barber, T. A. 1987. Limitations of light blockage particle counting in the analysis of parenteral solutions. In *Proc. PDA Conf. on Liquid Borne Particle Inspection and Metrology*, May 11–13, in Washington, DC.

Barber, T. A., and J. Williams. 1990. Analysis of dry powder antibiotics by light obscuration counting. In *Proc. PDA Intl. Conf. on Particle Detection, Metrology and Control*, May 11–13, in Arlington, VA.

Barnett, M. I. 1987. Resistance modulation particulate measurement. In *Proc. PDA Conf. on Liquid Borne Particle Inspection and Metrology*, May 11–13, in Washington, DC.

Bassoe, C. F., and R. Bjerknes. 1985. Phagocytosis by human leukocytes, phagosomal pH and degradation of seven species of bacteria measured by flow cytometry. *J. Med. Microbiol.* 19:15–125.

Bassoe, C. F., O. D. Laerum, J. Glette, G. Hopen, G. Haneburg, and C. O. Solberg. 1983. Simultaneous measurement of phagocytosis and phagosomal pH by flow cytometry: Role of polymorphonuclear neutrophilic leukocyte granules in phagosome acidification. *Cytometry* 4:254–262.

Bemer, D., J. F. Favries, and A. Renoux. 1990. Calculation of the theoretical response of an optical counter and its practical usefulness. *J. Aerosol Sci.* 21 (5):689–700.

Bjerknes, R. 1984. Flow cytometric assay for combined measurement of phagocytosis and intracellular killing of *Candida albicans. J. Immunol. Methods* 72:229.

Blanchard, J., J. A. Schwartz, and D. M. Byrne. 1977. Effects of agitation on size distribution of particulate matter in large volume parenterals. *J. Pharm. Sci.* 66:935–938.

Buettner, H. 1990. Measurement of the size of fine nonspherical particles with a light-scattering particle counter. *Aerosol Sci. Tech.* 12 (2):413–421.

Butera, M., J. H. Smith, W. D. Morrison, R. R. Hacker, F. A. Kains, and J. R. Ogilvie. 1991. Concentration of respirable dust and bioaerosols and identification of certain microbial types in a hog-growing facility. *Can. J. Anim. Sci.* 71 (2):271–278.

Caldow, R., and J. Blesner. 1989. A procedure to verify the lower counting limit of optical particle counters. *J. Parenteral Sci. and Technol.* 43:174–179.

Cambiaso, C. L., and J. N. Limet. 1989. Latex agglutination assay for human anti-Brucella IgM antibodies. *J. Immunol. Methods* (Netherlands), 122 (2): 169-75.

Cambiaso, C. L., J. P. Van Vooren, and C. M. Farber. 1989. Immunological detection of mycobacterial antigens in infected fluids, cells and tissues by latex agglutination: Animal model and clinical application. *J. Immunol. Methods* 129 (1):9–14.

Cantinieaux, B., P. Courtoy, and P. Fondu. 1993. Accurate flow cytometric measurement of bacteria concentrations. *Pathobiology* 61 (2):95–97.

Carver, L. D. 1969. Light blockage of particles as a measurement tool. *Ann. NY Acad. Sci.* 158 (3):710–721.

Clancy, T. P., and R. A. Hango. 1992. Response of laser particle monitor to bacterial challenges in high-purity water. *Ultrapure Water* 9 (4):51–54.

Coons, J. S., and R. S. Weinstein, eds. 1991. *Diagnostic flow cytometry.* Baltimore: Williams and Wilkins Co.

Cooper, D. W., and S. J. Grotzinger. 1989. Comparing particle counters: Cost vs. reproducibility. *J. Environmental Sci.* 35 (5):32–34.

DeLuca, P. P. 1977. Need for improved microscopic methods and understanding of correlations between microscopic and automatic methods. *Bull. Parenteral Drug Assoc.* 1:173–178.

DeLuca, P. P., B. Conti, and J. Z. Knapp. 1987. An overview of technical issues in particle detection. In *Proc. PDA Conf. on Liquid Borne Particle Inspection and Metrology,* May 11–13, in Washington, DC.

DiGrado, C. J. 1970. Liquid borne particle counting in the pharmaceutical industry: III. Method evaluation. *Bull. Parent. Drug Assoc.* 24:62–67.

Dodge, L. E. 1984. Calibration of the Malvern particle sizer. *Applied Optics* 23:2415–2427.

Drowart, A., C. L. Cambiaso, K. Huygen, E. Serruys, J. C. Yernault, and J. P. Van Vooren. 1993. Detection of mycobacterial antigens present in short-term culture media using particle counting immunoassay. *Am. Rev. Respir. Dis.* 147 (6 part 1):1401–1406.

Elias, H. G. 1972. The study of association and aggregation *via* light scattering. In *Light scattering from polymer solution,* edited by M. B. Huglin. London: Academic Press.

Gorman, S. P., D. F. McCafferty, and L. Anderson. 1986. Application of an electronic particle counter to the quantification of bacterial and candida adherence to mucosal epithelial cells. *Appl. Microbiol.* 2 (5):97–100.

Groves, M. J. 1969. The size distribution of particles contaminating parenteral solutions. *Analyst* 94:992–1001.

Hango, R., and T. Clancy. 1993. Laser particle monitor response observations in electronics-grade high-purity water. *Ultrapure Water* 10 (6):38–42.

Harfield, J. G., and W. M. Wood. 1990. Standard calibration materials in the Coulter counter. In *Particle size measurement,* edited by M. J. Groves, and J. Wyatt-Sargent. London: Society for Analytical Chemistry.

Hirleman, E. D. 1990. On-line calibration technique for laser diffraction droplet sizing instruments. *ASME* 83 (GT):232.

Hirleman, E. D. 1990. A general solution to the inverse near forward scattering particle sizing problem in multiple scattering environments: Theory. In *Proc. 2nd Int. Congr. Optimal Particle Sizing,* March 5–8, in Tempe, AZ.

Hopkins, G. H., and R. W. Young. 1974. Correlation of microscopic with instrumental particle counts. *Bull. Parenteral Drug Assoc.* 28:15–25.

Horvath, H., R. L. Bunter, and S. W. Wilkison. 1990. Determination of the coarse mode of the atmospheric aerosol using data from a forward scattering spectrometer probe. *AS&T* 12 (4):964–980.

Hovenac, E. A. 1990. Scattering from non-spherical particles. In *Proc. 2nd Int. Congr. Optical Particle Sizing,* March 5–8, in Tempe, AZ.

Jacangelo, J. G., E. M. Aieta, K. C. Carns, E. W. Cummings, and J. Mallevialle. 1989. Assessing hollow-fiber ultrafiltration for particulate removal. *Am. Water Works Assoc.* J8 (11):68–75.

Kastelein, J., M. T. Logtenberg, and P. G. M. Hesselink. 1992. Testing and evaluation of off-gas filters for bioreactor by a new bacterial aerosol challenge test method, TBAC. *Enzyme Microb. Technol.* 14 (7):553–560.

Kinsman, S. 1969. Electrical resistance method for automated counting of particles. *Ann. N.Y. Acad. Sci.* 158 (3):703–709.

Kleine, T. O., R. Hackler, and H. Meyer-Rienecker. 1990. Classical and modern methods of cerebrospinal fluid analysis. *Eur. J. Clin. Chem. Clin. Biochem.* 29 (10): 705–714.

Knollenberg, R. G. 1989. The measurement of latex particle sizes using scattering ratios in the Rayleigh scattering size range. *J. Aerosol Sci.* 20 (3):331–345.

Kogure, K., and I. Koike. 1987. Particle counter determination of bacterial biomass in sea water. *Appl. Environ. Microbiol.* 53 (2):274–277.

Lewis, C. M., E. E. Hargesheimer, and C. M. Yentsch. 1993. Application of particle counters to treatment of drinking water. *Amer. Laboratory* 25 (6):10–12.

Lines, R. W. 1967a. Particle counting in small containers. *Bull. Parent. Drug Assoc.* 21:118–123.

Lines, R. W. 1967b. An insertable orifice tube for *in-situ* contamination counts of solutions in opened ampoules and vials using the Coulter counter. *J. Pharm. Pharmacol.* 19:701–705.

Lines, R. W. 1981. Particle counting by Coulter counter. *Anal. Proc.* 18:514–519.

Liu, B. Y. H., W. W. Szymanski, and K. O. Ahn. 1985. On aerosol size distribution measurement by laser and white light optical particle counters. *J. Environ. Sci.* (May-June):19–24.

Maerz, F., R. Scheer, and E. Graf. 1990. Use of a laser diffraction particle counter to monitor integrity of reverse osmosis membranes and its inefficiency to detect bacteria in product water. *Int. J. Pharm.* 61 (1–2):57–66.

Masuko, M., S. Hosoi, and T. Hayakawa. 1991. A novel method for detection and counting of single bacteria in a wide field using an ultra-high-sensitivity TV camera without a microscope. FEMS, *Microbiology Letters* 65 (3):287–90.

McCoy, W. F., and B. H. Olson. 1987. Relationship among turbidity, particle counts and bacteriological quality within water distribution lines. *Water Res.* 20 (8):1023–1029.

Montague, W., and H. Sommer. 1990. Reliability and count accuracy of optical counters. *J. Environ. Sci.* May/June: 19–24.

Perez-Rosas, N., and T. C. Hazen. 1988. In situ survival of *Vibrio cholerae* and *Escherichia coli* in tropical coral reefs. *Appl. Environ. Microbiol.* 54 (1):1–9.

Philips, A. P., and K. L. Martin. 1988. Limitations of flow cytometry for specific detection of bacteria in mixed populations. *J. Immunol. Methods* 106:109–117.

Pinder, A. C., P. W. Purdy, S. A. Poulter, and D. C. Clark. 1990. Validation of flow cytometry for rapid enumeration of bacterial concentrations in pure cultures. *J. Appl. Bacteriol.* 69 (1):92–100.

Poole, R. K. 1982. Rapid estimates of sizes of microorganisms with the Coulter Nano-Sizer. *Microbiology Letters* 19 (75–76):109–118.

Rebagay, T., H. G. Schroeder, and P. P. DeLuca. 1977. Particulate matter monitoring II. Correlation of microscopic and automatic counting methods. *Bull. Parent. Drug Assoc.* 31:150–155.

Rhame, F. S., A. Streifel, C. McComb, and M. Boyle. 1986. Bubbling humidifiers produce microaerosols which can carry bacteria. *Infect. Control* 7 (8):403–407.

Salmeen, I., and A. M. Durisin. 1981. Some effects of bacteria population quantitation of Ames salmonella-typhimurium histidine reversion mutagenesis assays. *Mutat. Res.* 85 (3):109–118.

Schroeder, H. G., and P. P. DeLuca. 1980. Theoretical aspects of particulate matter monitoring by microscopic and instrumental methods. *J. Parent. Drug Assoc.* 34:183–191.

Seal, D. V., and R. P. Clark. 1990. Electronic particle counting for evaluating the quality of air in operating theatres: A potential basis for standards? *J. Appl. Bacteriol.* 68 (3):225–230.

Stramski, D., and D. A. Kiefer. 1991. Light scattering by microorganisms in the open ocean. *Prog. Oceanogr.* 28 (4):343–383.

Suzuki, A., Y. Namba, M. Matsuura, and A. Horisawa. 1984. Airborne contamination in an operating suite: Report of a five-year survey. *J. Hyg.* 93 (3):567–573.

Swithenbank, J., J. M. Beer, D. S. Taylor, D. Abbot, and G. C. McCreath. 1977. A laser diagnostic technique for the measurement of droplet and particle size distribution. *Prog. Astronaut. Aeronaut.* 53:421.

Tungate, D. S., and E. Reynolds. 1980. The MAFF on-line particle counting system. *Fisheries Research Technical Report No. 58.* Great Britain: Ministry of Agriculture, Fisheries and Food Research.

Weiner, B. 1979. Particle and spray sizing using laser diffraction. *Soc. Photoopt. Eng.* 170:53.

White, D. L., and R. W. Attwell. 1993. Use of the Elzone particle counter to monitor morphological changes in bacteria associated with endospore germination and direct viable counting. *J. Microbiol. Methods* 17 (1):7–83.

Whyte, W. A. 1983. Multicentered investigation of clear air requirements for terminally sterilized pharmaceuticals. *J. Parenter. Sci. Technol.* 37 (4):138–144.

Zhang, A., and M. J. Groves. 1988. Size characterization of mycobacterium bovis BCG (Bacillus Calmette Guerin) vaccine, Tice substrain. *Pharm Res.* 5 (9):607–610.

11
Quantitative ATP Analysis

Melvin E. Klegerman

Institute for Tuberculosis Research
University of Illinois at Chicago

MICROBIAL QUANTITATION

Different aspects of microbial biomass may be measured by various means, depending on the particular property examined.

Colony-Forming Units

The predominant method of quantitating viable microbes is the enumeration of colony-forming units (CFU). This may be done by spreading aliquots onto agar plates or suspending them in semisolid media and incubating for appropriate periods of time, after which the resultant colonies are counted (Shi et al. 1989; Rosenthal et al. 1962). Advantages of this method are the ability to verify the identity of colonies by visual inspection (contamination can usually be recognized) and the direct enumeration of viable particles. The usefulness of this method can be enhanced by subjecting colonies to various identification procedures. Extremely low cell concentrations can be detected by passing large volumes of fluid through sterile filters, which are then covered with a nutrient agar and incubated (Meltzer 1991).

Disadvantages of the CFU technique include the 24–48 hours incubation time and inaccuracies due to discrepancies between viable particle and cell numbers.

Considerable differences between CFU and viable cell number can occur because of cell aggregation or microcolonial morphology, such as in the case of metabolically stressed streptococci (Stewardson-Krieger et al. 1977). Several shortcomings of the CFU technique have converged in the case of BCG (*Bacillus Calmette-Guérin*) vaccine, which is an attenuated strain of *Mycobacterium bovis* used for the prophylaxis of tuberculosis. The organisms are extremely aggregated by secretion of an integument (Devadoss et al. 1991). Many cells are killed by autolysis and by mechanical shear forces used to partially disperse the organisms. Thus, each CFU represents an unknown, highly variable number of viable cells (Shi et al. 1989). Furthermore, as a slow-growing mycobacterium, BCG colonies require at least 3–4 weeks to become visible (Shi et al. 1989). This disadvantage can become extreme in the case of *Mycobacterium tuberculosis*, since patients with multiple drug-resistant tuberculosis have died before drug susceptibilities of infecting isolates could be determined (Hughes 1994).

Turbidity

A much faster, more convenient method for determining microbial mass has long been the quantitation of turbidity. Scattering of light is a function of particle size and number (Olson et al. 1990). Therefore, if the organisms comprise a relatively homogeneous suspension, this property fairly accurately reflects the microbial mass. On the other hand, if the suspension is heterodisperse, such as in the case of cell aggregation or microcolony formation, significant anomalies can occur (Olson et al. 1990). A major drawback of this method, when samples other than defined cultures are examined, is the possibility that turbidity might be due to factors other than biomass. Microbial turbidity is usually measured with a turbidimeter, which quantitates light scattered perpendicularly to the incident beam, or a colorimeter/spectrophotometer, since light scattered from the incident is operationally indistinguishable from absorbed light (Currie and Klegerman 1992, 175–204). In the latter case the "absorbance" is usually measured between 420 and 660 nm (Ingraham et al. 1983, 232).

Quantitation of Metabolites and Microbial Components

Viable cell mass can be determined by measuring metabolism. The best example of this method is the quantitation of $^{14}CO_2$ production from ^{14}C-labeled carbon sources, which is the basis of the BACTEC® radiometric technique (Laszlo et al. 1983). While a very sensitive method for relatively rapid detection and identification of isolates in a large-volume clinical laboratory, the BACTEC® system is expensive and not well suited for infrequent quantitation of viable cell number. Another way of quantitating biomass is to measure cell components or metabolites that are more or less characteristic of their parent life forms. For example, the measurement of DNA, RNA, or protein by established methods after disruption and/or extraction of cells has long been used to monitor microbial growth. Identification of specific DNA nucleotide sequences by the polymerase chain reaction (PCR) and other recombinant DNA techniques is a powerful new way to detect specific organisms that is finding widespread use for diagnostic applications, but can also be adapted to quantitate the organisms (Woodbury 1992).

MEASUREMENT OF ADENOSINE TRIPHOSPHATE

One metabolite that is particularly well suited to quantitation of vi
ber is adenosine triphosphate (ATP), which is universally present
(Shi et al. 1989). The quantity of ATP per cell has been found to be
stant among bacterial species and the nucleotide is rapidly lost from dead cells,
not only because of plasma membrane permeability, but also because of enzymatic conversion to other metabolites.

Spectrophotometry

Like all nucleotides, ATP can be quantitated spectroscopically, based on the fact
that heterocyclic bases absorb ultraviolet light maximally around 260 nm. The assay can be made specific by treating the sample with an ATPase, calculating the
ATP content from the decrease in optical density with correction for contribution
to the A_{260} of the added enzyme. The sensitivity of this method, however, is only
about 0.1–1.0 μg/ml (in the order of about 1 μM) (Bonner et al. 1968). Since bacteria contain about 0.5 fg of ATP per cell (Shi et al. 1989), ATP measurement relying on A_{260} determination cannot detect less than 2×10^8 cells/ml, which is an
order of magnitude worse than turbidity measurement (Davis et al. 1973, 96).

Coupling to Other Reactions

More sensitive assays for ATP could be devised by exploiting the numerous enzymatic reactions in which the nucleotide serves as a substrate or cofactor. Unless
the reaction product can be detected at far lower levels than ATP, however, this
strategy is hardly an improvement over direct ATP spectrophotometry since the
nucleotide is functionally destroyed during the reaction.

BIOLUMINESCENT MEASUREMENT OF ATP

McElroy (1947) discovered that firefly luminescence in the lanterns of the beetle
Photinus pyralis is dependent on ATP. Since then, it has been learned that, in this
system, the chemical energy contained within the terminal phosphoanhydride
bond of ATP is converted to a 2.2 ev photon ($\lambda = 562$ nm, having a yellow-green
color) by the compound luciferin, the structure of which is shown in Figure 11.1.
The reaction consists of two stages, both of which are catalyzed by the enzyme luciferase:

$$LH_2 + ATP \leftrightarrow AMP{\cdot}LH_2 + PP_i \tag{1}$$

$$LH_2{\cdot}AMP + {}^1/_2 O_2 \rightarrow L{\cdot}AMP^* + H_2O \rightarrow L{\cdot}AMP + h\nu + H_2O \tag{2}$$

Throughout the reaction, luciferin (L) is bound by the enzyme most tightly when
it is in its excited state (L·AMP*). Therefore, at higher concentrations of substrate
($>$nM), the enzyme will become saturated with the excited intermediate,
competitively inhibiting binding of reduced luciferin species and setting an
upper limit for the assay range (Strehler 1968; Chaplin 1991, 82–96).

Figure 11.1. Structure of firefly luciferin.

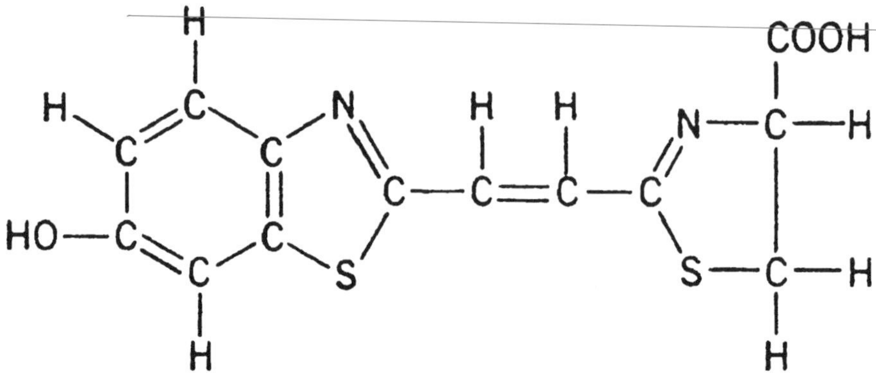

The reaction is magnesium and oxygen-dependent and has a pH optimum of 7.8–7.9, although ATP bioassay with this system can be performed from pH 6.5 to 8.0 (McElroy and Strehler 1949; Strehler, 1968). Other divalent metal cations can substitute for Mg^{++}, in the order of $Mn^{++} > Co^{++} > Fe^{++} > Ni^{++} > Zn^{++}$ (McElroy and Strehler 1949). When the reaction is carried out in vitro, the light produced can be quantitated by photodetection devices. Provided the enzyme, other substrates, and cofactors are present in excess, the amount of light produced is directly proportional to the concentration of ATP in the medium (Gheorghiu and Lagranderie 1979).

Applications

With the commercial development of relatively inexpensive photometers, or *luminometers*, featuring varying degrees of automation, bioluminescent measurement of ATP has emerged as a popular, convenient, and rapid means of quantitating microbial biomass for various purposes. The assays are generally linear from 10 pM up to about 1 μM, meaning that as little as 1×10^4 bacterial cells can be detected by this method, which is at least five orders of magnitude better than spectrophotometric measurement of ATP (Weeks et al. 1986). Currently available instruments can detect less than 1 fmole of ATP, corresponding to less than 1,000 bacterial cells.

Clinical microbiological applications include assessment of bacteriuria in suspected urinary tract infections (UTIs) and antibiotic susceptibility testing. Bacterial or microbial counts have been determined for raw milk, cream, raw meat, prepared foods, and food components. Sterility testing can be performed on beverages, cosmetics, and pharmaceuticals, especially parenteral products. Environmental applications include determining the effectiveness of sewage treatment and other bactericidal protocols, soil microbiology, and the assessment of microbial biomass in marine environments (*Lumac's* 1985; Stevenson et al. 1979). Of course, bioluminescent measurement of ATP could be used in any research protocol that requires quantitation of viable cells, whether prokaryotic or eukaryotic. For instance, this method is suitable for the determination of cytotoxicity in immunologic, toxicologic, or pharmaceutical investigations.

ATP-Related Applications

Exploitation of ATP's pivotal role in many biochemical reactions to quantitate other compounds becomes feasible with the bioluminescence technique. An appropriate enzyme, for which the compound of interest is a substrate and ATP is either a substrate or product, is chosen. Reagent concentrations are optimized so that the production or consumption of ATP is directly proportional to the compound's concentration. Compounds that have been assayed this way, using firefly bioluminescent measurement of ATP, include adenosine diphosphate (ADP), adenosine 3',5'-monophosphate (cyclic AMP), phosphocreatine, creatine, glucose, coenzyme A, nucleoside triphosphates, and phosphoenolpyruvate. Examples of other compounds that could be assayed by this method are pyrophosphate, nicotinamide adenine dinucleotide (NAD^+), and charged transfer RNA. If all substrates and cofactors are present, bioluminescent measurement of ATP could be used to assay enzymes catalyzing the reactions. Examples of this approach include apyrase, ATPase, myokinase, creatine kinase, nucleoside triphosphate phosphokinases, hexokinase, and pyruvate kinase (Strehler 1968; Gorus and Schram 1979; Whitehead et al. 1979).

Non-ATP Applications

The luciferases of several luminous bacteria have been found to require NAD^+ and flavin mononucleotide (FMN). Therefore, bacterial luciferases can be used to assay these cofactors and any component of reactions in which they are involved, greatly extending the applications of the bioluminescence technique. Species measured by this approach include long-chain aldehydes, oxygen, lactate dehydrogenase, alcohol dehydrogenase, malate dehydrogenase, glucose-6-phosphate dehydrogenase, aspartate aminotransferase, alanine aminotransferase, malate, oxaloacetate, glucose-6-phosphate, glucose, ammonia, ethanol, and urea (Strehler 1968; Gorus and Schram 1979; Whitehead et al. 1979).

Other Applications of Luminescence Reactions

Chemiluminescence

Reactive oxygen intermediates (ROI) generated by phagocytic cells, such as macrophages and neutrophils, after ingestion of foreign particles can emit light in the visible range when they decay to more stable species without reacting with a substrate (Allen et al. 1972). The intensity of the light emission can be greatly enhanced by the use of efficient photon-generating intermediates that are excited by ROI, especially hydrogen peroxide. Two such compounds that have found wide use are the cyclic hydrazide luminol (5-amino-2,3-dihydro-1,4-phthalazinedione, Figure 11.2) and the acridinium ester lucigenine (10,10'-dimethyl-9,9'-biacridinium nitrate, Figure 11.3) (Strehler 1968; Berthold 1991, 139–150).

Figure 11.2. Structure of luminol (5-amino-2,3-dihydro-1,4-phthali-dazinedione).

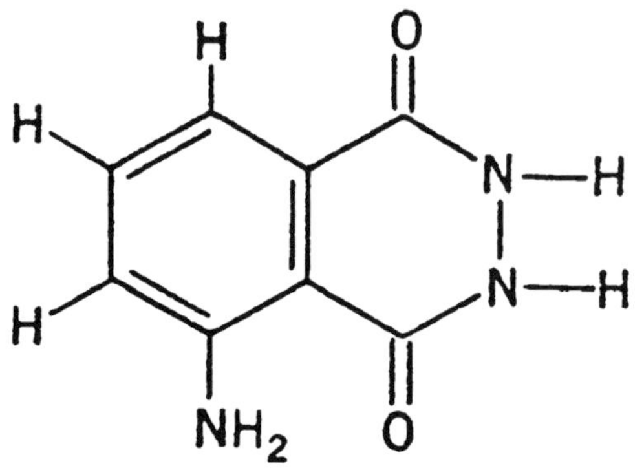

Figure 11.3. Structure of 10,10′-dimethyl-9,9′-biacridinium nitrate (lucigenine).

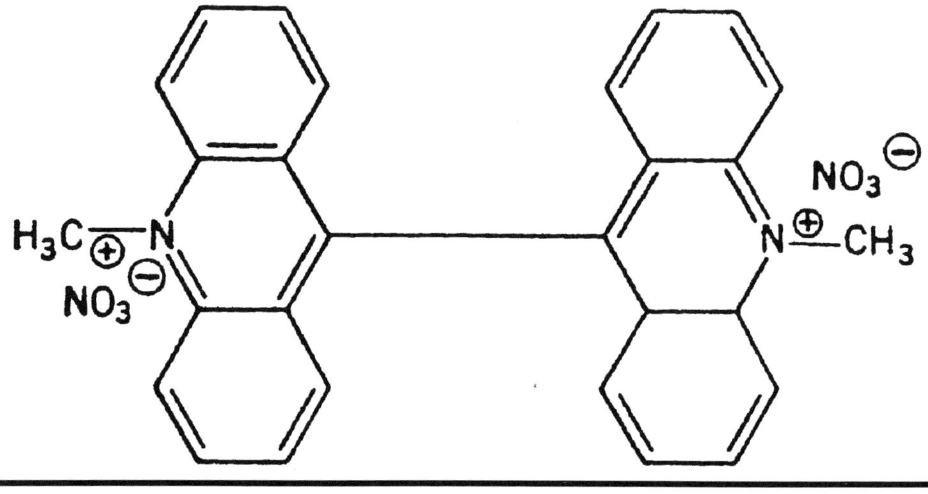

Chemiluminescence, as this process is called, has been used for more than 20 years to monitor the phagocytic function of neutrophils and monocytes in in vitro models of infectious diseases. More recently, these assays have been improved by the addition of luminol (De Chatelet and Shirley 1981; Stevens et al. 1978). The photon-generating intermediates can also be used to measure specific ROI, including oxygen itself. In addition, luminol is an effective indicator of small quantities of heme compounds (Strehler 1968).

Immunoassay

A major development in the area of high-sensitivity, quantitative immunoassay, which had been dominated by radioimmunoassay (RIA) for more than 10 years after its inception around 1960, was the introduction of enzyme immunoassays.

Radioimmunoassays rely on the competitive binding of radiolabeled and unlabeled antigen to specific antibodies, so that the amount of unlabeled antigen in a sample is inversely proportional to the amount of labeled antigen bound to antibody (which could be separated by various methods and then quantitated with radiation counters) in the test system (Currie and Klegerman 1992).

Soon after solid-phase RIA, in which antibodies were adsorbed to plastic test tubes, became widely available, enzyme immunoassays in which either antibodies or antigens were adsorbed to multiwell plastic microtiter plates were developed. This technique permitted the successive layering of immunoreactions, culminating with the specific binding of an immunoreagent covalently coupled to an enzyme. The assay is "developed" by addition of a colorless substrate that is converted to a colored product by the enzyme. After termination of the reaction, the plate is "read" using specialized colorimeters. The intensity of color development is directly proportional to the quantity of captured enzyme-linked ligand, which is ultimately proportional to the concentration of measured compound in the test sample (Voller and Bidwell 1985, 77–86).

Currently, there is a proliferation of analogous immunoassays that utilize fluorescence and luminescence to detect labeled ligands. Luminescent labels that have been used include luminol, isoluminol, acridinium esters, trioxylate compounds, various luciferases, and horseradish peroxidase. In principle, ligands can be labeled with any component of bioluminescent or chemiluminescent reactions. For development of the assay, the remaining components, (i.e., substrates, enzymes, and cofactors) are added and the light generated is measured. Specialized photometers for this purpose (e.g., in a microtiter well format) are being marketed. Compounds that have been assayed with this type of immunoassay include thyroxine, immunoglobulin G (IgG), insulin, digoxin, estradiol, estriol, progesterone, estrogen metabolites, biotin, and human chorionic gonadotropin (hCG) (Gorus and Schram 1979; Whitehead et al. 1979; De Boever et al. 1983; Weeks et al. 1986; Collins 1985; Van Dyke and Van Dyke 1990).

Recombinant DNA Applications

Detection of Nucleic Acid Hybridization.   By labeling oligonucleotide probes with a luminescent label, such as an acridinium ester, all the procedures used with radiolabeled probes could be employed, including radioautography. In this method electrophoretograms or dot blots of highly polymerized nucleic acid preparations are incubated with the labeled probe. After washing the remaining components of the luminescent reaction are added and a piece of X-ray film is positioned below the membrane or paper in a dark chamber. The film will be exposed where the probe has hybridized. A method commonly used to detect the presence of nucleic acid hybridizing to these kind of probes, such as in diagnostic and forensic applications, is to adsorb single-stranded nucleic acid to a suspended stationary phase, incubate with the probe, and elute double-stranded nucleic acid segments by changing the ionic strength of the medium. If the eluates are transferred to microtiter wells, the remaining components of the luminescent reaction are added, and the response determined with X-ray film. Otherwise, the eluates could be reacted and assayed with a luminometer (Granato and Roefaro 1989; Clyne et al. 1989; Nelson et al. 1990, 293–318).

Reporter Genes.   A particularly ingenious use of luminescence technology is the insertion of luciferase genes into hosts in order to assess the function of regulatory genes and to monitor cell viability in small populations. The firefly luciferase gene was inserted behind a promoter on a plasmid, which was then transfected into mammalian cells. In this way it was determined (by measurement of light generation with a luminometer) that the SV40 (simian virus) promoter was 10 times more efficient in rat pituitary adenoma cells than the rat growth hormone promoter (Matamoros and Strobl 1990, 311-318).

A well-known example of the cell viability monitoring application is the transfection of the firefly luciferase gene into clinical strains of *Mycobacterium tuberculosis*. Since dead bacteria no longer synthesize proteins and, therefore, fail to luminesce, they can be distinguished from living transfected bacilli (which glow) by simple microscopic examination or by assay with a luminometer. In this way drug susceptibilities of clinical isolates of tubercle bacilli in relatively small samples could be determined much more rapidly than by conventional culture methods that require at least three weeks to develop. This difference could be life-saving to someone infected with a multiple drug resistant strain of the bacterium (Jacobs et al. 1993).

METHODS OF BIOLUMINESCENT ATP MEASUREMENT

Extraction

Numerous methods have been employed for extracting ATP from bacteria prior to bioluminescent assay of the nucleotide. Extraction agents fall into the general categories of acids, organic solvents, hot aqueous buffers, and ionic surfactants. Perchloric acid, trichloroacetic acid, hot chloroform, n-butanol (with or without n-octanol to facilitate phase separation), acetone, dimethylsulfoxide, and boiling Tris-ethylenediaminetetraacetate (EDTA) buffer are all relatively effective means of extracting ATP from bacteria, but various studies have differed on the most efficient method in a particular circumstance. Shi et al. (1989), however, found that NRB, an "aqueous solution of surfactants" marketed by Lumac/3M for the release of ATP from microbial cells, was far less efficient in extracting the nucleotide from cells of *Mycobacterium bovis*-BCG than boiling buffers or n-butanol/ n-octanol. In general, the use of established extraction agents for acid-fast bacteria should be reevaluated in each case because of the relative impermeability of the complex cell wall.

Extraction efficiency can be determined by calculating the recovery of exogenous ATP standard added to a cell suspension, which is then subjected to the extraction procedure; a parallel extraction is carried out without added ATP.

$$\text{Percent recovery} = \frac{(\text{RLU of extracted cells} + \text{ATP}) - (\text{RLU of extracted cells})}{\text{RLU of ATP in extraction solution}} \times 100$$

where RLU is relative light units detected by the luminometer. Comparison to the ATP standard in the extraction solution eliminates the contribution of quench and yields extraction efficiency alone.

Manufacturers and distributors of luminometers, such as Lumac/3M, provide detailed protocols for extraction methodology appropriate for a number of applications. For instance, assessment of microbial contamination of meats requires homogenization of tissue, followed by removal of ATP released by muscle cells before extraction of microbial ATP. If digestion by an ATPase is used to remove ATP, then the extraction method should destroy the enzyme. Novel applications require design of protocols that avoid inaccuracies due to interfering factors and loss of ATP, which will be discussed later in this chapter. Several relatively recent reviews have addressed the issues of sample preparation and extraction for the detection and enumeration of microbes by bioluminescent ATP measurement (Stannard and Gibbs 1986; Kricka 1988; Stanley 1989; Griffiths 1993).

Measurement

Instrumentation

Potentially, any photometer that detects visible light, specifically between 390 and 630 nm, can be used to measure bioluminescence. Such instruments include fluorimeters and liquid scintillation counters (provided that the coincidence circuitry is disabled). An increasing array of specialized photomoters known as luminometers have become available for luminescence applications, however. These include instruments from Dynatech Laboratories (Chantilly, VA), Lumac BV (Landgraaf, The Netherlands), Amersham Corp. (Arlington Heights, IL), Turner Designs (Mountain View, CA), Biotrace, Ltd. (Bridgend, UK), Biolite, Ltd. (London, UK), Berthold Analytical, Inc. (Nashua, NH), Pharmacia LKB Nuclear, Inc. (Gaithersberg, MD), and Source Scientific Systems, Inc. (Garden Grove, CA).

All luminometers suitable for bioluminescent measurement of ATP employ photomultipliers for detection. Therefore, instruments configured for microtiter well plates scan the plates like enzyme immunoassay readers, since photodiode array technology does not yet provide the sensitivity appropriate for analytical luminescence protocols. These photomultipliers have a maximum quantum efficiency (ratio of photoelectrons released to number of photons impinging on the photocathode) of 20–28 percent in the wavelength range of 400–450 nm, which coincides with the emission wavelengths of luminol and acridinium esters (420–430 nm). The quantum efficiency is only 3–6 percent at 560 nm, which is the emission wavelength of luciferin, indicating that chemiluminescent measurements are inherently more sensitive than those made with bioluminescent technology (Berthold 1991).

Nevertheless, a lower limit of 0.1 pg ATP has been achieved, corresponding to about 100 average bacterial cells, but the practical limit is 500 bacteria (Berthold 1991). In a study of 10 different luminometers, Jago et al. (1989) found that the single-tube luminometers Lumac M2010A and Turner 20 TD were an order of magnitude more sensitive (0.20 and 0.27 pg ATP, respectively) than microplate readers from Amersham and Dynatech (1.79 pg ATP), probably because single-tube chambers can be better sealed from light.

ATP-dependent bioluminescence exhibits flash kinetics. An initial flash of light, which attains a maximum in 2–3 seconds, is followed by a rapid, then a gradual decrease in the emission (Stevenson et al. 1979). Therefore, it is necessary for luminometers to be outfitted with injectors, at least for the enzyme/substrate mixture, so that counting can begin immediately, and for the signal to be integrated over a period of at least five seconds. In fact, the instruments usually have multiple injection ports so that various operations could be carried out conveniently or automatically.

Examples.   *Lumac BV.*   The Biocounter® M 2500 (Figure 11.4), the successor to the M 2010, is a single-chamber instrument with three injection ports, which deliver 100 μℓ aliquots. The sensitivity for ATP is 0.1 pg, which is the best currently achievable. The digital output encompasses four decades with a maximum count rate of 50,000 counts per second. A rate mode enables determination of reaction kinetics and the instrument can be interfaced with a computer, allowing automatic data processing. The Biocounter® M 1500 Light (Figure 11.5) is a portable unit that is suitable for field work. It has the same sensitivity, range, and count rate as the M 2500, but the unit is powered by a rechargeable battery or line current, features a single integration time of 10 seconds, and reagents provided by the manufacturer alter the reaction kinetics sufficiently to obviate the need for an injection port.

Berthold Analytical.   The Lumat LB 9501, a single-sample luminometer, is apparently identical to the Monolight 2010 (Figure 11.6), which is marketed by Analytical Luminescence Laboratory (San Diego, CA). This unit has two injection ports and is programmable to perform up to 30 assay protocols. A thermal

Figure 11.4. The Lumac Biocounter® M 2500.

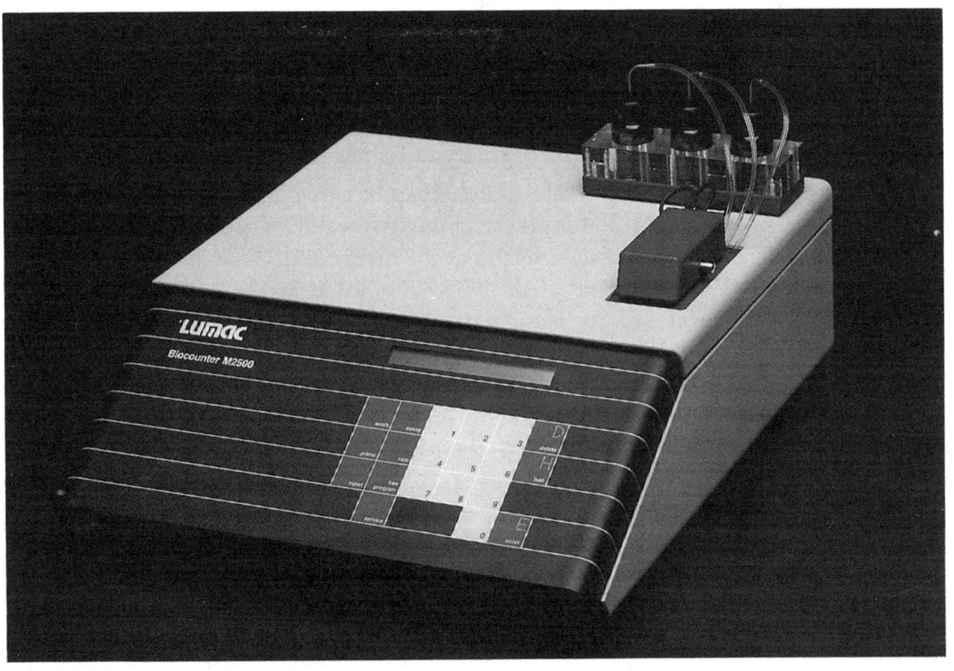

Figure 11.5. The Lumac Biocounter® M 1500 Light.

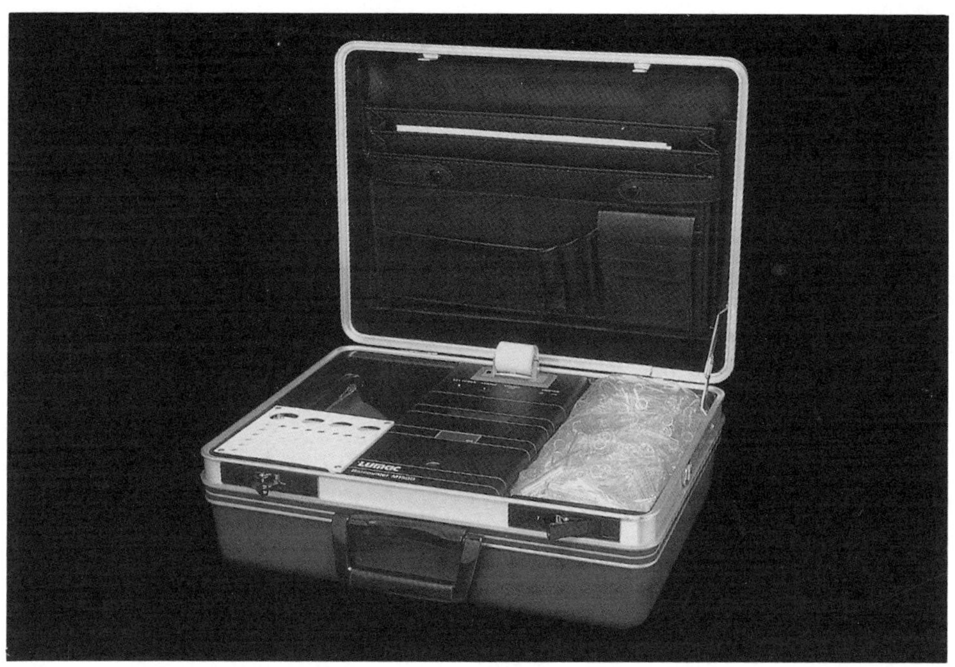

Figure 11.6. The Monolight 2010 Luminometer.

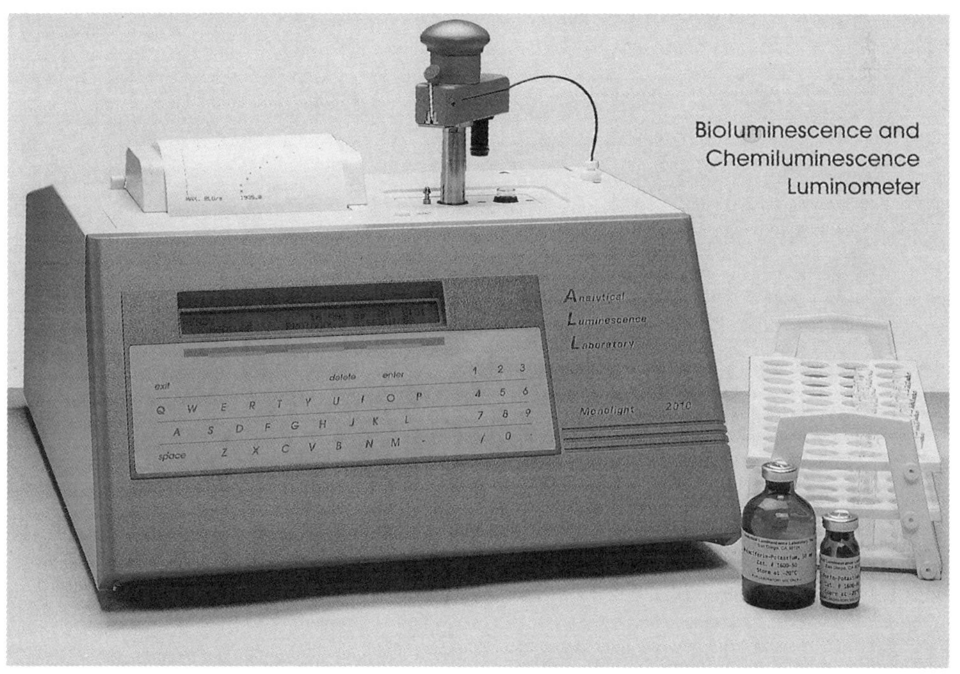

printout can also provide a kinetics profile in histogram form and print integral and peak RLU values.

Dynatech Laboratories.   The ML 1000 Microplate Luminometer is an automated instrument that measures the luminescent signal from up to 96 test samples in less than one minute. It has a dynamic range of 7 decades and a sensitivity of <1 fmole ATP, which is comparable to the Lumac instrument. The Microlite™ model can be interfaced with an optional printer and can be fitted with up to three single-reagent, syringe-type dispensers that dispense simultaneously with instantaneous detection. The dispensed volume can be varied between 10 $\mu\ell$ and 250 $\mu\ell$. A variable plate temperature control can elevate the sample temperature between 30°C and 45°C. The Model 2.4 unit has a built-in impact printer and includes a single dispenser. Both models can be interfaced with a computer. Dynatech markets a special opaque microtiter plate for use with these luminometers.

Lumac (U.S. outlet, Integrated BioSolutions, Monmouth Junction, NJ) and Analytical Luminescence Laboratory market a full line of bioluminescence reagents for their instruments. Tropix (Bedford, MA) provide reporter gene materials for the Dynatech instruments. Of course, these reagents can be used in any luminometer and individual components can be purchased from many different suppliers.

Figure 11.7. The Dynatech ML 1000 Microplate Luminometer, Model 2.4.

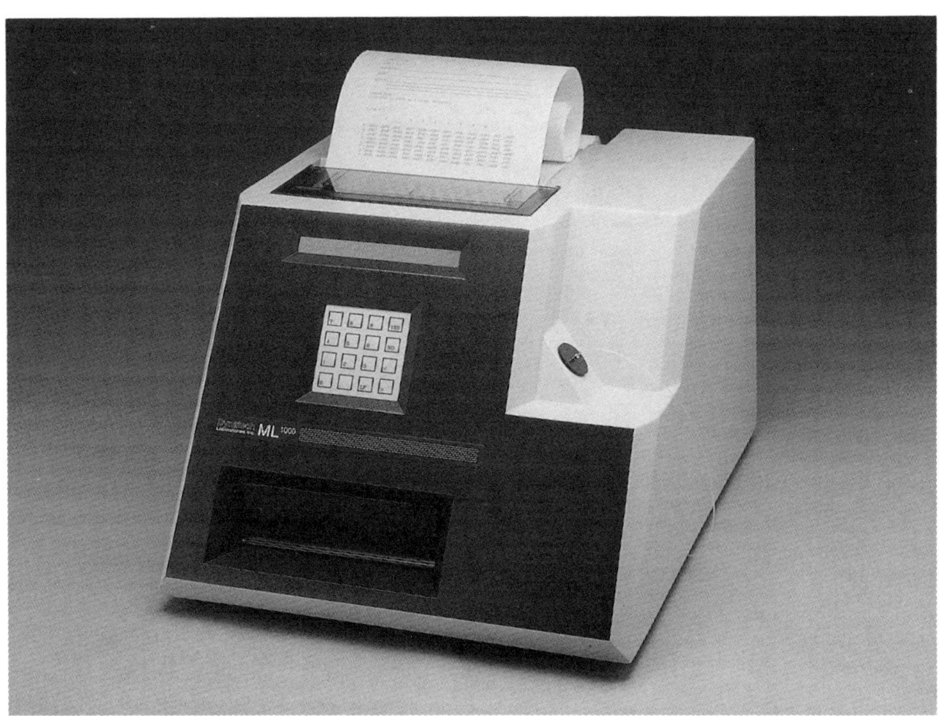

Automation.   Both Lumac and Berthold provide fully automated models that utilize cuvettes. The Lumac Autobiocounter M 4000 can handle sample volumes of up to 500 $\mu\ell$ and can process 240 samples in one hour. The sensitivity is the same as the other Lumac instruments and a fixed integration time of 5 seconds is utilized. The Berthold LB 953 handles up to 150 samples in tubes from 12 $\times$ 47 mm to 12 $\times$ 75 mm. Three injection ports are under microprocessor control and the sample compartment temperature can be varied up to 45°C (Berthold 1991).

Interference

Quench.   As in liquid scintillation spectroscopy, quench refers to any factor that attenuates photon detection. This can occur at any point between the generation of photons and their impingement on the photocathode. Color quench indicates that photons of a discrete wavelength can be absorbed by atoms or molecules excited at that wavelength. As a general rule, colored light will be absorbed by a substance having the complementary color. Therefore, the yellow-green light produced by the firefly luciferase/luciferin reaction will be strongly absorbed by red and purple compounds, so that hemolyzed serum and plasma samples will be particularly difficult to assay by this method.

Other substances, irrespective of their color, can quench the signal by interfering with the propagation of photons through the medium. A prime example of this is phosphate buffer, which was found to decrease the light detected from a luminescence reaction at pH 7.9 by about two-thirds (Strehler 1968). More recently, Shi et al. (1989) reported that boiling phosphate buffer extraction of *Mycobacterium bovis*-BCG exhibited a quench of 56 percent. In this study the quench factor found for boiling Tris-EDTA, n-butanol/glutamate-arsenate buffer and n-butanol/Tris-EDTA was 25, 3.3, and 4.6 percent, respectively. Other ions that can quench to some degree are chloride, cadmium, and zinc (*Lumac's* 1985).

Endogenous Enzymes.   Microbial cells contain a large variety of ATPases that can be detected by their rapid destruction of added ATP standard. These enzymes can be destroyed by heat, as in boiling buffer extractions, or inhibited chemically (e.g., by using Tris buffers), or by changing the pH (e.g., *Lumac's* Trans-NRS extraction reagent) (Lundin and Thore 1975; *Lumac's* 1985). Some ATPases and other enzymes, however, are capable of generating ATP from ADP and AMP, and falsely elevating ATP levels; the conditions for enzyme inactivation vary. Klegerman et al. (1991) found that, while boiling for two minutes destroyed ATPase activity in BCG vaccine filtrate, it did not prevent degradation with time of endogenous ATP extracted from the cells. The best policy is to assay extracts as quickly as possible after extraction or thawing. In the latter case extracts should be snap frozen immediately after extraction and they should not be allowed to stand at room temperature after thawing.

Other Factors.   A number of factors affect the photon-generating luciferase/luciferin reaction. Firefly luciferase shows sharp temperature and pH maxima, which occur at 25°C and pH 7.8, respectively (Strehler 1968). Assay buffers should be strong enough to counteract variant pH values of samples. Cofactors such as Mg^{++} and oxygen can be depleted by various means. Therefore, care should be taken to avoid excessive concentrations of EDTA in Tris-EDTA extraction buffers and to assure that samples are sufficiently aerated. Firefly luciferase can be reversibly inhibited by dissociation to monomers and by oxidation in the absence of substrate and irreversibly denatured, but factors promoting these processes are not well understood (Chaplin 1991).

In addition, sample turbidity, adsorption of ATP to surfaces or particles, molecular binding of ATP, and other sources of luminescence can all render the assay inaccurate (Strehler 1968).

Standardization and Quality Control

For many routine bioluminescent ATP assays, internal standardization is typically performed by doping replicate samples with a fixed quantity of ATP standard. The difference in instrument response between the doped and undoped samples is equated with the concentration of added ATP in the doped sample. ATP concentration in the unknown is determined by dividing the detected light output (RLU) by the RLU of the internal standard and multiplying by the ATP concentration of the standard (Shi et al. 1989).

This method relies on several assumptions, however. First, the detected light output must be directly proportional (a linear 45° slope) to the ATP concentration over the effective range of the assay, a condition that should be confirmed by performing internal standardization with several concentrations of ATP standard. The linearity of instrument response should also be determined with ATP standards in water. If the internal standardization curve is nonlinear, ATP standards should also be run in assay buffer to determine the effect of quench or other factors on the dose response. This approach allows recovery efficiency, as well as quench, to be assessed separately. Besides adsorption and binding of ATP, destruction of the nucleotide by extraction procedures and enzymatic conversion could contribute to reduced recoveries.

Second, internal standardization should be performed so as to correct for both quench and recovery factors. Therefore, the standard should be added to the bacterial suspension immediately before extraction, rather than to extracts. Important quality control parameters are intraassay variance (precision; determined by assaying the same sample repeatedly in a single assay), interassay variance (reproducibility; determined by assaying the same sample repeatedly in different assays), quantity of ATP per CFU (which should be assessed from a linear plot of ATP mass vs. CFU, a parameter that should remain relatively constant for a particular bacterial strain under defined conditions), and accuracy. The last attribute is assessed by serially diluting a sample containing high levels of endogenous ATP. If the dose response parallels the standard curve, then the method of standardization is believed to afford an accurate estimate of the actual ATP content of similar samples.

A single ATP standard in water can serve as a process control that also affords a cumulative interassay variance parameter. For instance, the minimum acceptable response of the Lumac M 2010 luminometer to 1.25 ng of ATP in water is 50,000 RLU. In addition, each luminometer requires checks of proper instrumental function. The Lumac instrument must show a dark count in the absence of all reagents of <30 RLU and a reagent blank in the absence only of ATP of <70 RLU. If the unit fails either of these tests, special cleaning routines are required.

CONCLUSIONS

Bioluminescent measurement of ATP using luciferin/luciferase systems is a very sensitive, specific, rapid, and convenient means of assaying the nucleotide that has much potential for the measurement of a wide range of compounds involved in ATP-dependent enzymatic reactions. The use of this method to quantitate viable microbial cells, however, requires the use of various validation procedures and safeguards, depending on the specific application. For several aspects of experimental research, the problem of accuracy is important. This is a problem common to all methods of ATP quantitation, however, since the potential inaccuracies arise from extraction and recovery deficiencies.

Problems of assay interference are the most important difficulties associated with bioluminescent ATP measurement, but appropriate internal standardization can compensate for such factors. Variations in operator technique can have large impacts on these assays. Time between extraction and assay, for instance, can make significant differences in recovery efficiencies unless this variable has been shown to be of minor importance in a given system experimentally. Appropriate quality control measures must be carefully and logically developed to anticipate possible systematic errors in a given application; these measures must be applied stringently and routinely in the monitoring process.

Most industrial applications of bioluminescent ATP measurement involve detection and quantitation of microbial contamination in a particular product, such as the rapid identification of a critical level of contamination in output material. The most important consideration for this purpose is whether the assay reliably detects the critical level of viable biomass. Validation will rely heavily on comparison of ATP assay results with other proven methods of microbial quantitation, especially CFU determinations, in the same sample. Positive samples should also be verified by CFU determination.

Another consideration is cost. The Lumac Autobiocounter M 3000, for instance, is currently priced at nearly $50,000. Materials, especially firefly luciferase, are also somewhat expensive, with the cost per sample being about $1.25 (based on the current price of Lumac reagents). Cloning of the firefly luciferase gene, however, promises to be a much more efficient, consistent, and inexpensive source of recombinant enzyme. With the realization of this potential and the marketing of automated luminometers that will undoubtedly be improved greatly in terms of throughput capacity, bioluminescent ATP measurement is becoming the method of choice for large-scale monitoring of viable microbial biomass.

REFERENCES

Allen, R. C., R. L. Stjernholm, and R. H. Steele. 1972. Evidence for the generation of an electronic excitation state(s) in human polymorphonuclear leukocytes and its participation in bactericidal activity. *Biochem. Biophys. Res. Comm.* 47:679–684.

Berthold, F. 1991. Luminescence instrumentation for microbiology. In *Physical methods for microorganisms detection,* edited by W. H. Nelson. Ann Arbor, MI: CRC Press.

Bonner, J., G. R. Chalkley, M. Dahmus, D. Fambrough, F. Fujimura, and C. H. Ru-Chih. 1968. Chromosomal nucleoproteins. *Meth. Enzymol.* 12:3–65.

Chaplin, M. F. 1991. The preparation and properties of immobilized firefly luciferase for use in the detection of microorganisms. In *Physical methods for microorganisms detection,* edited by W. H. Nelson. Ann Arbor, MI: CRC Press.

Clyne, J. M., J. A. Running, M. Stempien, R. S. Stephens, H. Akhavan-Tafti, A. P. Scharp, and M. S. Urdea. 1989. A rapid chemiluminescent DNA hybridization assay for the detection of *Chlamydia trachomatis. J. Biolumin. Chemilumin.* 4:357–366.

Collins, W. P., ed. 1985. *Alternative immunoassays.* New York: John Wiley & Sons.

Currie, B. L., and M. E. Klegerman. 1992. Analytical techniques commonly used in biotechnology. In *Pharmaceutical biotechnology: Fundamentals and essentials,* edited by M. E. Klegerman, and M. J. Groves. Buffalo Grove, IL: Interpharm Press, Inc.

Davis, B. D., R. Dulbecco, H. N. Eisen, H. S. Ginsberg, W. B. Wood, Jr., and M. McCarty. 1973. *Microbiology.* Hagerstown, MD: Harper & Row.

De Boever, J., F. Kohen, and D. Vandekerckhove. 1983. Solid-phase chemiluminescence immunoassay for plasma estradiol-17β during gonadotropin therapy compared with two radioimmunoassays. *Clin. Chem.* 29:2068–2072.

De Chatelet, L. R., and P. S. Shirley. 1981. Evaluation of chronic granulomatous disease by a chemiluminescence assay of microliter quantities of whole blood. *Clin. Chem.* 27:1739–1741.

Devadoss, P., M. E. Klegerman, and M. J. Groves. 1991. Surface morphology of *Mycobacterium bovis* BCG: Relation to mechanisms of cellular aggregation. *Microbios* 65:111–125.

Gheorghiu, M., and M. Lagranderie. 1979. Mesure rapide de la viabilité du BCG par dosage de l'ATP. *Ann. Microbiol.* 130B:147–156.

Gorus, F., and E. Schram. 1979. Applications of bio- and chemiluminescence: Its potential in the clinical laboratory. *Clin. Chem.* 25:512–519.

Granato, P. A., and M. F. Roefaro. 1989. Evaluation of a prototype DNA probe test for the noncultural diagnosis of gonorrhea. *J. Clin. Microbiol.* 27:632–635.

Griffiths, M. W. 1993. Applications of bioluminescence in the dairy industry. *J. Dairy Sci.* 76:3118–3125.

Hughes, S. 1994. The resurgence of tuberculosis. *Scrip Magazine* 24:46–48.

Ingraham, J. L., O. Maaløe, and F. C. Neidhardt. 1983. *Growth of the bacterial cell.* Sunderland, MA: Sinauer Asso.

Jacobs, W. R., Jr., R. G. Barletta, R. Udani, J. Chan, G. Kalkut, G. Sosne, T. Kieser, G. J. Sarkis, G. F. Hatfull, and B. R. Bloom. 1993. Rapid assessment of drug susceptibilities of *Mycobacterium tuberculosis* by means of luciferase reporter phages. *Science* 260:819–822.

Jago, P. H., W. J. Simpson, S. P. Denyer, A. W. Evans, M. W. Griffiths, J. R. M. Hammond, T. P. Ingram, R. F. Lacey, N. W. Macey, B. J. McCarthy, T. T. Salusbury, P. S. Senior, S. Sidorowicz, R. Smither, G. Stanfield, and P. E. Stanley. 1989. An evaluation of the performance of ten commercial luminometers. *J. Biolumin. Chemilumin.* 4:131–145.

Klegerman, M. E., J. Lajeune, M. Shi, and M. J. Groves. 1991. Adenosine triphosphatase activities in lyophilized Tice™-substrain BCG vaccine. *Biomed. Lett.* 46:213–217.

Kricka, L. J. 1988. Clinical and biochemical applications of luciferases and luciferins. *Anal. Biochem.* 175:14–21.

Laszlo, A., P. Gill, V. Handzel, M. M. Hodgkin, and D. M. Helbecque. 1983. Conventional and radiometric drug susceptibility testing of *Mycobacterium tuberculosis* complex. *J. Clin. Microbiol.* 18:1335-1339.

Lumac's rapid microbiology testing technology. 1985. Schaesberg, The Netherlands: Lumac/3M.

Lundin, A., and A. Thore. 1975. Comparison of methods for extraction of bacterial adenine nucleotides determined by firefly assay. *Appl. Microbiol.* 30:713–721.

Matamoros, M. C., and J. S. Strobl. 1990. Detection of gene promoter activity in mammalian cells using the firefly luciferase reporter gene. In *Luminescence immunoassay and molecular applications,* edited by K. Van Dyke, and R. Van Dyke. Ann Arbor, MI: CRC Press.

McElroy, W. D. 1947. The energy source for bioluminescence in an isolated system. *Proc. Natl. Acad. Sci. USA.* 33:342–345.

McElroy, W. D., and B. L. Strehler. 1949. Factors influencing the response of the bioluminescent reaction to adenosine triphosphate. *Arch. Biochem.* 22:420–433.

Meltzer, T. H. 1991. Pharmaceutical water: Generation, storage, distribution, and quality testing. In *Sterile pharmaceutical manufacturing: Applications for the 1990s,* Vol. 1, edited by M. J. Groves, W. P. Olson, and M. H. Anisfeld. Buffalo Grove, IL: Interpharm Press, Inc.

Nelson, N. C., P. W. Hammond, W. A. Wiese, and L. J. Arnold, Jr. 1990. Homogeneous and heterogeneous chemiluminescent DNA probe-based assay formats for the rapid and sensitive detection of target nucleic acids. In *Luminescence immunoassay and molecular applications,* edited by K. Van Dyke, and R. Van Dyke. Ann Arbor, MI: CRC Press.

Olson, W. P., M. J. Groves, and M. E. Klegerman. 1990. Cell mass of *Mycobacterium bovis* BCG estimated by gas chromatography. *Biologicals* 18:83–88.

Rosenthal, S. R., A. H. Kenniebrew, and N. Raisys. 1962. A modified semi-solid medium and technique for determining viability of BCG. *Acta Tuberc. Pneumol. Scand.* 42:167–172.

Shi, M., M. E. Klegerman, and M. J. Groves. 1989. Viability of freeze-dried Tice™-substrain BCG by bioluminescent measurement of adenosine triphosphate. *Microbios* 59:145–155.

Stanley, P. E. 1989. A review of bioluminescent ATP techniques in rapid microbiology. *J. Biolumin. Chemilumin.* 4:375–380.

Stannard, C. J., and P. A. Gibbs. 1986. Rapid microbiology: applications of bioluminescence in the food industry—a review. *J. Biolumin. Chemilumin.* 1:3–10.

Stevens, P., D. J. Winston, and K. Van Dyke. 1978. In vitro evaluation of opsonic and cellular granulocyte function by luminol-dependent chemiluminescence: Utility in patients with severe neutropenia, and cellular deficiency states. *Infect. Immun.* 22:41–51.

Stevenson, L. H., T. H. Chrzanowski, and C. W. Erkenbrecher. 1979. The adenosine triphosphate assay: Conceptions and misconceptions. In *Native aquatic bacteria: Enumeration, activity, and ecology*, edited by J. W. Costerton, and R. R. Colwell. Philadelphia: Am. Soc. Testing and Materials.

Stewardson-Krieger, P. B., K. Albrandt, A. R. Kretschmer, and S. P. Gotoff. 1977. Group B streptococcal long-chain reaction. *Infect. Immun.* 18:666–672.

Strehler, B. L. 1968. Bioluminescence assay: Principles and practice. *Meth. Biochem. Anal.* 16:100–181.

Van Dyke, K., and R. Van Dyke. 1990. *Luminescence immunoassay and molecular applications,* Ann Arbor, MI: CRC Press.

Voller, A., and D. E. Bidwell. 1985. Enzyme immunoassays. In *Alternative immunoassays*, edited by W. P. Collins. New York: John Wiley & Sons.

Weeks, I., M. L. Sturgess, and J. S. Woodhead. 1986. Chemiluminescence immunoassay: An overview. *Clin. Sci.* 70:403–408.

Whitehead, T. P., L. J. Kricka, T. J. N. Carter, and G. H. G. Thorpe. 1979. Analytical luminescence: Its potential in the clinical laboratory. *Clin. Chem.* 25:1531–1546.

Woodbury, C. P., Jr. 1992. Recombinant DNA technology. In *Pharmaceutical biotechnology: Fundamentals and essentials*, edited by M. E. Klegerman, and M. J. Groves. Buffalo Grove, IL: Interpharm Press, Inc.

12

Limulus Amebocyte Lysate (LAL) Assays

Thomas J. Novitsky
Associates of Cape Cod, Inc.

Endotoxins have been recognized as pyrogens since the early 1800s (Westphal et al. 1977, 221–238). Because of their ubiquitous nature and their resistance to steam sterilization, endotoxins are the most common contaminant of parenteral drugs, biologicals, and medical devices. The ability to detect endotoxins in pharmaceuticals is, therefore, of utmost concern not only to the FDA, but also to the manufacturer and various pharmacopoeias. The first pharmacopoeial pyrogen test appeared in 1941. The test was based on the earlier work of Florence Seibert (Seibert 1923). It was an animal test that took advantage of the ability of endotoxin and/or other pyrogens to cause a measurable increase in the temperature of a laboratory rabbit. This test is known as the Pyrogen Test, and is still in use today.

In 1968 Levin and Bang (1968) described an in vitro assay that was specific for endotoxins. The reagent for this assay was derived from an aqueous lysate of the blood cells, or amebocytes, of the North American horseshoe crab, *Limulus polyphemus.* The assay became known as the *Limulus* amebocyte lysate (LAL) test. Although used initially to detect endotoxin in blood as a diagnostic for endotoxemia and as a rapid test for gram-negative bacteremia, the FDA became interested in LAL as an alternative to the Pyrogen Test. In 1977 an article appeared in the *Federal Register* outlining the requirements for substituting the LAL test for the Pyrogen Test. At the same time the FDA licensed the first commercial manufacturer of LAL (Novitsky 1991). Since then the LAL test has been slowly replacing the rabbit as a release test for pyrogens (endotoxins) in parenteral drugs and medical devices. More importantly, however, the LAL test has seen widespread use as an "in-process test" to control endotoxin contamination. In 1980 a Bacterial Endotoxins Test (BET) chapter appeared in the USP, and in 1983 the USP replaced the Pyrogen Test with the BET for Water for Injection (USP 1993).

NATURE OF ENDOTOXINS/PYROGENS

Although there are many classes of pyrogens (i.e., fever-causing substances), including inorganic as well as organic molecules, and living as well as dead microbial entities, the most frequently encountered pyrogen, and arguably the most potent one, is endotoxin. The term *endotoxin* was coined by Richard Pfeiffer in 1892 and was meant to imply a toxin intimately associated with the body of the bacterium that produced it (Westphal et al. 1997, 221–238). Endotoxin is comprised of protein, lipid, and carbohydrate. It was found that the biologically active component of endotoxin included the lipid and carbohydrate portion; these were chemically linked to form a lipopolysaccharide (LPS). Further studies showed that the molecule was biphasic, with the biologic (including pyrogenicity and activation of LAL) activity confined to the lipid A region, while the antigenicity resided primarily with the carbohydrate, or O polysaccharide. A generalized chemical structure of LPS is shown in Figure 12.1.

In addition to pyrogenicity and activation of LAL, endotoxin exhibits a variety of biological activities. Unlike pyrogenicity and the activation of LAL (which in the horseshoe crab provides defense against invasion by gram-negative bacteria), most of the biological activity of endotoxin if it enters the circulation is considered harmful. A list of the most common biological activities of LPS are listed in Table 12.1.

NATURE OF THE LAL TEST

Biochemistry

The LAL test is enzymatic in nature. The essential components of the assay are Factor C, Factor B, proclotting enzyme (protein), and coagulogen. The first three components are enzymes of the serine protease type. The last component is a small protein with no enzymatic activity. Although the complete biochemistry of the reaction is not known, the reaction appears to be a cascade (Iwanaga et al. 1992). That is, Factor C, once activated by endotoxin or LPS, activates Factor B, which in turn activates the proclotting enzyme (becomes the clotting enzyme on activation). The clotting enzyme then acts on the coagulogen, cleaving it in two places. The cleaved coagulogen (coagulin) undergoes a configuration change becoming insoluble and forming a solid clot or gel. A schematic showing the LAL enzymatic cascade is shown in Figure 12.2. It should be noted that all of the commercially available LAL assays contain Factors C and B and the proclotting enzyme. As will be explained later, the coagulogen can be replaced by several different synthetic substrates.

Specificity

It was originally thought that the LAL assay was highly specific for endotoxin. It was found, however, that LAL contains another enzyme, Factor G, which when activated by β 1,3 glucans or by LAL reactive material (LALRM), will activate the proclotting enzyme. The result of this reaction is indistinguishable from that of

Figure 12.1. Schematic representation of lipopolysaccharide.

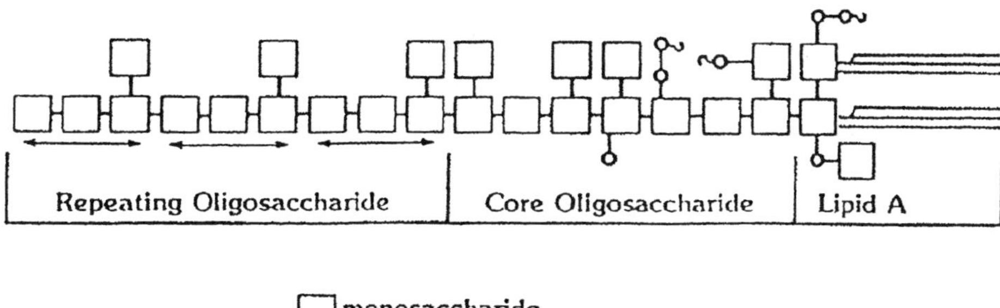

Table 12.1. Biological Activity of Lipopolysaccharide

Stimulation of lymphoreticular cells

    B-cell mitogenicity

    Polyclonal antibody synthesis

    Immunogenicity to polysaccharides

    Adjunctivity

    Macrophage activation

Suppression and tolerance

    Induction of endotoxin tolerance

    Suppression of thymic dependent responses

    Regulation of oral tolerance and IgA responses

Generalized effects

    Pyrogenicity (in man includes headache, chills, myalgia, and nausea)

    Abortion

    Tumor necrosis

    Lethality

    Disseminated intravascular coagulation

    Complement activation

    Activation of the fibrinolytic system

    Lung toxicity

    Gelation of LAL

Figure 12.2. Limulus amebocyte lysate enzymatic cascade.

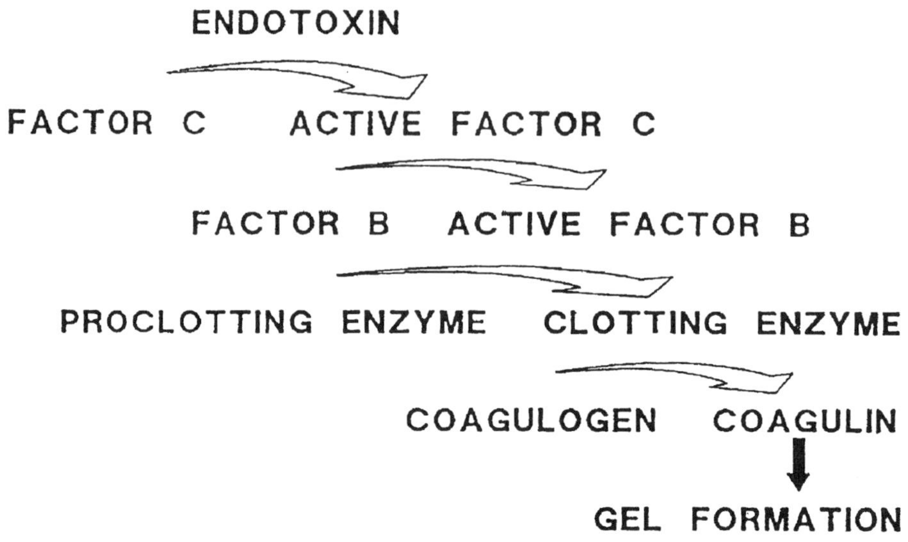

ENDOTOXIN

FACTOR C   ACTIVE FACTOR C

FACTOR B   ACTIVE FACTOR B

PROCLOTTING ENZYME   CLOTTING ENZYME

COAGULOGEN   COAGULIN

GEL FORMATION

endotoxin. Fortunately, about 1000 times more glucan than endotoxin is required for activation. In addition, although both endotoxin and glucan are highly resistant to destruction by heat, endotoxin is sensitive to mild alkaline hydrolysis while endotoxin is not. There is also a commercially available LAL that does not contain Factor G, making it absolutely endotoxin specific. While it may seem that LAL reactivity with glucan (false positive) would present a problem for pharmaceutical manufacturers, this has not been the case for two reasons:

1.   Positive tests on pharmaceutical products are produced under exacting (i.e., cGMP) conditions.

2.   1000 times more glucan than endotoxin is needed for a positive test.

In practice, a positive LAL test is considered an indication of contamination, whether glucan or endotoxin. On the rare occasion of a positive, it is easy to differentiate between the two, thereby helping to isolate the source of contamination. For example, endotoxin contamination usually arises from a contaminated water system, while glucans (which are mostly fungal constituents) enter the system from filters or raw materials (Roslansky and Novitsky 1991).

Sensitivity

The LAL assay is one of the most sensitive bioassays known. With certain modifications, the assay can detect as little as 0.001 Endotoxin Unit (EU)/ml, or the equivalent of the amount of endotoxin contained in 10 *Escherichia coli* bacteria. The sensitivity of LAL is now reported in Endotoxin Units, a measure of activity based on a reference standard endotoxin (RSE). The RSE was prepared by the FDA from *Escherichia coli* O113:H10:K(–), Westphal (i.e., phenol-water extracted) purified LPS (Hochstein et al. 1983; Westphal et al. 1977, 221–238). The original standard referred to as EC has gone through several lots—the current one is designated as EC-6. The endotoxin unit was selected since it was difficult to compare activities of different endotoxins on a weight basis. The RSE in use when the switch was made to EUs was EC-2. This RSE was arbitrarily assigned the potency of 5 EU/ng of *Escherichia coli* O113 LPS by the FDA. The current RSE, EC-6 however, has a potency of 10 EU/ng. Although it originated from the same source as EC-2, slight differences in handling, filling, nonendotoxin additives (stabilizers), most probably account for the difference in activity. It should be noted that when referring to endotoxin, the term *potency* is used, compared to the term *sensitivity* when referring to LAL. Potency is usually given as EU per vial or per ng of endotoxin, while sensitivity is given as EU/ml.

CONFIGURATION OF THE ASSAY

Gel-Clot Method

The original description of the LAL reaction included the phenomenon of a solid gel or clot formation. This so-called gel-clot test formed the basis for the first in vitro LAL assay. In this assay a small amount of LAL (0.1 ml) is added to an equal volume of sample in a 10×75 mm glass test tube. The mixture is then incubated for 60 ± 2 minutes at $37° \pm 0.5°C$. At the end of incubation, the tube is slowly inverted. If a solid gel-clot is formed (i.e., one that does not break on 180° inversion), the test is scored as positive. If the clot breaks, or the test remains liquid, it is scored negative. Usually standards in the range of 0.015–0.5 EU/ml (with twofold dilution intervals) are run as a control and to validate the label claim (sensitivity) of the LAL. If a positive sample is obtained, an approximation of the amount of endotoxin present can be made by repeating the test using further dilution of the sample. To calculate the endotoxin concentration, the dilution factor of the highest dilution with a positive result (i.e., firm gel-clot) is multiplied by the confirmed sensitivity of the LAL. Thus, a sample positive at a 1:4 dilution using an LAL of 0.5 EU/ml will have an endotoxin content of 2 EU/ml (4×0.5 EU/ml). Although the gel-clot test was the original LAL assay and alternative assays have appeared, the gel-clot method, used as a pass/fail test or limits test, is still the most widely used LAL technique.

Turbidimetric Method

In the original description of the clotting of LAL, the formation of turbidity that lead to the formation of a solid gel-clot was used to assess the enzymatic activity of the assay. Turbidity was subsequently used in an alternative to the gel-clot assay. In the first available turbidimetric test, LAL was incubated with sample or standard for a period of time. This was usually less than 60 minutes, since turbidity formed before a gel-clot. At the end of the incubation period, the mixture was read in a spectrophotometer or nephelometer. If the samples were a series of standards, a standard curve could be produced by plotting turbidity (as optical density) against endotoxin concentration. Optical densities of unknown samples could then be used to extrapolate the endotoxin concentrations of these samples. This test was less subjective than the gel-clot test and was more quantitative. Variations of this assay appeared as well as specialized equipment. In 1986 the LAL-4000 was introduced to incubate automatically and read turbidimetric results. Results could be tabulated automatically through a computer interface. Furthermore, this assay used the kinetics of turbidity formation, rather than the end point. This expanded the useful range of the assay from 1 log to over 5 logs. Several improvements on the 4000 were eventually introduced, including the Toxinometer, and the LAL-5000 series. The Toxinometer (mainly in Japan) and the LAL-5000 series (in the US and Europe) are still widely used today, especially by large pharmaceutical houses. A new machine, the ATi-6000, patterned after the LAL-5000 but with added sample capacity and the ability to use monochromatic light, has recently been introduced in Europe.

Chromogenic Method

In 1977 Japanese investigators described the use of synthetic substrates in the LAL assay (Nakamura et al. 1977). These substrates were the same ones used for human blood factor analysis. The ones that worked best contained an amino acid cleavage site similar to that in coagulogen. These synthetic substrates, when cleaved by the LAL clotting enzyme, released a chromophore that could be detected by a spectrophotometer. The most common of these substrates employ para-nitroanilide as the chromophore (405 nm absorbance). The chromogenic test originally described was a two-step endpoint type assay. In this assay LAL was first incubated with sample for some time, usually 15–30 minutes. Following this preliminary incubation, the chromogenic substrate was added. Following this addition, the assay was incubated a second time, usually for only a few minutes, then stopped with the addition of acid. During the second incubation, the chromophore was released. Standard curves could then be constructed, as with the turbidimetric assay (i.e., optical density [OD] versus endotoxin concentration). The advantage of the chromogenic method over the turbidimetric method was that the reaction could be stopped (acid addition), so that once the full color had developed, reading could be delayed for up to several hours. Turbidity, on the other hand, had to be read precisely at the end of the incubation period.

More recently the chromogenic method has been modified to a one-step assay, allowing the use of kinetics. The routine use of the kinetic chromogenic assay also relied on development of incubating microplate readers. There are several brands of these, with accompanying computer software now available.

Other Methods

Many other methods of performing the LAL test have been described. All of these, however, have as their starting point the *Limulus* enzymes. These methods and modifications have been reviewed and compared (Novitsky 1983, 1986, 1987a, 1987b, 1993). It should be noted, however, that generally only three basic methods are described and accepted by the FDA and the pharmacopoeias: gel-clot method, turbidimetric (endpoint and kinetic) method, and chromogenic (endpoint and kinetic) method.

REAGENTS/AVAILABILITY

Commercial Supply

Currently, there are six licensed manufacturers of LAL in the United States (one of these is a Japanese company).

- Associates of Cape Cod, Inc., Woods Hole, MA
- Biowhittaker, Inc., Walkersville, MD
- Endosafe/Charles River, Inc., Charleston, SC
- Haemachem, Inc., St. Louis, MO
- Seikagaku Corporation, Tokyo, Japan
- Baxter Healthcare, Deerfield, IL

There are several other LAL (includes lysate made from the Asian horseshoe crabs, *Tachypleus* and *Carcinoscorpius*) manufacturers in Asia, especially China; some European companies distribute LAL or LAL kits on an original equipment manufacturer (OEM) basis. Since horseshoe crabs only occur on the east coast of North America, and in the Far East (mainly in the South China Sea and Japan) and India, all of the world's supply comes from these areas.

Type of Kits Available

All of the methods described in the FDA guidelines or by the various pharmacopoeias are available commercially: gel-clot, endpoint turbidimetric, kinetic turbidimetric, endpoint chromogenic, and kinetic chromogenic. All commercially available LAL is supplied in freeze-dried form. The gel-clot reagents are usually packaged in multitest (50, 20, and 10) and single test sizes. The turbidimetric reagents are commonly packaged in multitest (e.g., 50) vials. The chromogenic reagents are packaged in multitest vials that usually correspond to volumes used in microplates (e.g., 96 and 60 [corresponds to all but perimeter wells of a 96-well microplate]). Several manufacturers provide multiple multitest vials in a kit (e.g., a 50 test vial $\times$ 5 vial configuration would be a 250 test kit). These kits often include LAL reagent water (LRW) used to reconstitute LAL, make dilutions, and as a negative control. Some kits also include an endotoxin standard referred to as a control standard endotoxin (CSE) to distinguish it from the RSE. Because no kit is ever universal, many of the manufacturers supply LAL, LRW, and CSE separately; customers compile their own "kits" from these components.

Supplies Needed to Perform the Assay Not Supplied with Kits

All LAL assays require liquid containment and transfer devices that are not contaminated with endotoxin and will not adsorb endotoxin or otherwise interfere with the LAL enzymes. Glass labware that has been depyrogenated by the user is always the best choice. Although plastics are widely available and have been successfully used with the LAL test, their unqualified use can lead to disaster (Novitsky et al. 1986). All plastics should, therefore, be validated for use in the LAL test. This process is made easier by consulting the LAL supplier for sources of acceptable plasticware. As a rule of thumb, polystyrene tubes often have undetectable levels of endotoxin and do not absorb endotoxin from solution. On the other hand, polypropylene tubes are often contaminated, absorb endotoxin strongly from solution, and can contain excess plasticizers and/or mold-releasing agents that can be potent enzyme inhibitors. Fortunately, the drawbacks to polypropylene tubes do not entirely apply to polypropylene pipet tips. Since liquid is only in contact with the tip a short period of time, this is not a problem. Contamination, however, and the possibility of other interfering substances should always be of concern with tips. It is recommended that the user validate any disposable labware (including glass [i.e., pipets]) used for the LAL assay by rinsing several representative samples and testing the rinse with LAL. Remember that sterile does not equal nonpyrogenic and nonpyrogenic does not equal endotoxin-free.

21 CFR

The LAL test is regulated in the United States by the Center for Biologics Evaluation and Research (CBER) of the FDA. As such, labeling and product circulars are controlled by 21 CFR. Thus, although the manufacturer's instructions for use may vary slightly from company to company, a U.S.–licensed pharmaceutical manufacturer must follow these instructions exactly in order to be in compliance with current good manufacturing practices. In light of the FDA guidelines, FDA guidances, and pharmacopoeial chapters relating to LAL, users often forget the regulatory authority of 21 CFR. Of course, apart from the individual manufacturers' particular requirements (e.g., reconstitution, storage, order of addition, etc.), the guidelines and pharmacopoeial chapters offer invaluable information on sample handling, replication, and required controls (see below). Although LAL for research purposes does not have these requirements, common sense dictates that the manufacturer's instructions be followed.

The latest version of the USP chapter <85> appears in the 8th Supplement to USP XXII (1993). This chapter covers the following topics:

- Reference standard and control standard endotoxins

- Preparatory testing

- Test for confirmation of labeled LAL reagent sensitivity

- Inhibition or enhancement test

- Maximum valid dilution (MVD)

- Test procedure, including preparation and procedure

- Calculation and interpretation, including geometric mean calculation, endotoxin content calculation, and interpretation

Each of these sections will be discussed briefly below. It is strongly suggested that the reader obtain and review a complete copy of this chapter from the USP.

Reference Standard and Control Standard Endotoxins

The first section describes the RSE—its preparation, storage, and comparison to the CSE. The RSE is supplied in a vial containing 10,000 USP EU per vial and is reconstituted with 5 ml of LRW. The RSE is available from the USP (as lot F) for a nominal fee. Reconstituted RSE can be stored refrigerated for no more than 14 days. Due to the relatively short supply of the RSE, users are allowed to use a CSE for day-to-day testing. The CSE must first be compared to the RSE. To accomplish this, the user should assay one vial of RSE and 1 vial of CSE with the LAL as described below under test procedure. Four replicates at each concentration level are needed. After the results of the test are recorded, the geometric mean end points of each standard are calculated and the ratio of the RSE to CSE is calculated. This result, in EU/ng, can then be used to assign EUs to the CSE. A CSE is deemed suitable if it has a potency of ≥ 2 to ≤ 50 EU/ng.

Preparatory Testing

Preparatory testing recommendations include using an LAL reagent of confirmed sensitivity, depyrogenated utensils, items to be tested that have been shown not to interfere with the test, and appropriate negative controls. In addition, validation of the test must be repeated if the brand of LAL or the method of manufacture of the article under test is changed.

Test for Confirmation of Labeled LAL Reagent Sensitivity

The labeled sensitivity of the LAL is confirmed by testing one vial of the reagent with concentrations of CSE or RSE, which range from 1/4 to 2 times the labeled sensitivity in twofold increments. The symbol for sensitivity is λ and implies EU/ml. The test is run in quadruplicate, including negative controls. The labeled sensitivity is confirmed if the result is within twofold of λ. Confirmation of label claim should be repeated if the lot of LAL reagent is changed.

Inhibition or Enhancement Test

The inhibition or enhancement test is a check on whether or not a test specimen interferes with the LAL assay (i.e., whether or not a particular sample can be validly tested). To perform this test, sample dilutions are prepared with and without endotoxin (final concentrations from 1/4 to 2 times λ in twofold increments) and tested. The test is done in quadruplicate with the diluted specimen together with a duplicate test of standard (diluted in LRW) and negative controls. It is also necessary that the specimen not be diluted past the point where the dose of the specimen, once administered, would be below the therapeutic level on a per kg body weight basis. To simplify this requirement, a function called the MVD is employed.

When running an inhibition/enhancement test, it is also important that the specimen under test does not contain endotoxin in the dilution used for the assay. Usually, concentrations above the MVD that do not cause inhibition/

enhancement are tested. This provides a margin of error or safety. Since the LAL test is so sensitive, there is usually quite a bit of dilution that can be used, if necessary, without exceeding the MVD to overcome inhibition/enhancement effects, and to dilute out small amounts of endotoxin.

Several "tricks" that can be used to facilitate the inhibition/enhancement assay include pH adjustment, filtration, and heating of the sample or dilution of sample.

Maximum Valid Dilution

Each drug product (injection or solution) has an MVD that can be calculated from the dose of the drug (mg, ml, or unit/kg) and the tolerance limit of humans/rabbits, which is defined as 5 EU/kg. Once the amount of drug per human (adult average weight taken as 70 kg) and the starting concentration of the drug is known, the amount the drug can be diluted and still provide a valid test can be calculated (i.e., the MVD). For most drugs a list of the endotoxin limits (i.e., how much endotoxin is allowed per mg, ml, or unit of a particular drug is allowed [also calculated from the dose and the tolerance limit] or the MVD) has been published. The MVD is considered a limit dilution factor and cannot be exceeded for a valid test.

Test Procedure

Preparation

In general, precautions in handling specimens, reagents, and so on should be used to avoid endotoxin contamination. It is not necessary to use a special hood or area, but areas containing a lot of dust or high traffic should be avoided. For each specimen and standard, two replicates at each level are recommended. It is also recommended that the reagent manufacturer's instructions be followed since these vary between brands of LAL. If the pH of the diluted specimen/LAL mixture is outside the range of 6.0–8.0, the pH should be adjusted on the diluted sample. Since all LAL has some buffering capacity (although the amount varies between brands), the pH of the mixture (LAL and sample) is more important than the sample alone. For example, Water for Injection (WFI) usually has a pH below 6, but reaction mixture of LAL plus WFI will be in the 6–8 range. On samples that have their own buffering capacity or whose pH overwhelms the LAL reagent, pH adjustment can be made with sodium hydroxide or HCl solutions. (Note: These can easily be made with low to no detectable endotoxin by starting with NaOH pellets or concentrated HCl and diluting in LRW using depyrogenated containers to a workable normality, e.g., 0.1 N).

It is not necessary to check the pH each time the assay is performed for routine testing of products which have been validated for the LAL procedure. The product positive control, included with each assay, serves, in part, as a check on the pH of the sample mixture. Thus, if the pH is out of range, the product positive control will be negative.

Procedure

The test is run by adding samples, standards, and negative and positive controls to 10 3 75 mm glass test tubes or single test vials (contain prepackaged LAL reagent). Since the composition of the test tube (flint or sodalime glass or borosilicate glass) can have an effect on the clotting reaction, it is important to follow the LAL manufacturer's instructions for the best type/brand of tube to use. Single test vials, however, have already been selected by the manufacturer for optimal reactivity. Once the sample has been added, add LAL and place all tubes in a 37° ± 1°C incubator (preferably a nonagitating water bath, although a block heater will do, but not a convection oven/incubator) for 60 ± 2 minutes. The test should be allowed to incubate undisturbed, as jarring the clot can result in a negative test. At the end of the incubation period, a test scores positive if a solid gel-clot is formed (i.e., one that can withstand a 180° inversion). Negative tests are those that remain liquid or have semisolid clots which break on inversion. Since the gel-clot method is a limit test, endotoxin presence below the LAL sensitivity may result in a semi-solid gel. However, since this test is highly subjective, it is not wise to record ± results. Some gel material (floculation) in LRW negative controls, however, should alert the user to a potential contamination source. A test is valid if the standard is within 1 ± twofold dilution of the label claim, the negative control is negative, and the product positive control (i.e., product to which 2 ℓ endotoxin has been added) is positive. If the product results in a negative test, the product passes at the limit of detection. If the product shows a positive result, the product fails. If a product contains endotoxin, an approximation of its content can be made by repeating the tests on higher dilutions of the sample. The endotoxin content can then be calculated.

Calculation and Interpretation

Geometric Mean Calculation

Endotoxin content is always calculated using the geometric mean end point (GME) calculation. To find the geometric mean, record the value in EU/ml of the highest dilution positive in a series (i.e., the last positive test in a series of decreasing concentrations of endotoxin). This value is referred to as the endpoint, E. Convert E to its logarithm (e), add all e values, and divide by the number of e values to obtain the logarithmic mean. The geometric mean end point is the antilog of the logarithmic mean. Thus, GME = antilog ($\Sigma e/f$) where f is the number of replicates (end points).

Endotoxin Content Calculation

To calculate the endotoxin concentration in a sample, it is necessary to calculate the endpoint concentration, E, for each series of dilutions by multiplying the reciprocal of the dilution of the highest dilution positive in each series by λ, the label claim of the LAL used. The GME is then calculated as above. (Note: Endotoxin content can only be calculated if there is a clear end point in each series [i.e., a negative dilution must be present following the last positive dilution]; in other words, a valid end point must be present).

Interpretation

It should be quite obvious to the user that if all the controls are in place and the results in order, an article under test will pass if the limit of the test or the endotoxin content of the article does not exceed the endotoxin limit set by the USP (i.e., the limit indicated in the article monograph).

FDA GUIDELINES/GUIDANCE

Technician/Laboratory Validation

The FDA guidelines recommend that manufacturers assess the variability of their testing laboratory before performing any official tests (FDA 1987). This usually includes confirmation of label claim and inhibition/enhancement tests on the article(s) under test. It is no longer necessary to compare pyrogen test results to LAL results to validate the LAL test. In order to be validated for performing the LAL test, each technician should show proficiency in confirming label claim with at least one lot of LAL and a single lot of endotoxin standard. Any methods approved for use by the FDA (i.e., from a licensed supplier) can be used. These currently include the gel-clot, turbidimetric, and chromogenic methods. Although the FDA allows use of gel-clot LAL in turbidimetric and chromogenic applications, it is much safer to use LAL particularly suited for the method. Since these are now readily available from several sources, this should not be a problem. If the gel-clot method is used for the turbidimetric or chromogenic methods, however, it is up to the user to validate the method and to use the appropriate criteria (i.e., specific to the method, for interpreting the results).

Products are validated by running the inhibition/enhancement test as described above on at least three production batches of each finished product considered for testing.

Products Applicable to LAL Testing

Human and Animal Drugs and Antibiotics

For drugs subject to New Drug Applications (NDAs), or abbreviated NDAs (ANDAs), antibiotic drug applications, and animal drugs subject to New Animal Drug Applications (NADAs) and abbreviated NADAs (ANADAs), manufacturers are required to submit a supplemental application for LAL testing. The supplement should contain the initial quality control data, inhibition/enhancement testing, and the minimum endotoxin limit (or proposed limit) for the drug product. Biological products require an approved license amendment before the LAL test can be substituted for the pyrogen test. Manufacturers of drugs not subject to premarket approval can use the LAL test without notifying the agency. However, it is recommended that the manufacturer have LAL validation data on file for each product for which the test is used.

Medical Devices

If a manufacturer of medical devices wishes to label its product as "nonpyrogenic," the LAL product must be validated for the particular device. If any regulated manufacturer wishes to use LAL test procedures that differ significantly from the guidelines, a premarket notification under section 510(k) of the Federal

Food, Drug, and Cosmetic Act or a Premarket Approval Application (PMA) must be submitted noting the deviations. Manufacturers who wish to reduce their volume of endproduct testing (i.e., not all lots tested), must demonstrate adequate control of endotoxin using routine testing at key stages of production. A premarket notification or PMA is also required.

FOREIGN REGULATIONS

European Pharmacopoeia

The European Pharmacopoeia (EP) covers LAL in Chapter V.2.1.9. Bacterial Endotoxins (1987). In general, the EP roughly follows the USP, but on the whole is not as detailed. Please note the EP differs from the USP as follows:

- The endotoxin standard used is the Biological Reference Preparation (BRP), which is calibrated in International Units through comparison with the International Standard.

- pH adjustment is tighter than the USP, being 6.5–7.5 in the EP.

- The incubation period can be flexible, from a minimum of 20 minutes to a maximum of 60 minutes.

Japanese Pharmacopoeia

The Japanese Pharmacopoeia (JP) (English Version) covers LAL in the section *General Tests*, section 7. Bacterial Endotoxins Test (1991). As with the EP, the JP generally follows the USP, and is about as detailed as the EP. Please note that the JP differs from the USP as follows:

- No particular endotoxin reference standard is specified in this chapter, although a standard containing 10,000 EUs/ml is recommended for use.

- pH adjustment is 6.0–7.5.

- The JP also covers LAL in the Reference Standards section under LAL. This section describes a method for determining the sensitivity of the LAL reagent and provides a statistical formula for assessing the accuracy of the assay.

Health Protection Branch—Canada

The Canadian Health Protection Branch's (HPB) requirements for using the LAL test roughly follow those of the FDA and reference the test protocol of the USP. In particular, the HPB requires manufacturers using the LAL procedure to have the following data (on three production batches of product) available:

- Results of inhibition and enhancement studies

- Results of studies demonstrating the sensitivity and reproducibility of the LAL procedure

- Results of the rabbit pyrogen test for drugs amenable to rabbit testing in order to confirm the absence of nonendotoxin pyrogens

Substitution of the rabbit test by the LAL test for drugs in schedules C and D to the Food and Drugs Act requires authorization by the Bureau of Biologics. Substitution of the rabbit test by the LAL test for drugs in New Drug Status requires the filing of a Supplemental New Drug Submission. Substitution of the rabbit test by the LAL test for other drugs does not require authorization; however, it is recommended that validation data be forwarded to the HPB for evaluation prior to changing tests.

DOMESTIC AND INTERNATIONAL STANDARDS

United States

The RSE of the United States is referred to as EC. EC was produced from *Escherichia coli* O113:H10:K(-) lipopolysaccharide. The LPS was purified from cells in the logarithmic phase of growth by the phenol-water purification method. Gram quantities of this LPS (EC) were produced by Dr. J. A. Rudbach under contract from the FDA (Rudbach et al. 1976). The original lot that defined the EU was EC-2. The current lot is EC-6 (although the USP refers to this as lot G). EC-6 was produced with the cooperation of the National Institute of Biological Standards and Control (NIBSC) in the United Kingdom, the United States Pharmacopeial Convention, and the FDA. EC-2 had a potency (by definition) of 5 EU/ng. In a collaborative study, the replacement standard EC-5 was found to have a potency of 10 EU/ng. Results with EC-6 indicate a potency similar to EC-5. EC-2 differed from EC-5 and EC-6 in its nonendotoxin constituents. EC-2 contained normal human serum albumin as a filler, while EC-5 and EC-6 contain lactose and polyethylene glycol (PEG) (9.9 mg and 1.0 mg respectively in each vial of EC-6). These fillers or additives are used to facilitate reconstitution, prevent aggregation, and possibly retard adsorption of the endotoxin by the container. It is possible that the difference in fillers is responsible for the difference in potency, but this has not been proven. The reconstituted stability of EC-5 was only 14 days, although data exists to indicate the shelf life is much longer. Although not recommended, this standard was often diluted and frozen to extend its reconstituted shelf life. Instructions for the new standard, EC-6, allow dispensing of one ml portions into glass vials with screw caps. These vials can then be stored at -20°C for several months. Unfortunately, in the instructions for use of EC-6, there is no indication of the reconstituted shelf life at 2°–8°C.

World Health Organization/National Institute of Biological Standards and Control

The World Health Organization (WHO) under the guidance of NIBSC produced an international reference standard (Poole and Mussett 1989). This standard consists of *Escherichia coli* O113:H10:K(-), from the same EC bulk material. The preparation consisted of 4000 ampoules containing 2 μg endotoxin plus 0.3 percent trehalose in lyophilized form. The vials were hermetically sealed after secondary desiccation and backfilling with dry nitrogen. Vials are stored at -20°C in the dark. The international standard was found to have a potency of 14,000 International Units (IU) (7 IU/ng), determined in a collaborative study with EC-5. For convenience, the WHO set the IU equal to the EU.

European Pharmacopoeia

The EP's standard is referred to as the BRP. The original BRP was known as lot NP-2, which was found to be contaminated with several species of bacteria (EP 1992). Once a vial of NP-2 was opened, the bacteria grew and affected the potency. A new lot of BRP was made, NP-3, and was found to be sterile and was introduced in 1992 as the 2nd European Reference Preparation. NP-3 was compared to the international standard and was found to contain 800 IU/ml. Unlike the other governmental or pharmacopoeial standards, the BRP is a liquid, packaged sterile in hermetically sealed glass ampoules. This form provides numerous advantages over the other standards, including the following:

- Liquid form does not require reconstitution, thus avoiding errors in choice of water, measurement, inadvertent contamination, and so on.

- Ampoules hermetically sealed eliminate moisture degradation problems encountered with freeze-dried standards.

- Liquid preparation allows production of unlimited numbers of vials with little variability (freeze-dried lots subject to size of lyophilizer, etc.).

In addition, the BRP is made from highly purified (phenol-water extracted and electrodialyzed) LPS from *Salmonella abortus equi*. Of all the standards available, this is the most well characterized. It is unfortunate other governments/ agencies have not adopted a technology similar to the one used to prepare the BRP, as it is clearly superior.

Japanese Pharmacopoeia

The JP has adopted an *Escherichia coli* UTK-B LPS as their standard (Kanoh 1982). This standard was compared to EC-5 in a Japanese collaborative study and was found to have a potency of 16,000 Japanese EU per vial. Like the WHO, the JP set their endotoxin units equal to the USP's EUs.

General Notes on Standards

Too many standards cause confusion. In equating IUs and other units of potency to EUs, it was intended that these other standards could serve as direct substitutes for the RSE and avoid user comparisons or EU/IU conversions. Unfortunately, collaborative studies assign "average" potencies to standards, making it difficult to compare results of one standard to another unless the confidence intervals are understood and employed. Users should note, therefore, that with particular lots of LAL it will not be unusual, in fact it is quite common, not to get label claim when using the international standard with an LAL whose potency was directly determined with EC-5 (i.e., IUs may not equal EUs). In an attempt to rectify this problem, the WHO, the USP, and the FDA in a rare collaborative effort, have jointly produced and have agreed to consider designating (pending the outcome of a collaborative study) the first truly international standard, EC-6. These groups are to be commended for attempting to simplify the interpretation of LAL results.

APPLICATIONS

Endproduct Testing

As described above, endproduct testing in lieu of the pyrogen test is allowed, provided the manufacturer's instructions (21 CFR), the FDA guidelines/guidance, and the USP (or applicable foreign regulations) are followed.

In-Process Control

Historically, the LAL test has always been an in-process control test. Although LAL use as an endproduct release test (as a substitute for the pyrogen test) has enjoyed more publicity than the more mundane in-process test, the latter has accounted for and still accounts for >99 percent of LAL tests performed by licensed pharmaceutical manufacturers.

Water

Water testing is LAL's easiest application (Novitsky 1987a, 77–96). Water (low endotoxin, of course) not only serves to reconstitute freeze-dried LAL, but is used as a negative control; as a diluent for samples and standards; and as an aid in removing endotoxin from filling lines, labware, filters, and so on. There is a circular argument here, however: How can one find water to serve as LRW, if the LAL test (which uses LRW as a reconstitution agent) is used to define LRW in the first place? This is one instance where technique and process control have played key roles. First of all, it is known from chemistry and physics that distillation is an effective means of eliminating organic impurities from water or other liquids. Second, endotoxins (pyrogens) and other organic materials can be destroyed by high heat (pyrolysis). Third, the hemolymph of the horseshoe crab is generally free of endotoxin, as long as the circulatory system is intact. Thus, starting with distilled water collected in depyrogenated vessels, it should be possible to collect amebocytes from intact horseshoe crabs and make LAL that has not been in contact with endotoxin. This was exactly what was done by Levin and Bang (1968) on their initial discovery of LAL and is done today by the commercial manufacturers of LAL. It is easy, therefore, to test water for endotoxin content by simply reconstituting a vial of LAL. Water so tested that does not clot or otherwise cause a reaction is called LRW. Since most LAL manufacturers provide LRW with LAL kits or as a separate item, the user can confirm the quality of their own water by comparison with the commercially available LRW. It should be noted that other commercially available water (e.g., sterile WFI), while designated as "non-pyrogenic," may not meet the standards set for LRW. This is because the pass/fail limit for WFI is only 0.25 EU/ml. Thus, batches of this water may actually contain LAL-detectable levels of endotoxin and would not be suitable for use as LRW.

Raw Materials

More and more manufacturers are now testing the raw materials used for parenteral formulations with the LAL test. The reasoning is simple—if the raw material is low in endotoxin and the manufacturing process (including water) is controlled, the likelihood of the final product containing a failing level of endotoxin is unlikely. Procedures used to test raw materials are similar to those used for final product testing. Since raw materials are often tested more concentrated than the final product, the inhibition/enhancement test becomes more critical. While final product allowable endotoxin levels are set by the USP or the FDA, endotoxin content of raw materials is the choice of user. Often acceptance criteria for the raw material is based on endotoxin content; thus, many raw material suppliers are now using the LAL test.

Clinical/Diagnostic

Sepsis

One of the original recommendations of Levin and Bang (1968) when they first described the LAL test was to use it as a diagnostic for endotoxemia and gram-negative bacterial sepsis. To date this use has seen only research studies. The FDA has yet to approve the assay as a clinical diagnostic for endotoxemia, although at least one manufacturer has completed formal clinical trials for an LAL-based diagnostic kit and another manufacturer sells one "for research only." The literature in this area is quite extensive and has recently been reviewed (Hurley 1995; Novitsky 1994; Pearson 1990, 52–62).

Bacteriuria

Bacteriuria is another area where much research has occurred, but no FDA–approved kit has appeared on the market. Gram-negative bacteria on the order of 10^5 CFU/ml can be detected easily in urine using the LAL assay. Although many studies have proven the accuracy of this test, other, less expensive and simple tests for bacteriuria have effectively prevented commercialization of an LAL–based diagnostic. LAL applied to bacteriuria has been extensively reviewed (Nachum 1990b, 81–96).

Spinal Meningitis

Once again, LAL has been studied extensively for use as a rapid diagnostic for gram-negative spinal meningitis. At least one LAL manufacturer was approved by the FDA for this application, but the assay was never commercialized. One drawback to the assay for this application is the lack of specificity for the type of endotoxin detected (e.g., *Neisseria* vs. *Haemophilus*). The test is rapid, however, and has been used routinely by some hospitals, especially in children with suspected meningitis. This literature has also been extensively reviewed (Nachum 1990a, 67–80).

Renal Dialysis

Since renal dialysis employs a lot of water in the preparation and processing of artificial kidneys, the potential for endotoxin contamination is well recognized. Currently, only the water used to dilute sterilant has an LAL test requirement (42 CFR Part 405, *Federal Register*, October 2, 1987, pp. 36926–36935; and AAMI *Recommended Practice*, July 1986). These include recommendations that the water used to dilute the sterilant have less than 1 nanogram (ng)/ml of endotoxin. As this industry evolves and new dialyzers and procedures are introduced, it is likely that tests on the dialysate and on reusable devices will become necessary. LAL use in this industry has been reviewed (Gould 1988).

Other

Any disease caused by gram-negative bacteria or any substance colonized or infected by gram-negative bacteria has the potential for an LAL application. Since endotoxins are often shed by growing bacteria and are left behind after gram-negative bacteria die, they are often present in amounts more amenable to detection than other methods for detecting bacterial presence. It should be recognized, however, that correlation of endotoxin content with bacterial numbers may be purely coincidental and that endotoxin presence cannot be equated with certainty to the presence of viable bacteria.

Food

As with the clinical applications, LAL can be used as a surrogate for detecting the presence of gram-negative bacteria in food. Most studies have been done with milk and ground beef, although the quality of certain fish has also been ascertained from an LAL test. It has been found that the keeping quality of ultra-high temperature (UHT) processed milk is directly related to its endotoxin content (i.e., its historical bacterial burden). In fact, several countries in Europe and Australia now have laws or recommendations for testing UHT milk (or milk intended for UHT preservation) with LAL. Other food applications have only been topics of research (Mikolajcik and Brucker 1983; Sullivan et al. 1983).

Semiconductor Process Water

Since the LAL test is one of the most sensitive bioassays available, it has been applied to the monitoring of water used to manufacture semiconductors. Studies have shown a correlation to organic carbon and bacterial content of semiconductor process water (Gould et al. 1991). Due to the relatively slow assay time (no online LAL test yet exists), only a few semiconductor manufacturers are actively using this assay.

Environmental

One of the first nonclinical applications of the LAL test was as an indirect assessment of the gram-negative bacterial number in the open ocean. This test confirmed the presence of relatively large numbers of bacteria (10^4 cells/ml) in the deep ocean (4000 m). It was also found that the LAL test correlated well with bacterial biomass (as carbon) and could be used in lieu of tedious direct counting procedures (Watson et al. 1977). The test has also been used in fresh water (Novitsky 1987a, 77–96).

Another application of the LAL test has been for air quality. It has been shown that certain kinds of dust are laden with endotoxin. The endotoxin component of the dust has been implicated in a variety of occupational lung diseases. Methods have been worked out to collect dust and extract the endotoxin for quantitation by the LAL assay (Reynolds and Milton 1983).

VALIDATION OF DEPYROGENATION

Next to the testing of water, validation and monitoring of depyrogenation is one of the most frequent applications of the LAL test. Validation of depyrogenation is quite straightforward (Dawson 1993; Novitsky 1988). Once a heating cycle is chosen and the hot and cold spots determined (by thermocouple placement), endotoxin is added to a typical load configuration and the cycle repeated. In a typical validation at least 1000 EU is added per container, replicates of which are placed at the thermocouple locations. Similar containers are retained as controls and the others are run through the cycle. Endotoxin recovery is effected by adding LRW to the containers and testing the resulting solution with LAL. If, after testing, the added endotoxin has been reduced by at least three logs, the cycle is considered validated.

Heat labile items can also be tested in a similar manner, the depyrogenation procedure of choice being a NaOH solution and/or water rinse with or without elevated temperature.

The most common problem associated with depyrogenation validation is complete recovery of the control endotoxin. Since endotoxin is readily adsorbed to container surfaces, it is difficult if not impossible to recover 100 percent of the control. Unfortunately, substances added to some endotoxin standards, which would otherwise help to recover the endotoxin, adversely influence the kinetics of endotoxin destruction and therefore cannot be used. It is possible to increase recovery by adding high concentrations of endotoxin in very small volumes, freeze-drying the added endotoxin rather than air drying, and using vigorous mixing or sonication to help recovery. It is not necessary to recover 100 percent of the added endotoxin, however, only to show that a three-log reduction has occurred. Thus, if 1000 EU's of endotoxin were added to a container and only 500 EUs were recovered in the control, then for a valid test less than 0.5 EU must remain after depyrogenation.

POTENTIAL NEW APPLICATIONS OF TECHNOLOGY

While not exactly new, several aspects of the LAL test continue to evolve and improve. These include the potential for automation exemplified by the LAL 5000 Automated Endotoxin Detection System and the various microplate readers and accompanying software. With the advent of antiendotoxin therapy, the demand for a clinical test for endotoxin will grow. An LAL–based test, currently the most sensitive bioassay for endotoxin, will most likely be the first licensed endotoxin diagnostic. The same substances (i.e., monoclonal antibodies and endotoxin binding proteins), may also lend themselves to LAL alternative assays or even assays for gram-negative bacteria which possess endotoxin on their surfaces. It remains to be seen whether the much desired "in-line" endotoxin assay can be developed. Perhaps a binding protein, or even the enzyme(s) from LAL itself, can be utilized in a "biosensor" device to realize this goal.

REFERENCES

Dawson, M. E. 1993. Endotoxin standards and CSE potency. *LAL Update* 11 (4):2–5.

European Pharmacopoeia Commission. 1992. 2nd endotoxin reference preparation. *Pharmeuropa* 4 (3):162.

EP. 1987. Bacterial Endotoxins. In *European Pharmacopoeia*, V.2.1.9. European Pharmacopoeia Commission.

FDA. *Guideline on validation of the Limulus amebocyte lysate test as an end-product endotoxin test for human and animal parenteral drugs, biological products, and medical devices*. Rockville, MD: U.S. Department of Health and Human Services, Public Health Service.

Gould, M. J. 1988. Performing the LAL gel-clot test in facilities. *Nephrol. News Issues* (Nov):26–29.

Gould, M. J., M. E. Dawson, and T. J. Novitsky. 1991. A review of the particulate nature of lipopolysaccharide and its measurement by the *Limulus* amebocyte lysate test. *Part. Sci. Technol.* 9:55–59.

Hochstein, H. D., D. F. Mills, A. S. Outschoorn, and S. C. Rastogi. 1983. The processing and collaborative assay of a reference endotoxin. *J. Biol. Stand.* 11:251–260.

Hurley, J. C. 1995. Endotoxemia: Methods of detection and clinical correlates. *Clin. Microbiol. Rev.* 8:268–292.

Iwanaga, S., T. Miyata, F. Tokunaga, and T. Muta. 1992. Molecular mechanism of hemolymph clotting system in *Limulus*. *Thromb. Res.* 68:1–32.

Kanoh, S., Y. Ogawa, T. Komura, and H. Kawasaki. 1982. Preparation and properties of a reference endotoxin (LPS) for Limulus and pyrogen tests. *Jpn. J. Med. Sci. Biol.* 35:112–113.

Levin, J., and F. B. Bang. 1968. Clottable protein in Limulus: its localization and kinetics of its coagulation by endotoxin. *Thromb. Diath. Haemorrh.* 19:186–197.

Mikolajcik, E. M., and R. B. Brucker. 1983. Limulus amebocyte lysate assay—A rapid test for the assessment of raw and pasteurized milk quality. *Dairy Food Sanit.* 3:129–131.

Nachum, R. 1990a. Detection of gram-negative meningitis. In *Clinical applications of the limulus amoebocyte lysate test*, edited by R. B. Prior. Boston: CRC Press.

Nachum, R. 1990b. Detection of gram-negative bacteriuria. In *Clinical applications of the limulus amoebocyte lysate test*, edited by R. B. Prior. Boston: CRC Press.

Nakamura, S., T. Morita, S. Iwanaga, M. Niwa, and K. Takahashi. 1977. A sensitive substrate for the clotting enzyme in horseshoe crab hemocytes. *J. Biochem.* 81:1567–1569.

Novitsky, T. J. 1983. Choosing a method. *LAL Update* 1 (1):1–4.

Novitsky, T. J. 1986. Methodology update. *LAL Update* 4 (4):1–2.

Novitsky, T. J. 1986. Validation controls for depyrogenation. *LAL Update* (March):1–3.

Novitsky, T. 1987a. Bacterial endotoxins (pyrogens) in purified waters. In *Biological fouling of industrial water systems: A problem solving approach*, edited by M. W. Mittelman, and G. G. Geesey. San Diego: Water Micro Associates.

Novitsky, T. J. 1987b. Methodology update (continued). *LAL Update* 5 (1):1–2.

Novitsky, T. J. 1988. Validation of dry heat depyrogenation in a small forced air oven. *LAL Update* 6 (2):1–3.

Novitsky, T. J., J. Schmidt-Gengenbach, and J. F. Remillard. 1986. Factors affecting the recovery of endotoxin adsorbed to container surfaces. *J. Parent. Sci. Technol.* 40:284–286.

Novitsky, T. J. 1991. Discovery to commercialization: The blood of the horseshoe crab. *Oceanus* 27:13–18.

Novitsky, T. J. 1993. Comparison of various LAL methods. *LAL Update* 11 (1):2–4.

Novitsky, T. J. 1994. Limulus amebocyte lysate (LAL) detection of endotoxin in human blood. *J. Endotoxin Res.* 1:253–263.

Pearson, F. C. 1990. Detection of endotoxemia. In *Clinical applications of the limulus amoebocyte lysate test*, edited by R. B. Prior. Boston: CRC Press.

Poole, S., and M. V. Mussett. 1989. The international standard for endotoxin: Evaluation in an international collaborative study. *J. Biol. Stand.* 17:161–171.

Reynolds, S., and D. K. Milton. 1993. Comparison of methods for analysis of airborne endotoxin. *Appl. Occup. Environ. Hyg.* 8 (9):761–767.

Roslansky, P. F., and T. J. Novitsky. 1991. Sensitivity of *Limulus* amebocyte lysate (LAL) to LAL–reactive glucans. *J. Clin. Microbiol.* 29:2477–2483.

Rudbach, J. A., F. I. Akiya, R. J. Elin, H. D. Hochstein, K. R. Thomas, M. K. Luoma, and E. C. B. Milner. 1976. Preparation and properties of a national reference endotoxin. *J. Clin. Microbiol.* 3:21–25.

Seibert, F. B. 1923. Fever-producing substance found in some distilled waters. *Am. J. Physiol.* 64:90–104.

Sullivan, Jr. J. D., P. C. Ellis, R. G. Lee, W. S. Combs, Jr., and S. W. Watson. 1983. Comparison of the *Limulus* amoebocyte lysate test with plate counts and chemical analyses for assessment of the quality of lean fish. *Appl. Environ. Microbiol.* 45:720–722.

The Society of Japanese Pharmacopoeia. 1991. *JP XII The Pharmacopoeia of Japan, Twelfth Edition, English Version.* Tokyo: Yakuji Nippo, Ltd.

USP. 1993 Bacterial endotoxins test <85>, *United States Pharmacopeia, XXII,* Eighth Supplement. Rockville, MD: United States Pharmacopeial Convention, Inc.

Watson, S. W., T. J. Novitsky, H. L. Quinby, and F. W. Valois. 1977. Determination of bacterial number and biomass in the marine environment. *Appl. Environ. Microbiol.* 33:940–946.

Westphal, O., U. Westphal, and T. Sommer. 1977. The history of pyrogen research. In *Microbiology,* edited by D. Schlessinger. Washington, DC: American Society for Microbiology.

13

Microbial Quantitation by Gas Chromatography

Wayne P. Olson

Oldevco

Axenic cultures of bacteria are required for the production of viable vaccines (e.g., *Mycobacterium bovis* strain BCG, Rosenthal 1937); for the production of parenterally administered therapeutic proteins from recombinant organisms (e.g., insulin produced in *Escherichia coli*; Goeddel et al. 1979); in the production of rapid slide agglutinins (e.g., *Bacillus anthracis* and *Salmonella typhimurium*); and myriad industrial applications (Marshall 1986; Volk and Wheeler 1988). The vaccine strains are maintained as seed lots from which a working seed lot is prepared for a given year. Any given ampoule is subcultured a small number of times, or only once, before a fresh ampoule of the working seed lot is opened for use.

One wants to know the approximate quantity of cells in an inoculum and at the end of a fermentation, and purity of the cultures at the seed lot, working seed lot, and production fermentation levels. Gas chromatography (GC) is almost a real-time method for estimating inoculum concentration (Olson et al. 1990) and for detecting microbial contaminants. Alternatives to GC include light scattering, particle analysis and related methods, standard plate counts (Greenburg et al. 1981), HPLC of extracted phospholipids (Henson et al. 1985), quantitation of muramic acid, and ATP quantitation. Of these, only standard plate counts provide for the quantitation of organisms *and* the detection of microbial contaminants, but the information is available a day or more after the inoculum has been used.

GAS CHROMATOGRAPHY AS A RAPID METHOD OF CHOICE

The identification of bacteria from fatty acid profiles has been documented profusely (Abel et al. 1963; Drucker 1976; Lechevalier 1977; Moss 1981; Welch 1991). The method is best based on dendrograms that separate species first on the presence or absence of certain fatty acids, then on fatty acid ratios. The fatty acid profile of *Mycobacterium bovis* BCG substrain Tice™, developed by GC of the methyl esters, is shown in Figure 13.1 (from Olson 1992). The typical profile of any bacterium shows a relatively small number of fatty acids (5 to 9 is common). Many strains or varieties of bacterial species can be identified by this method.

Milligram quantities of inoculum are required for the GC method presented at the end of this chapter. For inoculum resuspended from a working seed lot or the suspension in a fermenter, mg amounts are readily available and sample preparation can be completed in a maximum of 3 or 4 hours. In contrast, a streak plate requires a minimum of 18 to 24 hours incubation time. If a contaminant is a slow grower (usually the case if the host for a particular gene is *Escherichia coli*), then it may not be visible on a streak plate for 48 to 72 hours and *then* may be overgrown with the host. So streak plates, although they require a minimum of cellular material, have some drawbacks in speed and sensitivity.

MEMBRANE METHODS

The only culture method that seems likely to be competitive with GC for quantitating cells and detecting/quantitating contaminants is the membrane filter variation of the plate count method used widely in the quantitation of environmental microbes (see items 907 and 909A in Greenberg et al. 1981). One does logarithmic

Figure 13.1. Fatty acid methyl esters (FAMES) for *Mycobacterium bovis* BCG, substrain Tice™, grown as a pellicle on Sauton medium, from Olson (1992).

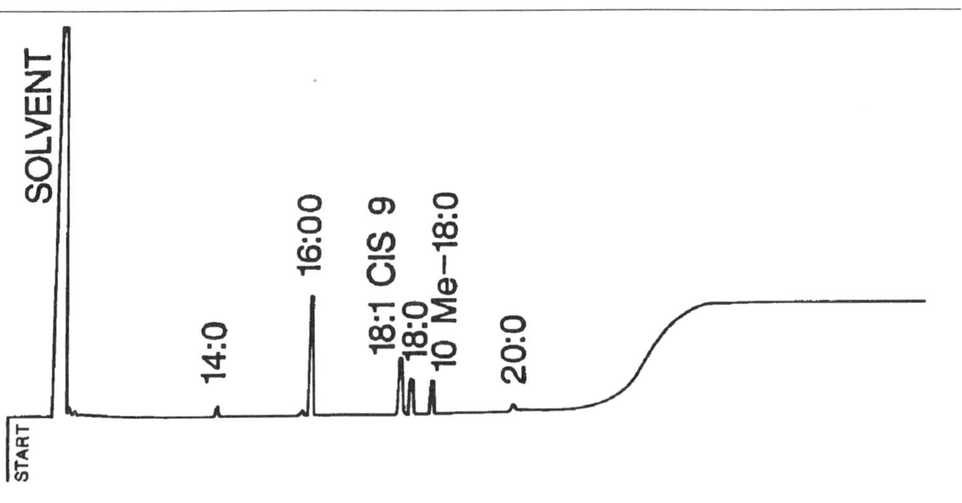

dilutions of 1 ml of cell mass in sterile saline (preferably phosphate-buffered saline) and passes a fixed volume (usually between 10 and 100 ml) of the dilutions through 0.45 micrometer-pore size sterile membrane filters (usually 47 or 50 mm diameter and gridded [e.g., HAWG 047 from Millipore]), then plates the filter onto an appropriate solid medium (e.g., nutrient agar in a Petri dish). This is preferable to plating out 1 ml of the dilutions on solid medium because

- The cells usually are well dispersed on a membrane filter.

- The membrane method is at least 20–50 percent more sensitive than plating on solid medium (W. Olson; Frederick Dawson—unpublished).

- Spreaders are less of a problem on membranes.

The major advantage of the plate count and membrane methods is that viables are counted. However, the investment in time and effort is significantly greater than for the GC technique.

BASIS OF MICROBIAL QUANTITATION BY GAS CHROMATOGRAPHY

The amount of each of the fatty acids recovered in a microbial sample is proportional to the cell mass, viable and nonviable. For purposes of quantitation, one selects a representative fatty acid and determines the area under the curve (AUC) for that fatty acid as a function of cell mass. This has been shown for the $C_{16:0}$ of BCG (Figure 13.2, from Olson et al. 1990) and the $C_{15:0\ iso}$ of *Bacillus cereus* (Olson 1992). One of the characteristic fatty acids for *Bacillus subtilis* is the $C_{15:0\ anteiso}$ (Figure 13.3, from Olson 1992) and the AUC for $C_{15:0\ anteiso}$ is a function of the cell mass and viable cell number under a given set of conditions (Figure 13.4). The function for Figure 13.4 is

$$Y = 1.13 \times 10^4 + 1.7102 \times 10^3 X$$

where Y = area under the 15:0 anteiso curve and X = *Bacillus subtilis* wet weight in mg. Similarly,

$$Y = 0.987 \times 10^3 + 6.085 \times 10^{-5} X'$$

where X' = *Bacillus subtilis* colony-forming units $\times 10^9$. For $n = 10$, $r^2 = 0.9656$. A second-order equation for AUC as a function of wet weight has an insignificantly better fit ($r^2 = 0.9674$).

A Validation of the GC Method

Groves et al. (1991) estimated numbers of BCG cells in vaccine lots by means of a Coulter Multi-Sizer™ and the $C_{16:0}$ AUC data for the same BCG lots were determined by Olson (1992). The BCG cell number in batches of Tice™ BCG vaccine, as a function of $C_{16:0}$ AUC for the same lots, is shown in Figure 13.5. For this function, $r^2 > 0.999$, indicating the equivalence of the methods and that the accuracy of the GC *and* the Coulter methods approaches unity.

Figure 13.2. Area under the curve (AUC) for the $C_{16:0}$ FAME of BCG grown on Sauton medium as a function of cell mass in mg, from Olson et al. (1990).

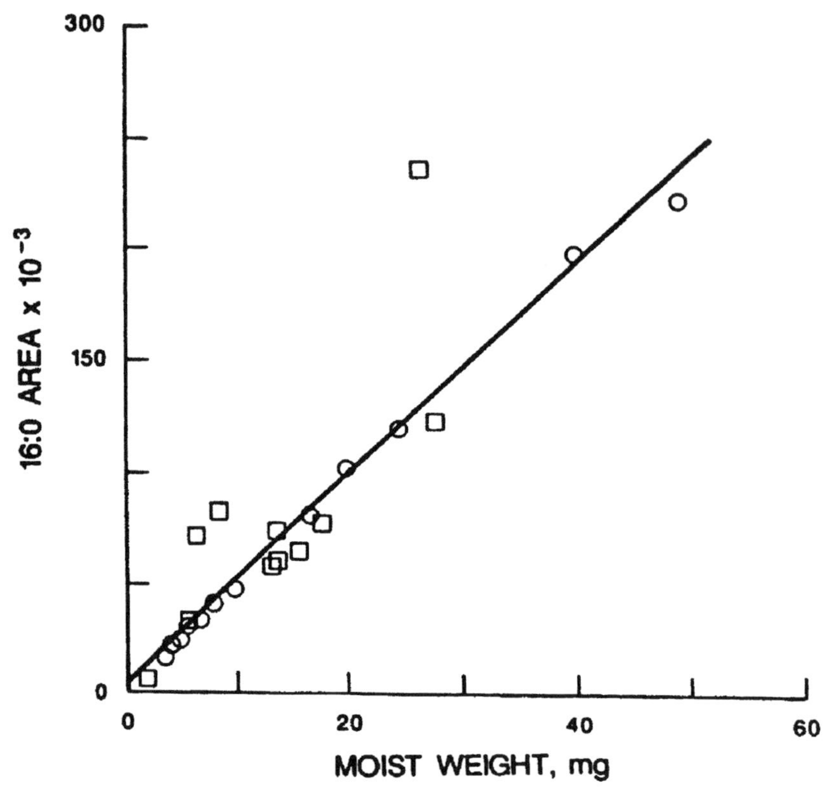

Potential Errors

Errors of the GC approach to microbial quantitation are as follows:

1. Cell mass or cell number estimates are based on viable plus nonviable cells.

2. Fatty acid composition of the cells can vary with cultural conditions.

Therefore, for the GC quantitation of an organism recovered directly from the inoculum or the fermenter, the standard curve must be prepared from the same medium and inoculum and *not* from the organism grown under markedly different laboratory bench conditions (e.g., on nutrient agar or in nutrient broth at 22°C). For example, it would be unwise to apply the functions of Figure 13.4 directly to the AUC for *Bacillus subtilis* grown and harvested under conditions different from ours.

Figure 13.3. Fatty acid methyl ester profile for *Bacillus subtilis*, from Olson (1992).

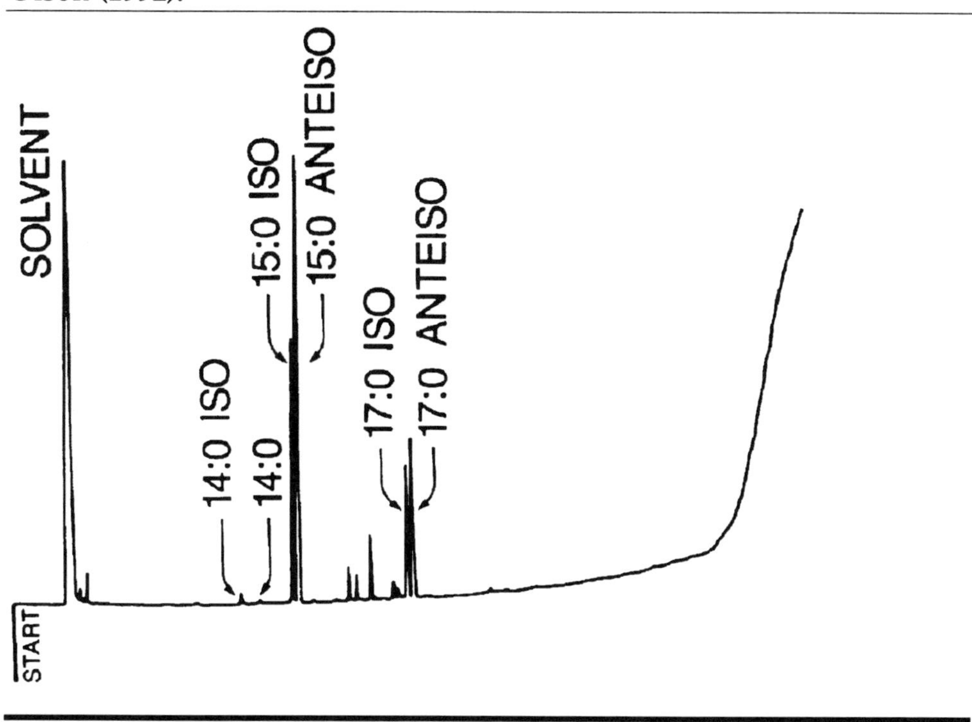

Microbial Contaminants in Axenic Cultures

Most recombinant peptides are produced in *Escherichia coli*, which has a very short doubling time. In practice, a low level of microbial contamination is tolerable, provided products of high toxicity or extremely high turnover proteases are not released or secreted by the contaminating microbe(s).

We estimate that the membrane variation on plate count methods may detect one contaminant in 300 *Escherichia coli*, and that the GC method detects (as contaminating fatty acid peaks), on average, one contaminant in 100–200 organisms, depending on the organisms (unpublished results). For a fast-growing bacterium like *Escherichia coli*, that level of contamination probably is trivial since, by the end of the fermentation, the *Escherichia coli*:contaminant ratio likely will be of the order of 10^5:1 or more. In such an instance, the detection of a microbial contaminant by any method is unlikely.

Where the organism of interest and the contaminant do *not* grow at very disparate rates, either the standard plate count/membrane method *or* the GC method should be acceptable for the quantitation of the desired organism and for the detection of microbial contaminants. An example of a *Bacillus subtilis* contaminant in BCG is presented in Figure 13.6.

Figure 13.4. AUC for the $C_{15:0\ anteiso}$ FAME of *Bacillus subtilis* as a function of cell mass and colony-forming units.

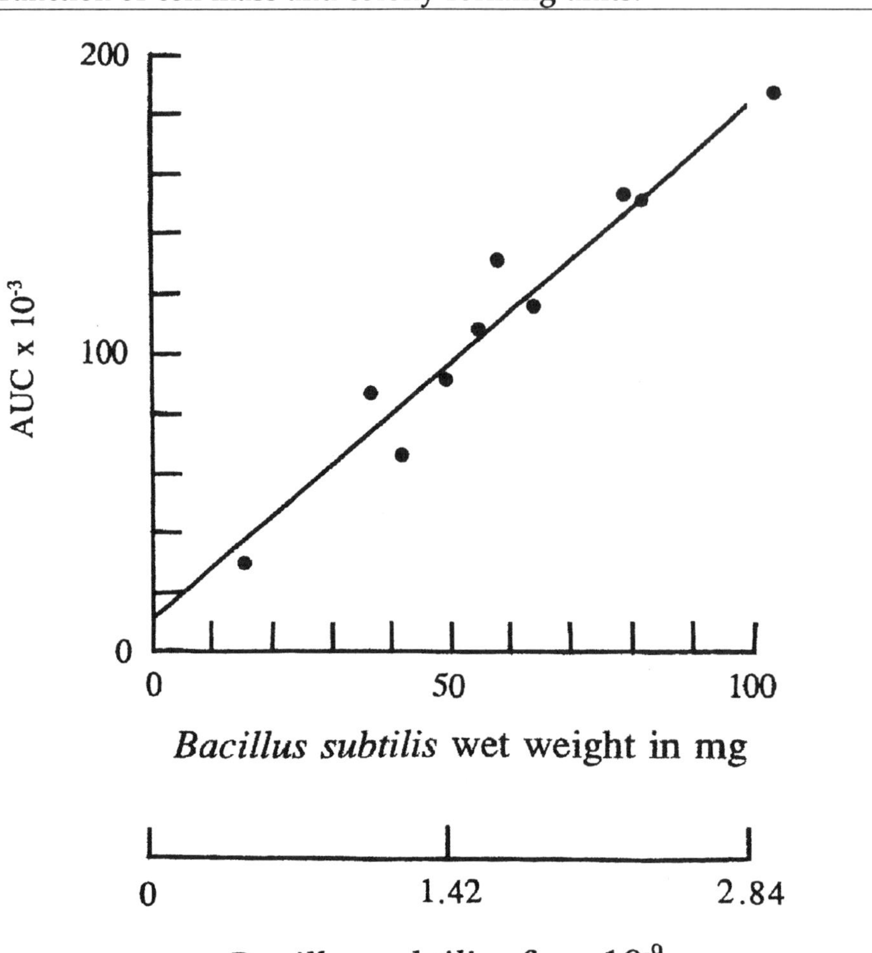

Sample Preparation for Bacterial Mass/Numbers Estimation by GC

Methods are as follows:

A cell concentrate is reconstituted in, for example, 1 ml saline or phosphate-buffered saline (PBS; the aqueous makeup solution need not be sterile), vortexed, and 5–50 μl of the suspension transferred in an Eppendorf or other suitable micropipette into a 13 × 100 mm Pyrex® screwtop culture tube containing 1 ml of 15 g percent NaOH in 50 percent aqueous methanol and a boiling bead (saponification step). The screwtop should have a Teflon® lining and the sample is heated for 2 hours at 100°C. At the completion of saponification, the tube is air

Figure 13.5. Cell number of various lots of BCG estimated by the Coulter method, as a function of AUC for the $C_{16:0}$ FAME of the same BCG lots, from Olson et al. (1990).

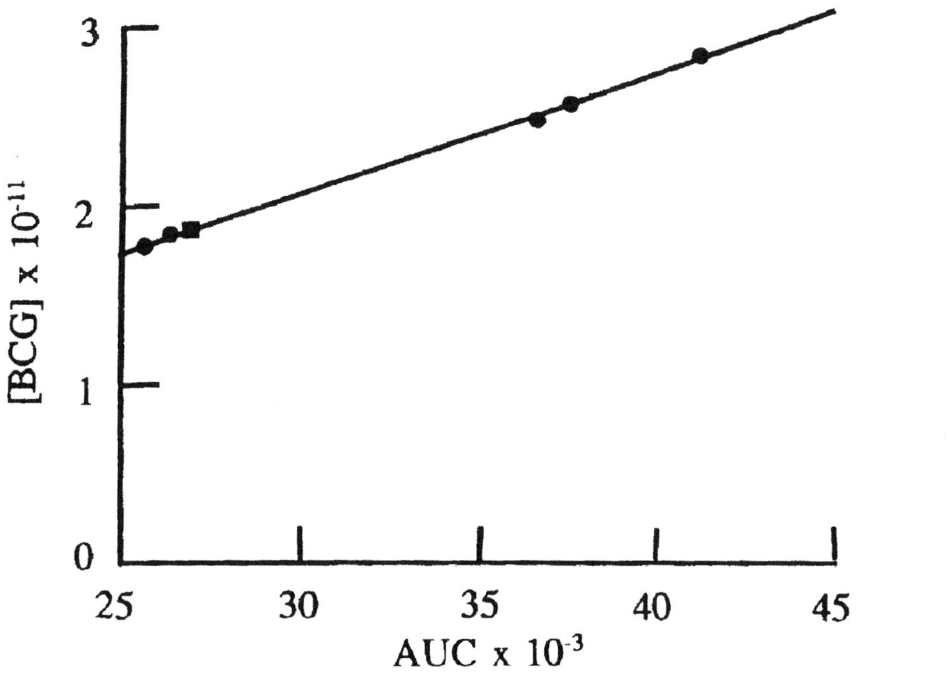

cooled and 4 ml of 3.25 N HCl in 45.8 percent aqueous methanol (methylation reagent) is added, vortexed, and the mixture incubated at 80°C for 1 hour. The tube is air or water cooled and 1.25 ml of *n*-hexane/methyl-*t*-butyl ether (1:1, v/v) is added and end-over-end rotated for 10 minutes. The *bottom* phase is removed with a Pasteur pipette and 3 ml of 1.2 g percent NaOH is added and end-over-end mixed for 5 minutes. The top phase is removed and placed in a GC vial for analysis. The solvent phase need not be recovered quantitatively since a GC with an autoinjector takes a finite sample of the solvent.

Even with a hard-to-extract cell mass like BCG, the precision of the estimation of the AUC, using the saponification, methylation, and extraction procedure given immediately above, is about ±3 percent. Using the methods in Moss (1981) with BCG, the precision is about ±15 percent (Olson 1992).

Automation

The MIDI MIS (Microbial Identification System), marketed by MIDI Inc. (Newark, DE) is an automated GC system that can provide and record AUC data, and

Figure 13.6. Fatty acid methyl ester profile for *Bacillus subtilis* seeded into BCG. Peaks 3, 4, 6, and 7 are from *Bacillus subtilis* and peaks 5, 8–11 are from BCG. Peak 1 is solvent and peak 2 is from the growth medium.

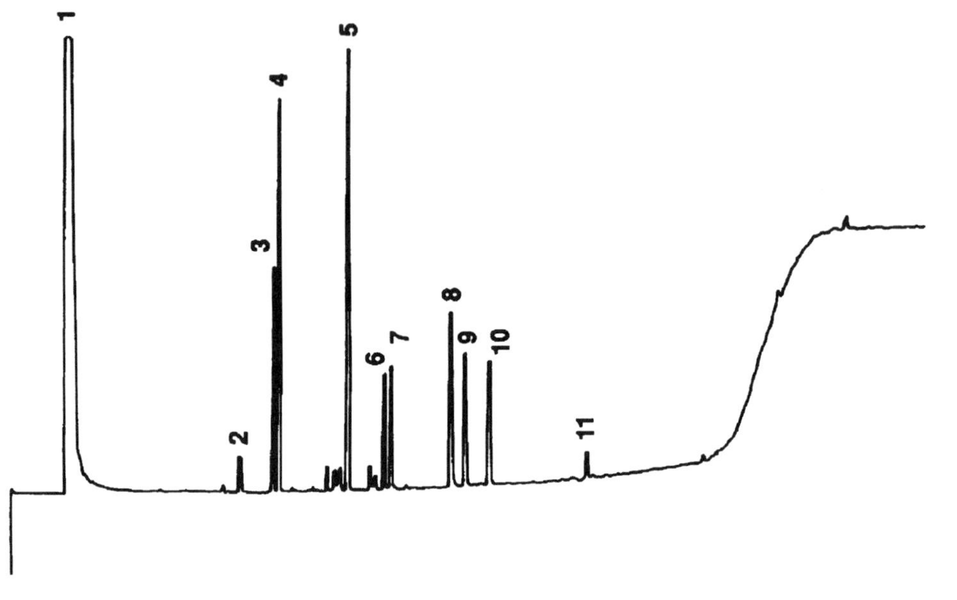

concentration data if desired. Microbial fatty acids can be extracted and esterified by a robotic system (MIRA), which also is marketed by MIDI.

ACKNOWLEDGMENT

The data for Figures 13.4 and 13.6 were developed by the author at the Institute for Tuberculosis Research (Univ. of Illinois at Chicago), Dr. Michael J. Groves and Dr. Melvin E. Klegerman.

REFERENCES

Abel, K., H. deSchmertzing, and J. J. Peterson. 1963. Classification of microorganisms by analysis of chemical composition: I. Feasibility of using gas chromatography. *J. Bacteriol.* 85:1039–1044.

Drucker, D. B. 1976. Gas-liquid chromatographic chemotaxonomy. *Meth. Microbiol.* 9:51–125.

Goeddel, D. V., D. G. Kleid, F. Bolivar, H. L. Heyneker, D. G. Yansura, R. Crea, T. Hirose, A. Kraszewski, K. Itakura, and A. Riggs. 1979. Expression in

Escherichia coli of chemically synthesized genes for human insulin. *Proc. Natl Acad. Sci. USA* 76:106–110.

Greenberg, A. E., A. A. Connors, D. Jenkins, and M. A. H. Franson. 1981. *Standard methods for the examination of water and wastewater,* 15th ed. Washington, DC: American Public Health Association.

Groves, M. J., M. E. Klegerman, P. O. Devadoss, O. N. Singh, and Y.-Y. Zong. 1991. Characterization of BCG vaccine by Coulter counter and scanning electron microscopic techniques. *Part. Part. Syst. Charact.* 8:194–199.

Henson, J. M., P. H. Smith, and D. C. White. 1985. Examination of thermophilic methane-producing digesters by analysis of bacterial lipids. *Appl. Environ. Microbiol.* 50:1428–1433.

Lechevalier, M. P. 1977. Lipids in bacterial taxonomy—a taxonomist's view. *CRC Crit. Rev. Microbiol.* 5:109–210.

Marshall, V. M. E. 1986. The microflora and production of fermented milks. In *Progress in industrial microbiology,* Vol. 23, *Microorganisms in the production of food,* edited by M. R. Adams. Oxford: Elsevier.

Moss, C. W. 1981. Gas-liquid chromatography as an analytical tool in microbiology. *J. Chromatog.* 293:337–347.

Olson, W. P. 1992. Targeted drug delivery based on the mode of action of Bacille Calmette-Guerin in bladder cancer. Ph.D. diss., University of Illinois at Chicago.

Olson, W. P., M. J. Groves, and M. E. Klegerman. 1990. Cell mass of *Mycobacterium bovis* BCG estimated by gas chromatography. *Biologicals* 18:83–88.

Rosenthal, S. R. 1937. Studies with BCG. I. The present method of BCG cultivation and vaccine production as practised at the Pasteur Institute and the Tice laboratories. *Am. Rev. Tuberc.* 35:678–684.

Volk, W. A., and M. F. Wheeler. 1988. *Basic microbiology,* 6th ed. New York: Harper & Row.

Welch, D. F. 1991. Applications of cellular fatty acid analysis. *Clin. Microbiol. Rev.* 4:422–438.

14

Quantitative Analysis of DNA in Protein Products and Fermentation Media

R. J. Cook

O. E. Khorkova

Regeneron Pharmaceuticals, Inc.

Recombinant proteins for pharmaceutical use are produced in increasing quantities every year. Microbial or cell line DNA copurifies with the target protein in the majority of industrial processes. The presence of relatively large amounts of residual cellular or contaminating microbial DNA (more than 100 pg per dose of recombinant protein, as stated in "Points to Consider . . ." (FDA 1993), particularly in nonorally administered agents, raises concerns about potential oncogenic or transforming activity of this DNA. Results of experiments that introduced large amounts of viral DNA into animals led a World Health Organization (WHO) study group to conclude that viral DNA in 1–10 µg doses is tumorigenic (Twomey and Krawetz 1990). It is unclear whether these results are relevant to recombinant proteins that contain trace amounts of DNA mostly of bacterial origin (Briggs and Panfili 1991), but since there is a possibility that these trace amounts could put the recipient at risk, safety issues must be based on very conservative theoretical assumptions. Therefore, ensuring that final products contain minimal levels of residual DNA is an important task in manufacturing biopharmaceuticals to the current cGMP standards.

DNA testing of recombinant proteins, testing for bacteriophages in fermentation media, and plasmid stability assays represent key elements of good manufacturing practices and regulatory evaluation (FDA 1993; 21 CFR 1992). The level of residual DNA must be determined in order to demonstrate control over the manufacturing process and comply with cGMP. The absence of adventitious agents (e.g., bacteriophages) is of concern because of obvious health and safety issues. Furthermore, upstream analysis of fermentation samples for bacteriophages is a cost-effective measure, because it allows one to abort the process at an early stage, avoiding purification and filling expenses. Plasmid stability assays ascertain the overall stability and consistency of the bacterial strain, as well as demonstrate cost-effective analysis, cGMP compliance, and good science.

The FDA policy regarding testing for residual DNA and other impurities and contaminants is expressed not in codified laws but in a series of opinions titled "Points to Consider . . ." (FDA 1987a,b). These opinions represent the current consensus at the Center of Biologics Evaluation and Review (CBER); they are periodically updated with input from industry, academia, and other regulatory agencies throughout the world.

THEORETICAL ASPECTS OF DNA ASSAYS

The process of manufacturing proteins, including the development of adequate purification processes and analytical methods for process validation and quality control, require fast, reliable, and sensitive quantitative assays for DNA impurities (FDA 1987b). Copurifying residual DNA is of primary concern because of regulatory requirements, potential health risks, and process control.

The required assay should be able to detect all kinds of DNA, irrespective of origin. The threshold assay (Briggs and Panfili 1991) solved this problem by using anti-DNA antibodies with very low sequence specificity; in hybridization-based assays, probe DNA is extracted from the cells in the fermentation media and theoretically represents all sequences that might be present in the resulting protein product. Both approaches have their limits: Antibodies have some sequence specificity and hybridization assays will not detect DNA from contaminating microorganisms. Hybridization assays should be complemented by sterility and bacteriophage testing, which should be able to detect contaminating microorganisms.

In testing for potentially oncogenic DNA the FDA is taking the most conservative position to minimize health risks (FDA 1987a, 1987b, 1993). Clearly, the analytical goal should be even more stringent than the actual standard for purity. In terms of assays, the FDA promotes analytical goals that require cutting-edge technology. Also, in most processes amounts of less than 10 pg DNA per dose of protein should be achievable. The required limit of detection is at picogram levels of DNA.

Testing the DNA content of the in-process samples is important because the ability of the industrial process to consistently remove specific contaminants, such as DNA, can be utilized as a demonstration of process stability, reproducibility, and the manufacturer's control over the process. DNA assays are also important in the course of process development and validation. Testing is necessary for both in-process samples and final product, as well as in the course of process development and validation. Because replicate runs are done for each sample, the sample load is significant. The assay should be automated. The threshold assay is automated and to a great extent has addressed problematic issues involving replicate analysis. Historically hybridization assays have been very lengthy and cumbersome. In this chapter we describe several modifications that helped us to make hybridization assays faster, more reproducible, and easier to perform (use of a slot-blot apparatus, heat deproteinization, and simplified hybridization buffer, which allows for shorter hybridization times, etc.).

A quantitative assay is always preferred over a semiquantitative or qualitative assay, because it eliminates subjectivity in estimating assay results, allows quantitative determination of accuracy, precision, and limits of detection; and quantitation, range, and linearity of the assay. A quantitative assay sets reliable standards for DNA testing, which leads to fine control over the process, and makes the assay validation procedure straightforward. Quantitative DNA assays allow trend analysis by obtaining statistically valid information. With the threshold assay, quantitative data are obtained by measuring pH changes caused by an antibody-conjugated enzyme (Briggs and Zuk 1988; Kung et al. 1990). With hybridization assays the level of radioactivity can be quantitated by densitometry of an autoradiogram, by direct counting of radioactivity, or by phosphorimaging, as described later in the chapter. Quantitative DNA assays are very useful in the course of purification process development and validation (21 CFR 1992).

DNA assays should be rugged enough to handle different types of samples. Purified products contain a significant excess of protein that causes interference. In-process samples may contain high concentrations of salts, urea, or other substances that interfere with the assay. DNA assays that are not rugged are time consuming because of time spent repeating assays that did not work. This repetition indicates that the assay is not reproducible or reliable.

The existing methods for detecting picogram amounts of DNA, which we will discuss here, are dot- and slot-blot hybridization using radioactive and digoxigenated probes, PCR, and the Threshold Total DNA System (Molecular Devices Corp., Menlo Park, CA).

APPLICATIONS

Polymerase Chain Reaction

Although the polymerase chain reaction (PCR) is the most sensitive technique, it is sequence-specific and not suitable for the detection of contaminating DNA. Using random primers with PCR is less specific, but is still questionable for determining the total contaminating DNA content (Pepin et al. 1990).

Threshold System

The Threshold System for determining total DNA content is fully automated. As supplied by Molecular Devices Corporation (Menlo Park, CA), it includes an instrument, computer, reagents, and disposables for running assays.

The system is based on two single-stranded DNA-binding proteins: biotin-conjugated ssDNA-binding protein from *Escherichia coli* (protein 1) and urease-conjugated monoclonal antibody against ssDNA (protein 2). Both proteins have very low DNA sequence specificity.

The test sample is heat denatured and incubated with both proteins and streptavidin for 1 hr at 37°C. The DNA-protein complex is then filtered through a biotinylated membrane, to which it becomes bound by the biotin-streptavidin linkage. Each capture site on the membrane contains an amount of urease that is quantitatively related to the amount of DNA in the sample. The membrane with a fixed DNA-protein complex is then placed into a special reader containing urea solution (Briggs 1991; Briggs and Panfili 1991; Briggs and Zuk 1988).

The reader detects pH changes resulting from the enzymatic transformation of urea caused by urease using a silicon-based light-addressable potentiometric sensor (LAPS) (Hafeman et al. 1988; Kung et al. 1990). This detector has a dynamic range greater than 3 logs of signal, which corresponds to greater than 2 logs of DNA. The rate of change of the surface potential is a direct measure of the enzyme activity. Enzyme activity is determined by the number of enzyme molecules, since the substrate is in great excess and reaction conditions are the same for all samples. About 10^8 urease molecules are required to produce, in 1 min, a response significantly higher than the background (Briggs and Panfili 1991).

Zero and positive calibrators on every membrane are used to compensate for membrane-to-membrane variability, increasing precision and reproducibility of the test. Calf thymus DNA is used as a positive calibrator. The standard curve ranges from 2 to 200 pg of DNA. Briggs and Panfili (1991) estimated that the mean quantitated level at 2 pg is separated from 0 by more than 2 standard deviations.

Although the threshold assay is not sequence specific, and is almost fully automated, highly sensitive, and quantitative, it has several drawbacks. The assay is limited by the ranges of ionic strength and pH, and susceptible to interferences and inhibitors in the samples (McKnabb et al. 1989). The *Threshold Application Notes* by Molecular Devices Corporation (1989) state that some samples might not be appropriate for direct assay in the threshold system, especially those containing greater than 0.01 M guanidine chloride or urea, greater than 0.1 percent trifluoracetic acid (TFA) and acetic acid, greater than 0.2 percent sodium dodecyl sulfate (SDS) or greater than 0.3 M salts (in many cases even lower).

After passing through all of the different steps in the purification process, the contaminating DNA is likely to be sheared. The threshold assay demonstrates a reduced response to fragments of less than 800 bases due to the decreasing probability that a small fragment can bind both protein conjugates (Briggs and Panfili 1991). This means that it is likely that threshold data would show lower amounts of residual DNA in protein than are actually present.

The threshold method is rather inconvenient for detecting picogram amounts of DNA in protein samples. Since the limit of detection is picograms, milligram amounts of protein should be used. This necessitates the use of proteinase K digestion and organic extraction because the excess protein may clog the biotinylated membrane and interfere with the assay. But incorporation of proteinase K digestion/organic extraction into the process may cause variability in results (Per et al. 1993; Scheinker in press). The presence of protein digestion products reduces recovery of DNA spikes introduced into protein samples (Briggs and Panfili 1991). Multiple repeats and organic extraction are time consuming and labor intensive, which defeats the purpose of the automated assay.

Hybridization

Hybridization under standardized conditions of temperature, ionic strength, and reaction stringency may be performed to determine the amount of residual DNA in the test sample that is complimentary to the probe used. DNA hybridization is a technique based on the phenomenon of base-pairing between two complimentary DNA strands one of which is labeled with radioactive nucleotides (Meinkoth and Wahl 1984; Wahl n.d.). The labeling reaction is based on the method in which a mixture of random sequence primers initiates DNA synthesis by Klenow fragment of DNA polymerase I from any single-stranded template (Feinberg and Vogelstein 1983, 1984). In the assay described the probe is prepared by extracting total DNA from the same strain of cells from which the protein has been purified (see Appendix 1).

Stages of the Hybridization Assay

Hybridization protocols for the detection of trace amounts of DNA in protein samples normally include the following steps, which we will consider in detail below:

1. Deproteinization of the sample

2. Application of samples and standards to the membrane

3. Labeling of the probe DNA

4. Incubation of the membrane in the solution of labeled DNA

5. Washing of the membrane

6. Quantitation of results

Deproteinization of Samples.   Deproteinization of samples, especially those containing milligram amounts of protein, is necessary because most proteins interfere with DNA hybridization. Most hybridization protocols include proteinase K digestion followed by organic extraction of DNA-contaminated protein samples. This method is lengthy, cumbersome, and labor intensive. It also involves the use of hazardous organic compounds, such as phenol and chloroform.

The heat deproteinization method proposed by Dr. Scheinker (in press) for the Threshold System has been successfully implemented in our hybridization assay. This method cuts the time needed for sample preparation from 3–4 hours to 15 minutes and does not use organic solvents. This method also does not use proteinase K buffer, which contains a high concentration of SDS. High concentrations of SDS in the sample increase the loss of DNA from nylon membranes and, as such, decrease hybridization signal intensity (Cannon et al. 1985).

For heat deproteinization protein samples are mixed with dilution buffer (for example, 100 mM Tris-HCl pH 8.0; 150 mM NaCl), incubated for 5 min in a boiling water bath and spun down in a microcentrifuge for 5 min at 14,000 rpm. The cleared supernatant is then applied to the membrane. DNA content estimated by this method is identical to the amounts obtained by proteinase K/phenol method (Figure 14.1) and by the threshold method (Table 14.1). We tried to use this method for different proteins in different buffers. In some cases the spike

Figure 14.1. Residual DNA content determined in the same protein sample deproteinized (A) by heating and (B) by proteinase K digestion/organic extraction. 1—protein, 2—protein + spike (50 pg).

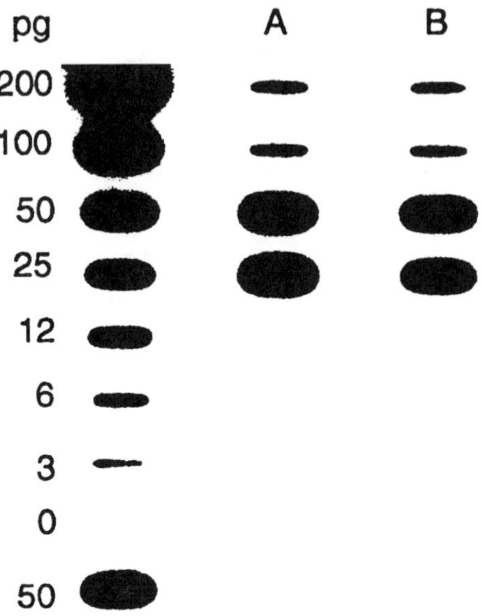

Table 14.1. Comparison of DNA Content in Different Protein Samples Estimated Using Threshold and Hybridization Methods

Sample	Threshold Data (DNA/protein)	Hybridization Data (DNA/protein)
4001 DL	0.44 µg/mg	0.5 µg/mg
4002 DL	0.46 µg/mg	0.7 µg/mg
4002 DP	<13.3 pg/mg	1.81 pg/mg
4003 DP	<16 pg/mg	6 pg/mg
4001 PB	<4 pg/mg	0 pg/mg

recoveries were 20–1000 percent (data not shown). This might be due to buffer-protein-DNA interactions. Normally we were able to find dilution buffer compositions suitable for a particular protein sample by means of varying the concentration of NaCl and Tris-HCl. It requires some extra effort, but this initial time investment pays off in the long run if many samples of this type are to be analyzed.

The presence of contaminating bacterial DNA in buffers, reagents, pipetters and so on can also cause unacceptably high spike recovery. This problem is exacerbated if probe DNA incorporates a broad range of bacterial sequences. Furthermore, ubiquitous contamination can exist if the bacterial strain from which the probe DNA originated is being used in the same lab where the assay is performed. We recommend storing the dilution buffer frozen and using sterile disposable supplies and aliquoted reagents where possible.

Application of Samples and Standards to the Membrane.   The dot- and slot-blotting techniques provide a rapid means of detecting DNA and RNA in unfractionated nucleic acid samples. Blotting can be performed by hand or using special apparatuses. Dot and slot blotters allow the application of larger sample volumes, are faster than manual application, and produce more uniformly shaped spots. The volume in which the sample was introduced to the membrane did not influence the results (data not shown). However, volumes larger than 1.5 ml are not practical for this assay, because it takes too long for the liquid to pass through the membrane. This type of blotting technique facilitates the subsequent quantitation on the phosphorimager.

Slot blotters, like Schleicher and Schuell's Minifold II, allow application of samples over a smaller surface area than dot blotters, eliminating variations in results caused by uneven sample immobilization and resulting in higher signal intensities. The Minifold II also does not use rubber O-rings, which are frequently the source of DNA cross-contamination. The unique design of the manifold in Schleicher and Schuell's Minifold II promotes even spotting of liquid on the filter without lateral migration of the solution that can distort results. Regular house vacuum can be used as a vacuum source (Wahl n.d.).

The choice of membrane is very important in the development of a hybridization assay. Nitrocellulose (NC) membranes were first used for hybridization, but now they are gradually being replaced by nylon membranes. Nitrocellulose membranes are very fragile, their handling is more labor intensive, and their sensitivity is consistently lower (Twomey and Krawetz 1990). It is believed that the mechanism of attachment of nucleic acids to NC membranes is noncovalent; on the other hand, primary amine groups of nylon can form a covalent cross-link with ultraviolet (UV)- or heat-activated thymidine (Khandjian 1987). Nylon membranes can also bind DNA in low-ionic strength buffers, which makes handling even less labor-intensive.

Observations in the literature indicate that various nylon membranes under optimal conditions bind DNA with varying degrees of efficiency. Parameters that determine binding efficiency include the type of nylon membrane, the method of DNA transfer to the membrane, and the method of fixing DNA to the membrane (Nierzwicki-Bauer et al. 1990; Twomey and Krawetz 1990; Meinkoth and Wahl 1984; Reed and Mann 1985). Using reported data and our own observations, we chose Bio-Rad's Zeta-Probe membrane (Twomey and Krawetz 1990). It has very good performance and comes with a certificate of analysis that is required for GMP methods.

Many problems of the hybridization assays are caused by incompatibility of the membrane, hybridization solution, and sample pretreatment procedures. It has been shown earlier that sensitivity of the nylon membranes can be increased by briefly denaturing the DNA prior to application to the membrane and by treating the membranes with alkali following or preceding transfer (Twomey and Krawetz 1990). However, this might not be true in all cases. For example, we used Zeta-Probe Membrane (Bio-Rad) presoaked in 0.4 N NaOH and NaOH-denatured probe with both old (cat #RPN 1518) and new (cat #RPN 1636) versions of Amersham's Rapid Hybe Buffer and obtained dramatically different results (Figure 14.2 A,D). The composition of the buffers is proprietary, but Amersham claims that the new buffer is a complete functional equivalent of the old one. Soaking Zeta-Probe Membrane in 0.4 N NaOH prior to sample application enhanced the hybridization signal with the old Rapid Hybe buffer from Amersham, and adding NaOH denatured probe did not have any adverse effects in this case. High background, high nonspecific binding, and low signal-to-noise ratios were observed with the new version of this buffer. The only way to accommodate the modified hybridization solution was to omit presoaking of the membrane in 0.4 N NaOH and use heat-denatured probe, but even with those modifications the method was not as reproducible and reliable as with the old version of Rapid Hybe. The intensity of the hybridization signal also considerably decreased. Similar problems were encountered when formamide-containing buffer was used with NaOH-soaked membrane and NaOH-denatured probe (Figure 14.2 B).

We tried to use a buffer, consisting of 0.25 M Na_2PO_4, 7 percent SDS and the different pretreatment systems. The results were acceptable, but the levels of nonspecific binding were high (Figure 14.2 C) (note apparent high amount of DNA in water control at "0" slot). This problem was solved by adding 1 percent of Blocking Reagent (Boehringer Mannheim, cat #1363514) to the hybridization buffer (Figure 14.3).

The following membrane-treatment combination proved to give the most consistent results and the highest hybridization signal intensity: Zeta-Probe Membrane soaked in water for at least 5 min, and NaOH-denatured probe (Figure 14.3). After application of the samples, the wells were washed with 0.5 ml of 0.4 N NaOH followed by 0.5 ml of water. Washing of the wells seemed to enhance the hybridization signal slightly, but it can be omitted to speed up the procedure without significantly affecting hybridization efficiency (data not shown).

It was shown that UV cross-linking in combination with alkali-hydrolyzed samples gives the best results (Twomey and Krawetz 1990). On the other hand, it has been noted that exposure of Zeta-Probe Membranes to high energy UV light leads to significant loss of hybridization signal (Reed and Mann 1985). In our experiments the membrane, while still wet, was UV cross-linked at 30 mJ for 20 sec and immediately transferred into the prehybridization solution. Absence of UV cross-linking did not affect the intensity of hybridization signal (data not shown), but might have some influence on reprobing of the membrane.

Assay controls were used in order to determine assay validity. Water, dilution buffer, and protein + spike samples for every protein sample used were run on each filter (Figure 14.3). It is widely accepted that there is contaminating DNA in the picograms per ml range, of presumably bacterial origin, in commercially available highly purified waters, such as Water for Injection (WFI) and water for processing semiconductors. Filtering water used for the DNA assays through positively charged nylon filters will reduce the level of contamination. In the case of hybridization, assays adding blocking reagents to the hybridization buffer and

Figure 14.2. Hybridization of DNA standard curves in different hybridization solutions with different pretreatment procedures. Hybridization conditions in all cases: prehybridization 30 min, hybridization overnight, both at 65°C with shaking, final wash in 0.2 × SSC/0.5 percent SDS at 65°C. A—new version of Rapid Hybe hybridization buffer from Amersham (cat #RPN 1636), NaOH-treated membrane, NaOH-denatured probe; B—hybridization buffer containing 50 percent formamide, 5 × SSC, 1 percent SDS, NaOH-treated membrane, NaOH-denatured probe; C—hybridization buffer containing 0.25 M Na$_2$HPO$_4$/ 7 percent SDS, membrane wetted in water, NaOH-denatured probe; D—old version of Rapid Hybe hybridization buffer from Amersham (cat #RPN 1518), NaOH-treated membrane, NaOH-denatured probe.

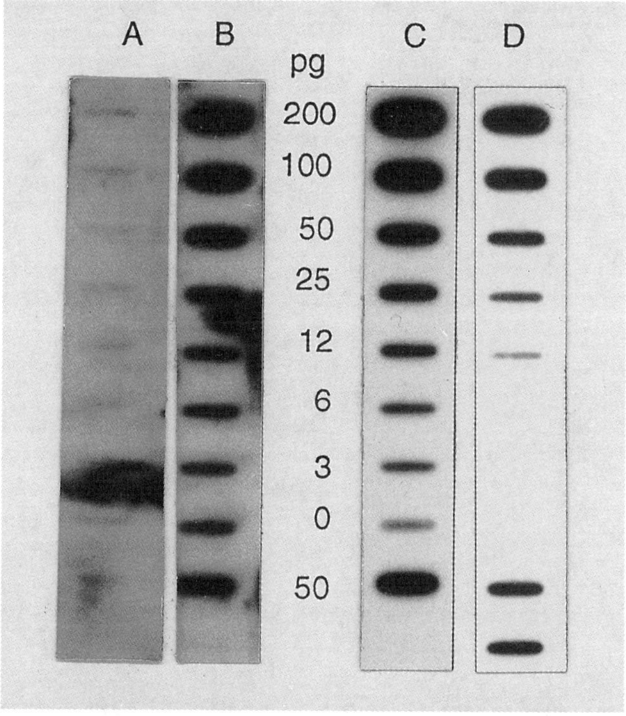

using high stringency washing conditions will also reduce the levels of nonspecific hybridization with background DNA.

The protein sample adopted as a standard in the quality control (QC) department was run periodically to ascertain the uniformity of results. Many problems associated with hybridization assays arise from the vials in which very diluted (1 pg/µl) DNA solutions, used to prepare standard curve, are stored. Glass and polycarbonate vials readily absorb DNA. Polypropylene vials are generally better, but not all types of polypropylene are entirely safe or DNA free. In our experience, Sarstedt screw-cap polypropylene tubes have the lowest affinity towards DNA. We also try to avoid storing solutions with concentrations of less than 1 ng/µl DNA.

Figure 14.3. DNA hybridization assay. Conditions: hybridization solution containing 0.25 *M* Na$_2$HPO$_4$/7 percent SDS/1 percent Blocking Reagent, water-soaked membrane, NaOH-denatured probe, 15 min prehybridization, overnight hybridization at 65°C with shaking, final wash in 0.2 × SSC/0.5 percent SDS at 65°C. A—heat-deproteinized protein samples + spike (50 pg), B—heat-deproteinized protein samples.

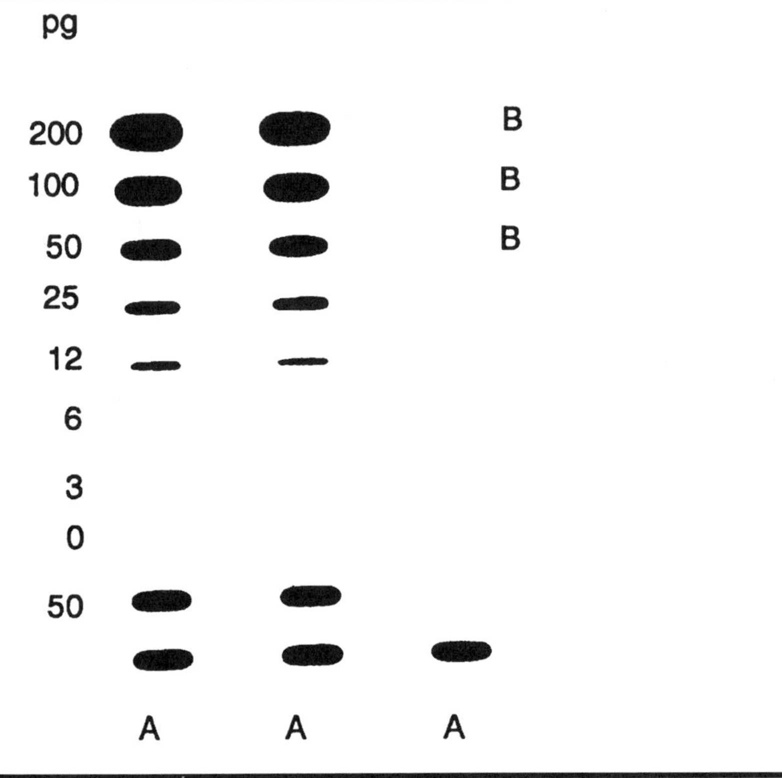

Labeling of the Probe.   The probe has been labeled by random primed hexanucleotide synthesis according to the method developed by Feinberg and Vogelstein (1983, 1984) using commercially available kit by (Amersham cat #RPN. 1600Z). The reaction was designed to obtain DNA probes labeled to high specific activity with ^{32}P. Two labeled deoxynucleoside triphosphates (dNTPs); deoxycytosine triphosphate and deoxyadenosine triphosphate (dCTP and dATP) with specific activities of 6,000 Ci/mmol were used. Labeling was carried out at ambient temperature for 3–5 hours. It should be noted that specific activities of greater than 3×10^9 disintegrations per minute (dpm)/µg of DNA were routinely obtained (data not shown).

Pharmacia Ready-to-Go labeling kits also were used. They are a complete premixed lyophilized labeling reaction mixture capable of producing DNA probes that have been radioactively labeled from 5×10^8 to 5×10^9 dpm/µg in

20 minutes. Labeled probe DNA was purified from unincorporated nucleotides by passing through a G-50 column to ensure that those nucleotides would not contribute to high background. The intensity of the hybridization signal strongly depended on the specific activity of the probe. High signal intensity, in its turn, allowed shorter hybridization and film exposure times, such that the assay could be completed in one day.

The actual DNA content of the sample is determined by comparing the signal intensity of the sample with a standard curve. DNA in the standard curve was from the same preparation as probe DNA. It was isolated from a sample of protein product at an early stage of purification and, thus, contained all sequences that might potentially be present in the purified protein. Since it has been highly processed, the isolated DNA contained a substantial amount of low molecular weight fragments, which contribute to UV absorption but have low hybridization efficiency. To avoid complications that might be caused by artificially high DNA concentration determined by UV absorption, we passed the isolated DNA through 1.5 × 30 cm column packed with Bio-Gel A-5 (Bio-Rad). Fractions with high UV absorbance were analyzed by agarose gel electrophoresis and those containing DNA fragments of 100–2000 base pairs (bp) were used as probe and standard DNA in the assay.

Incubation of the Membrane in the Solution of Labeled DNA.   Both prehybridization and hybridization were conducted in reusable polyethylene trays. Trays allow free movement of the membranes in the hybridization solution and easier handling of membrane, which are beneficial for reducing the background, especially if nonradioactive detection systems are used. Trays are easily decontaminated by soaking in Contrad (Curtin Matheson Scientific, Inc.) overnight.

Short prehybridization periods (15–20 minutes) in sodium phosphate/SDS/ blocking reagent buffer produced very acceptable results (Figure 14.3). Longer periods (up to 4 hr) did not have any adverse effect. Hybridization was conducted for 30 min or as long as overnight at 65°C with shaking. Longer hybridization periods (up to 5 hr) increased the intensity of the signal; after 5 hours signal intensity remained largely unchanged (data not shown). Care should be taken, especially with long hybridization periods, to prevent evaporation of the hybridization solution from the tray. This is the main cause of high background. We covered the trays with Saran Wrap™ before closing the lid. This seemed to produce an evaporation-proof closure.

Concentration of DNA in hybridization solution was crucial for obtaining low background levels. In our case it had to be about 3 ng probe DNA/ml of hybridization solution. If we increased it at least 2–3-fold, the background was significantly increased (data not shown).

Washing of the Membrane.   Washing is necessary to eliminate nonspecifically bound probe, which otherwise would create high background and make quantitation of target DNA difficult. In our assay we finished with a high stringency wash (0.2 × SSC/0.5 percent SDS at 65°C). High stringency washes always removed nonspecifically bound probe to levels sufficient to produce consistent signal-to-noise ratios.

Quantitation. The main problem with DNA hybridization assays is quantitation. Most frequently used methods are visual, densitometry of autoradiograms, scintillation counting of the samples cut out of the membrane, flatbed scintillation counting, and phosphorimaging. Visual assessment is highly subjective. Although it can provide a good estimate of DNA content in the sample, the numbers are very approximate and could always be argued. Densitometry of autoradiograms is traditionally used to estimate radioactivity of the samples tested by hybridization assay. It has two major limitations: poor sensitivity (less than 5 percent of beta-particles emitted by a ^{32}P-labeled sample actually interact with the emulsion of a standard X-ray film to form a latent image center) and limited, nonlinear dynamic range (Johnston et al. 1990). The problem with dynamic range is further complicated by the characteristic sigmoidal density vs. log exposure response curve. Scintillation counting of individual samples cut out of the membrane is laborious and prone to technical error, especially when sample load is high. In flatbed scintillation counting the intact membrane is placed into a sample bag with a small amount of scintillation liquid. The photomultiplier tubes are mounted close together on a plate over the sample and each counts a separate rectangular area. The sensitivity of ^{32}P-labeled probe in this system was determined as 100 pg DNA, which makes the method unsuitable for assessing residual DNA content in recombinant protein preparations (Hyypia et al. 1990).

In our assay quantitative image analysis has been obtained with a phosphorimager (Fuji BAS 2000). The phosphorimager takes advantage of storage phosphor technology to detect and quantify radioactivity. Storage phosphorimaging plates can detect ^{32}P-labeled sample at an exposure level of 1 disintegrations/ mm^2 (Johnston et al. 1990). After suitable incubation time (5–15 min), phosphor screens are scanned by a neon-helium laser at 633 nm. Energy from the beta particles returning to the ground state is released at 390 nm. Luminescence intensity is measured and stored digitally. The result is a quantitative representation of the image formed on the storage phosphor screen by the original radiolabeled membrane (Figure 14.4).

Phosphorimaging plates store energy in a linear capacity over 5 orders of magnitude, while X-ray film is linear over about 1.5 orders of magnitude (Johnston et al. 1990). Sensitivity of imaging plates for ^{32}P at the midpoint of the film response is about 250 times higher than direct autoradiography and 15 times higher than autoradiography using an intensifying screen (Reichert et al. 1992; Zouboulis and Tavakkol 1994). Resolution is in the 0.3 mm range and therefore suitable for gel and blot applications (Johnston et al. 1990).

The imaging plates commercially available from Fuji Photo Film Co. are composed of fine crystals of BaFBr:Eu^{+2} in an organic binder. High energy radiation (e.g., beta particles) can excite an electron of the Eu^{+2} ion into the conduction band. This electron is then trapped in the "F-center" of the BaFBr complex with the resultant oxidation of Eu^{+2} to Eu^{+3}. The excited complex has a distinct absorption band around 600 nm. By exposing the excited complex to light from a neon-helium laser (633 nm), the electrons are liberated back to the conduction band, reducing Eu^{+3} to Eu^{+2}. Eu^{+2*} then releases a photon at 390 nm as it returns to the ground state. The intensity of the luminescence is measured and stored digitally in relation to a position of the scanning laser beam. Before and after use

Figure 14.4. Printout of the phosphorimager screen showing 5 membranes. Exposure time 15 min.

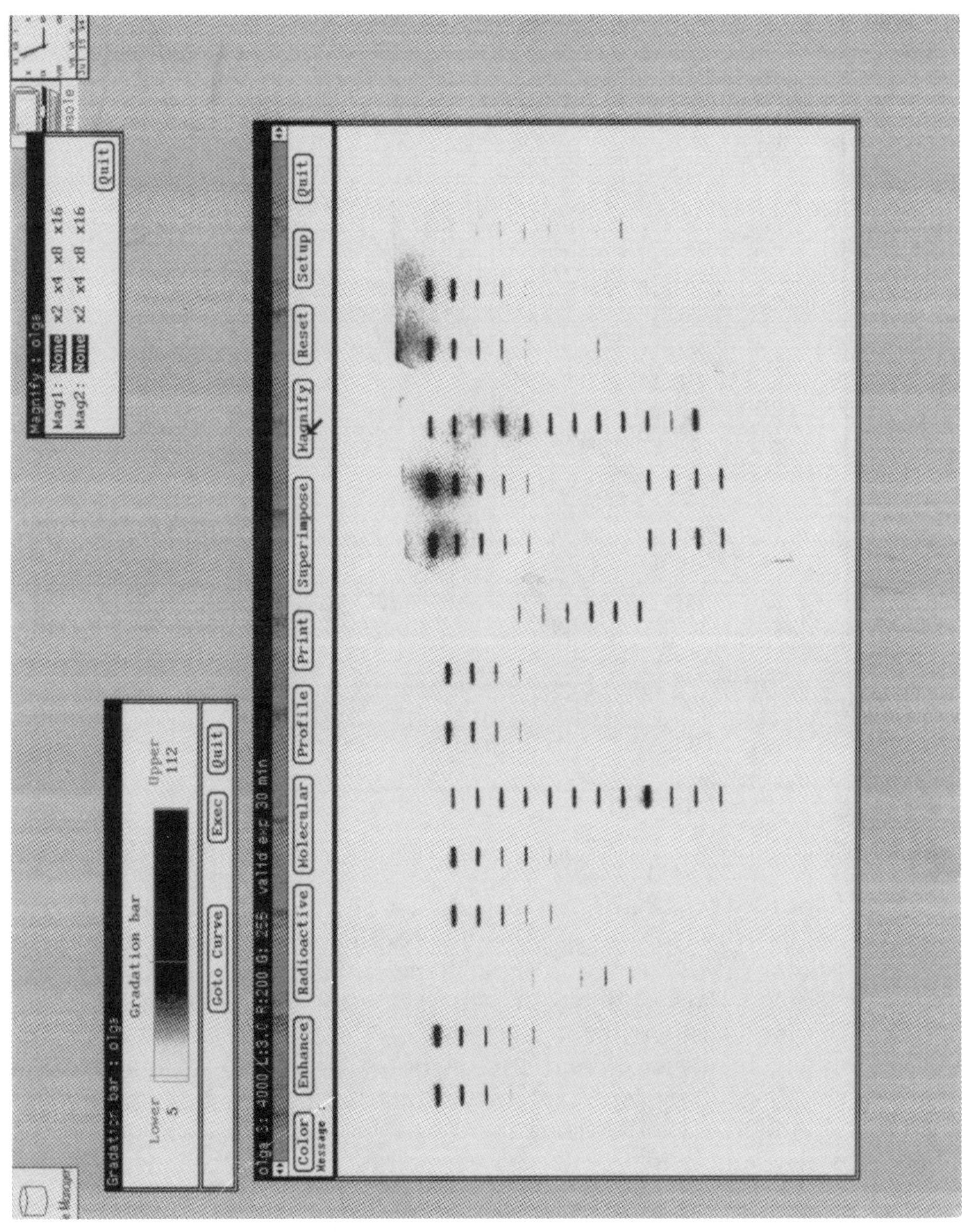

remaining images and background noise are erased by exposing imaging plates to bright visible light using Image Eraser. Thus, the imaging plates are reusable (Johnston et al. 1990).

Figure 14.5. Curve fit for the phosphorimager values obtained from the standard curve. Exposure time 5 min.

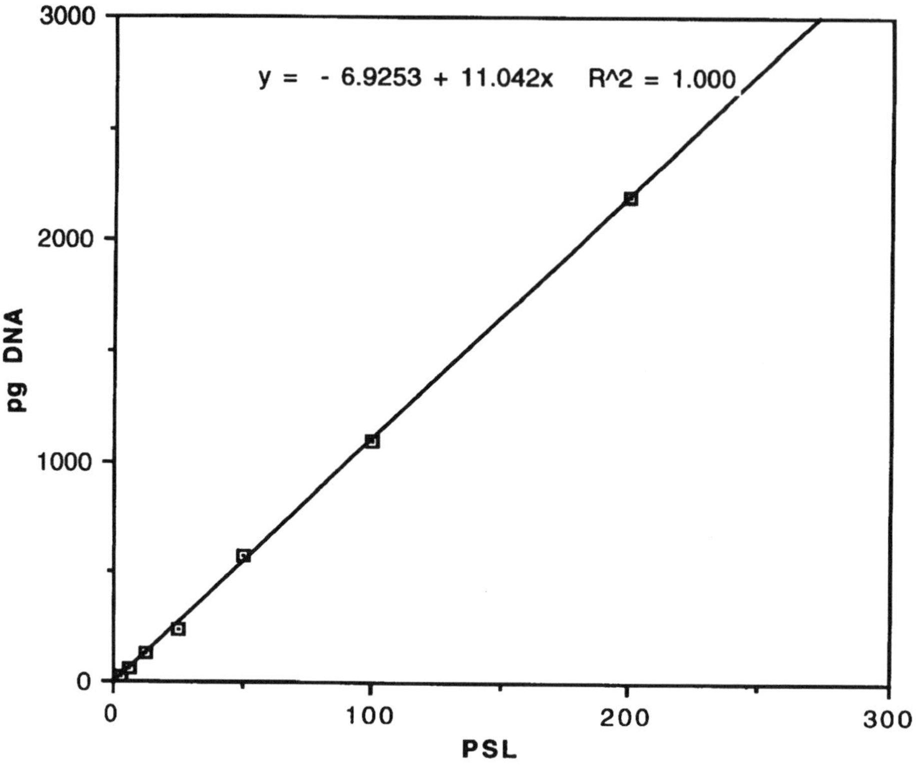

Signal imprinted on the imaging plate decays over time. Short exposures are almost half as stable as longer ones. About 20 percent of the signal in long exposures is lost in the first 3 hours, and about 40 percent over 3 days. However, exposures of several days and weeks will not necessarily give proportional detection. This is because environmental radiation is also detected with great sensitivity, resulting in a background that increases over time (Zouboulis and Tavakkol 1994). In addition, long exposures are affected by a gradual decay of the stored signal. Another factor affecting the efficiency of detection is absorption of radiation by the sample matrix itself. Detection of ^{14}C and ^{35}S is reduced by 50 percent in dried gels compared to the same radioactivity spotted on filter paper; the more energetic ^{32}P radiation is not significantly affected (Johnston et al. 1990).

Data generated after integration of chemiluminescence signal per slot area of the standard curve can be plotted (e.g., in CricketGraph) against DNA amounts applied per slot (Figure 14.5). The data obtained allows the determination of the linearity of the standard curve using linear regression analysis. Within the range used in our experiments (1–200 pg), the phosphoimager response was very close to linear ($r^2 = 0.99 \pm 0.009$).

Percent spike recovery for sample spikes can be quantitated by using the linear regression equation. In order to estimate the accuracy of the assay, the closeness of the expected amount of DNA introduced into a protein sample to the apparent level of DNA detected by the assay was determined. Spikes of 100, 50, and 25 pg of DNA were run on 5 separate blots with standard curves and blots counted in a phosphorimager (Table 14.2). Spike recovery ranged from 94 percent to 98 percent and did not depend on the amount of DNA in the spike.

In order to determine the precision of the assay, equal amounts of DNA (6, 50, and 100 pg) were applied each to one blot (n = 20) (Table 14. 3). The coefficient of variation was 16–25 percent. It was lower for the 100 pg, spike which could be due to the relatively lower input from the background. A comparison of different membranes shows that the extent of variability depends mostly on different background levels in different areas of the filter. If higher precision is needed, samples should be placed as close to the standards and each other as possible. Using the threshold method, McKnabb et al. (1989) obtained similar coefficients of variation for these DNA amounts.

The lowest amount of DNA that we were able to detect on an autoradiogram was 0.5 pg (after an overnight exposure). This amount could not be accurately quantitated by the phosphorimager because it did not significantly differ from the background. However, 3 pg are routinely quantitated, and 1.5 pg can be quantitated on blots with higher signal-to-noise ratios.

Table 14.2. Recovery of DNA Spikes from Protein Samples Estimated Using Phosphorimaging Techniques

Amount of DNA in the Spike (pg)	% Spike Recovery	SD	CV (%)
25	98	7.4	7.5
50	94	5.2	5.5
100	95	10.9	11.4

Table 14.3. DNA Hybridization Assay Precision Estimated Using Phosphorimaging Techniques

Amount of DNA (pg)	Average PSL	SD (PSL)	CV (%)
6	86.5	21.79	25
50	625.1	146.1	23
100	353.2	58.5	16

Conclusions

We were able to develop a fast, reliable, quantitative method for determining residual DNA in recombinant proteins for pharmaceutical use. The hybridization-based method was chosen because this is the method accepted by the FDA (FDA 1987a), but also because it is sensitive and versatile. The developed assay may be performed in one day because of the use of an improved deproteinization technique, labeling the probe to high specific activity, using buffer for rapid hybridization and using phosphorimaging techniques for quantitation. At the same time the assay is very flexible in terms of the time frame and could be adjusted to different working schedules. The robustness and ruggedness of this method are acceptable in terms of high reproducibility and low number of required assay repeats. Most of the subjectivity in calculating and reporting the data for this proposed method has been removed; the assay relies on objective quantitative data obtained from phosphorimager analysis.

THEORETICAL ASPECTS OF BACTERIOPHAGE DETECTION

Bacteriophage contamination can be harmful to the fermentation process and potentially dangerous to the patient (Primrose 1990). Therefore, it is an FDA requirement to test for bacteriophage in fermentation samples. Testing for bacteriophage as well as for other contaminating microorganisms is a necessary complement to the total DNA hybridization assays because the DNA hybridization assay would not detect the DNA of nonrelated organisms due to the specificity of the probe.

Bacteriophage burst size (number of particles released from one host cell) varies from 30 to 300, and the time from infection to lysis varies from 20 to 100 min depending on the particular host-phage system and environmental conditions (Primrose 1990). Advanced bacteriophage infection can be recognized by changes in fermentation parameters, such as decrease in cell metabolic rate, indicated by an increase in dissolved oxygen and a decrease in CO_2 concentration. The rate of alkali addition and turbidity of the culture are also decreased, and the consumption of antifoam agents is increased because of cell lysis which liberates a lot of protein (Hongo et al. 1973). Eventually, the cell culture becomes completely clear. However, at the earlier stages, when only 1–5 percent of cells are lysing, it is difficult to recognize the infection. Since phage particles could be both introduced (Rapp et al. 1992) or mutated to a new virulent variety at any stage of fermentation (Primrose et al. 1982, 66), the FDA requires that a bacteriophage assay be performed for each fermentation run (FDA 1987b).

The ideal assay should detect all kinds of bacteriophages present in the smallest quantities. However, due to the natural diversity of bacteriophages and the constant mutation process, it is extremely difficult to develop a test capable of detecting all possible species with equal efficiency. Diversity and constant generation of new bacteriophage strains render all kinds of DNA/RNA related assays useless.

At the current state of knowledge, the microbiological approach is probably the most appropriate, although it is also plagued with many difficulties. To begin with, not all phages infect all bacteria; even if they do infect a particular strain they might need special conditions (i.e., pH, temperature, media composition, etc.) (Smith and Ritchie 1980). This results in an infinite number of combinations of strains and conditions and also makes the possibility of having a valid positive control for this assay questionable. The working positive control would indicate that the infection conditions were right for the particular bacteriophage chosen as a positive control, but not for all possible phage contaminants. On the other hand, maintaining a bacteriophage culture in the QC laboratory in close proximity to fermentation facilities, creates a danger of infecting the fermenter. In our mind the danger of infection outweighs the doubtful value of the positive control and our decision was not to use it in the assay.

For practical purposes it seems most appropriate to select as indicator cells the current production strain. Although the assay in this case will not detect phages that do not infect this particular strain, it will still be of use, because it will cover the species that are most likely to show up in a particular process and cause the most damage. While developing the assay, we tried to plate out cells from the fermenter, incubate them overnight, and check the plates for bacteriophage plaques. The assay would detect only phages capable of infecting a particular production cell line under the fermentation conditions used. However, since the cells at harvest had very poor viability, they produced an uneven lawn (data not shown). This made the interpretation of results extremely difficult.

The procedure in which samples from the fermenter were shaken with chloroform to facilitate cell lysis and potential release of phage particles was eventually developed (see Appendix 2). Lysed cells were centrifuged down. Supernatant, presumably containing phage particles, was mixed with indicator cells. Indicator cells were obtained from the manufacturers' working cell bank (i.e., representing the same cell line as used in production). The conditions of infection in this case had to be chosen arbitrarily. We currently use a 20 minute incubation of a mixture of indicator cells and supernatant from a fermentation sample prior to addition of soft agar. After the soft agar solidifies, the plates are incubated at 37°C overnight and then scored for the number of plaques (Figure 14.6). If plaques are detected, a more specific assay for the phage should be developed (Primrose et al. 1982, 66).

THEORETICAL ASPECTS OF PLASMID STABILITY

Although naturally occurring plasmids are normally very stable, this is not true for hybrid multicopy plasmids used for the production of recombinant proteins (Margaritas and Bassi 1991). They often are lost from the cell and may also lose the introduced protein gene because of deletions and rearrangements in their DNA. This is a significant problem in large-scale fermentations, because the overall yield of protein product may be reduced raw materials and time would be wasted, which would result in significant nonproductive costs. This also brings up the FDA issue of being in control of the process.

Plasmid instability can be of two types: segregational and structural. Segregational instability occurs because of unequal distribution of plasmids

Figure 14.6. Photograph of plates as usually submitted with the results of the bacteriophage assay showing no bacteriophage contamination.

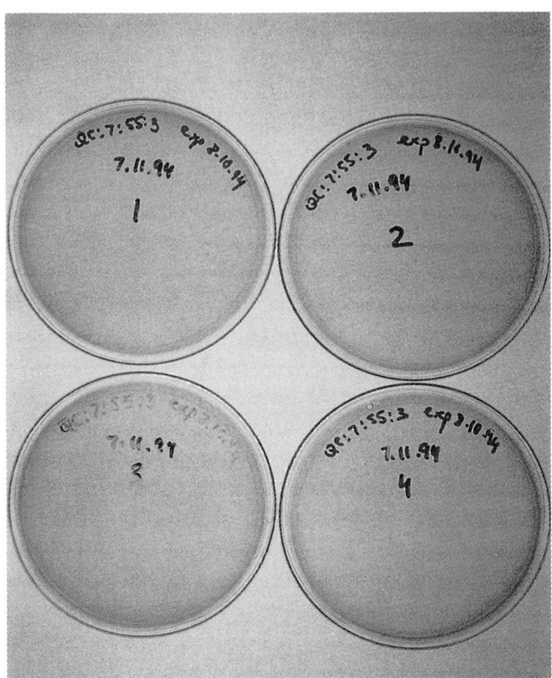

during cell division. Some plasmids carry a partition locus that ensures even segregation of plasmids, and removal of this locus increases plasmid instability (Nilsson and Skogman 1986). High copy number plasmids are more segregationally stable than low copy number plasmids (Margaritas and Bassi 1991). Jones et al. (1980) found that segregational instability depends primarily on the genetic makeup of the plasmid rather than on the host cell. However, other factors such as environmental conditions, especially limitations in nutrients essential for DNA replication, phase of growth, and plasmid-host compatibility also may influence plasmid stability (Balows et al. 1991; Kim and Meyer 1984; Bron and Luxen 1985).

Structural instability occurs as the result of deletions, insertions, or rearrangements in plasmid DNA. This type of instability is especially frequent in sequences with short direct repeats. It has been noted that plasmids carrying strong promoters are more prone to structural instability. Environmental and plasmid-host interaction may also contribute to this type of instability.

Currently used microbiological methods for the detection of plasmid stability are applicable only to plasmids with selectable markers, such as resistance to antibiotics (Balows et al. 1991). Cells from fermenters are diluted and plated on a nonselective media to obtain separate colonies. Cells from a certain number of these colonies are then transferred onto two plates: one with selective (e.g., containing the appropriate antibiotic) and one with nonselective media. Plates are incubated overnight at 35°C and the number of colonies that survived both on selective and nonselective media are compared. The nonselective medium is used for assurance that the transfer of cells actually occurred. If no growth appears on the selective plate and there is growth in the corresponding point on the nonselective plate, it is assumed that the colony is sensitive to the selective agent and, therefore, has lost the plasmid. The limitations of this method are that it will only work with plasmids carrying resistance genes; it will not detect cells that have lost a portion, but not all of the multiple copies of a plasmid; and it will not detect structural instability.

We developed a method for PCR quantitation of the amount of plasmid released into fermentation media at a particular stage of fermentation (see Appendix 3). The cells are spun down and an aliquot of supernatant is introduced into the PCR reaction as a template. After 30 cycles of synthesis, the obtained amount of DNA is compared to a standard (a known amount of DNA amplified in the same conditions). However, the significant tube-to-tube variability characteristic of PCR reactions affects the accuracy and reproducibility of direct comparison. The problem of variability is minimized by introducing into the same tube a competitive DNA template that uses the same primers as the target DNA but that is easily distinguished from it after amplification.

A competitive template can be prepared by introducing a point mutation that, for example, creates a new restriction enzyme site. cDNA can be used vs. a genomic copy of the same gene containing a small intron (Gilliland et al. 1990). Competitive templates also can be prepared by using PCR with low stringency primer annealing (Forster 1994a). The resulting multiple products are then separated by electrophoresis, eluted, and used as competitive templates. Another method by Forster (1994), which we used, employs a linker primer to generate a shorter competitive template.

Target DNA is coamplified with a dilution series of competitor DNA of known concentration. After amplification, relative amounts of target and competitive DNA can be determined visually, measured by direct scanning of ethidium-stained gel, or by incorporation of radiolabeled dNTPs. The target and the competitor DNA are amplified in exactly the same conditions without tube-to-tube variability; hence, the initial ratio of these templates should be maintained. Since the starting concentration of the competitive template was known, the concentration of target DNA can be calculated (Gilliland et al. 1990). This method can be used to quantitate less than 0.5 pg of target DNA accurately (Figure 14.7).

To establish if released plasmids will be digested in the media by nucleases, we incubated samples of media and cells from the same stage of fermentation with a molecular weight standard. The amount of DNA as estimated by gel electrophoresis did not significantly decrease, even after an overnight incubation at 37°C (data not shown). This indicates that there are no significant amounts of active nucleases in fermentation media at the preinduction stage; the plasmid loss estimated by a PCR reaction would not be artificially low due to plasmid digestion.

For the given fermentation conditions calculation of plasmid stability or retention should be straightforward. The known fermentation conditions are cell

Figure 14.7. 1.5 percent agarose/TBE (Tris Borate EDTA) gel showing the results of plasmid retention assay. Amounts of competitor fragment of 221 bp (fragment B) taken in each PCR reaction shown at the bottom. Fragment A is a target sequence amplified from the plasmid present in the media (3 μl of media at preinduction stage were taken into each 100 μl reaction). Lane 6 contains Boehringer Mannheim Molecular Weight Marker IV, 7-100 μl reaction with 3 μl media and no competitive fragment, 8–100 μl reaction with no templates (see Appendix 3 for details).

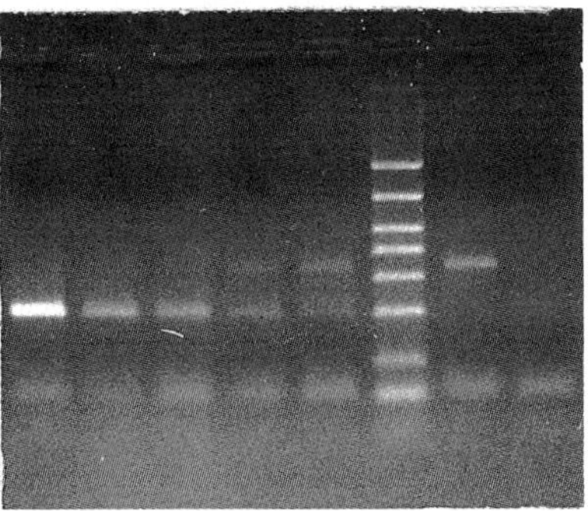

concentration, stage of fermentation, and plasmid copy number. An important factor is that the fermentation sample must be collected at the same stage of fermentation each time. This allows the comparison of different fermentation runs while maintaining consistent fermentation conditions. The stage that we have chosen is preinduction, because at harvest optical density (OD) cell viability may likely be decreasing and plasmid loss increasing.

The following is an example of how we have been able to determine plasmid stability in our fermentation production runs:

$$\% \, PR = 100 - (Nm/Nc \times 100)$$

where % PR = plasmid retention in %; Nm = number of plasmids in media; and Nc = theoretical number of plasmids in cells.

The number of plasmids in cells (Nc) can be calculated in the following way:

$$Nc = Ncopies \times C$$

where Ncopies = plasmid copy number per cell and C = cell concentration.

Ncopies can also be directly determined by using the PCR method described here. In this case washed cells should be used as a template in the PCR reaction or, if the total number of plasmids in a given volume must be determined, an aliquot of fermentation media with cells should be the template.

The number of plasmids in media (Nm) can be calculated using the following equation:

$$Nm = \frac{C_{PCR}}{M_W} \times 6.02 \times 10^{23}$$

where C_{PCR} = concentration of plasmid DNA in media determined by PCR and M_W = molar weight of the amplified fragment.

For example, the copy number per cell is a characteristic of a given bacterial strain and plasmid or is determined separately by the same PCR method. In our example it will be 300. Cell concentration is estimated from the optical density of the sample (e.g., 7×10^{13} cells/ℓ). Concentration of plasmid DNA lost into the media is determined by the PCR test (e.g., 10^5 pg/ℓ). The molar weight of the amplified fragment is calculated from its length (e.g., 2.2×10^{17} pg/mol); 6.02×10^{23} is Avogadro's number.

$$Nm = 10^5 \ (pg/\ell)/2.2 \times 10^{17} \ (pg/mol) \times 6.02 \times 10^{23} = 3 \times 10^{11} \text{ copies,}$$
$$\text{where } Nm = \text{number of plasmids in media}$$

$$Nc = 7 \times 10^{13} \text{ cells}/\ell \times 300 \text{ copies/cell} = 2.1 \times 10^{16} \text{ copies}$$

$$\% \text{ Plasmid Retention} = 100 - (3 \times 10^{11}/2.1 \times 10^{16} \times 100) =$$
$$100 - 0.001 = 99.999\%$$

Conclusion

Molecular biology and biochemical methods can solve many of the issues and problems concerning plasmid stability. Many of these solutions will involve optimizing media components and conditions, operating conditions, and fermenter construction and design (Margaritas and Bassi 1991). The ability to determine plasmid stability values before fermentation harvest predicts product yield and lower fermentation cost. The aforementioned assay that we describe here gives quantitative values such that measurements are objective and production decisions can be made on a solid basis.

ACKNOWLEDGMENTS

We thank Dr. J. Fandl for his help in developing the plasmid retention assay and for reading the manuscript critically.

REFERENCES

Balows, A., W. J. Hausler, K. L. Herrmann, H. D. Isenberg, and H. J. Shadomy, eds. 1991. *Manual of clinical microbiology.* Washington, DC: American Society for Microbiology.

Briggs, J. 1991. Sensor-based system for rapid and sensitive measurement of contaminating DNA and other analytes in biopharmaceutical development and manufacturing. *J. Parent. Sci. Technol.* 45:7–12.

Briggs, J., and P. R. Panfili. 1991. Quantitation of DNA and protein impurities in biopharmaceuticals. *Anal. Chem.* 63:850–859.

Briggs, J., and R. Zuk. 1988. Fast biosensor-based system for measuring total DNA in protein solutions. Symposium on Continuous Cell Lines as Substrates for Biologicals, in Arlington, VA. *Develop. Biol. Standard.* 70:245–253.

Bron, S., and E. Luxen. 1985. Segregational instability of pUB110-derived recombinant plasmids in *Bacillus subtilis*. *Plasmid* 14:235–239.

Cannon, G., S. Heinhorst, and A. Weisbach. 1985. Quantitative molecular hybridization on nylon membranes. *Anal. Biochem.* 132:6–13.

Code of Federal Regulations. 1992. Title 21, Part 211, *Good manufacturing practices for finished pharmaceuticals*, 160. Washington, DC: U.S. Government Printing Office.

FDA. 1987a. *Points to consider in the characterization of cell lines used to produce biologicals*. Rockville, MD: Office of Biologics Research and Review.

FDA 1987b. *Points to consider in the manufacture and testing of monoclonal antibody products for human use*. Rockville, MD: Office of Biologics Research and Review.

FDA. 1993. *Points to consider in the characterization of cell lines used to produce biologicals*. Rockville, MD: Center for Biologics Evaluation and Research.

Feinberg, A. P., and B. Vogelstein. 1983. A technique for radiolabeling DNA restriction endonuclease fragments to high specific activity. *Anal. Biochem.* 149:229–237.

Feinberg, A. P., and B. Vogelstein. 1984. A technique for radiolabeling DNA restriction endonuclease fragments to high specific activity: Addendum. *Anal. Biochem.* 137:266–267.

Forster, E. 1994a. Rapid generation of internal standards for competitive PCR by low stringency primer annealing. *Biotechniques* 16:1006–1008.

Forster, E. 1994b. An improved general method to generate internal standards for competitive PCR. *Biotechniques* 16:18–20.

Gilliland, G., S. Perrin, and H. F. Bunn. 1990. Competitive PCR for quantitation of mRNA. In *PCR protocols: A guide to methods and applications*, edited by M. A. Innis, D. H. Gelfand, J. J. Sninsky, and T. J. White. New York: Academic Press.

Hafeman, D. G., J. W. Parce, and H. M. McConnell. 1988. Light-addressable potentiometric sensor for biochemical systems. *Science* 240:1182-1185.

Hongo, M., T. Oki, and S. Ogata. 1973. Phage contamination and control. In *The microbial production of amino acids*, edited by Yamada et al. New York: John Wiley.

Hyypia, T., E. Auvinen, S. Kovanen, and T. H. Stahlberg. 1990. Rapid quantitation of DNA spot hybridization by flatbed scintillation counting. *J. Clin. Microbiol.* 28:159–162.

Jones, I. M., S. B. Primrose, A. Robinson, and D. C. Ellwood. 1980. Maintenance of some ColE1-type plasmids in chemostat culture. *Molec. Gen. Genet.* 180:579–581.

Johnston, R. F., S. C. Pickett, and D. L. Barker. 1990. Autoradiography using storage phosphor technology. *Electrophoresis* 11:355–360.

Khandjian, E. W. 1987. Optimized hybridization of DNA blotted and fixed to nitrocellulose and nylon membranes. *BioTechnology* 5:165–167.

Kim, K., and R. J. Meyer. 1984. Novel system of recognizing and eliminating foreign DNA in *Pseudomonas putida. J. Bacteriol.* 678:158–161.

King, R. S., and P. R. Panfili. 1991. Influence of fragment size on DNA quantitation using DNA-binding proteins and a sensor-based analytical system: Applications in the testing of biological products. *J. Biochem. and Biophys. Methods* 23:83–89.

Kung, V. T., P. R. Panfili, E. L. Sheldon, R. S. King, P. A. Nagainis, B. Gomez, Jr., D. A. Ross, J. Briggs, and R. Zuk. 1990. Picogram quantitation of total DNA using DNA-binding broteins in a silicon sensor-based system. *Anal. Biochem.* 187:220–227.

Margaritis, A., and A. S. Bassi. 1991. Plasmid stability of recombinant DNA microorganisms. In *Recombinant DNA technology and applications,* edited by A. Prokop, R. K. Bajpai, and C. Ho. New York: McGraw-Hill, Inc.

McKnabb, S., R. Rupp, and J. L. Tedesco. 1989. Measuring contaminating DNA using monoclonals. *Biotechnology* 7:343-347.

Meinkoth, J., and G. Wahl. 1984. Hybridization of nucleic acids immobilized on solid supports. *Anal. Biochem.* 138:267–284.

Nierzwicki-Bauer, S. A., J. S. Gebhardt, L. Linkkila, and K. Walsh. 1990. A comparison of UV-cross-linking and vacuum baking for nucleic acid immobilization and retention. *BioTechniques* 9:472–478.

Nilsson, J., and S. G. Skogman. 1986. Stabilization of *E. coli* tryptophan production vectors in continuous cultures: A comparison of three different systems. *Biotechnology* 4:901–903.

Noack, D., M. Roth, R. Geuther, G. Muller, K. Undisz, C. Hoffmeier, and S. Gaspar. 1981. Maintenance and genetic stability of vector plasmids pBR322 and pBR325 in *E. coli* K12 grown in a chemostat. *Mol. Gen. Genet.* 184:121–125.

Pepin, L. D. J., R. B. Lang, N. Lee, M.-J. Liao, and D. Testa. 1990. Detection of picogram amounts of nucleic acids by dot blot hybridization. *Biotechniques* 8:628–632.

Per, S. R., C. R. Aversa, P. D. Johnson, T. L. Kopasz, and A. F. Sito. 1993. Determining residual DNA levels: Comparison of hybridization with an alternative method. *BioPharm* 6:34–40.

Primrose, S. B. 1990. Controlling bacteriophage infections in industrial bioprocesses. In *Advances in biochemical engineering/biotechnology*, vol. 43, edited by A. Feichter. Berlin-Heidelberg: Springer-Verlag.

Primrose, S. B., N. D. Seeley, and K. B. Logan. 1982. Methods for the study of virus ecology. In *Experimental microbial ecology*, edited by R. G. Burns, and J. H. Slater. Oxford: Blackwell Scientific Publications.

Potter, C. G., G. T. Warner, T. Yrjonen, and E. Soini. 1986. A liquid scintillation counter specifically designed for samples deposited on a flat matrix. *Phys. Med. Biol.* 31:361–369.

Rapp, M. L., T. Thiel, and R. J. Arrowsmith. 1992. Model system using coliphage X174 for testing virus removal by air filters. *Appl. Environ. Microbiol.* 58:900–904.

Reed, K. C., and D. A. Mann. 1985. Rapid transfer of DNA from agarose gels to nylon membranes. *NAR* 13:7207–7221.

Reichert, W. L., J. E. Stein, B. French, P. Goodwin, and U. Varanasi. 1992. Storage phosphor imaging technique for detection and quantitation of DNA adducts measured by the ^{32}P-postlabeling assay. *Carcinogenesis* 13:1475–1479.

Scheinker, V. S. In press. Heat protein coagulation as a pre-treatment procedure for DNA quantitation.

Sverdlov, E. D., G. S. Monastyrskaya, L. I. Guskova, T. L. Levitan, V. J. Sheishenko, and E. J. Budowsky. 1974. Modification of cytidine residues with a bisulfite-o-methylhydroxylamine mixture. *Biochem. Biophys. Acta* 340:153–165.

Smith, K. M., and D. A. Ritchie. 1980. *Introduction to modern virology*. London: Chapman and Hall.

Threshold application notes. Total DNA assay series. 1989. Molecular Devices Corp.

Twomey, T. A., and S. A. Krawetz. 1990. Parameters affecting hybridization of nucleic acids blotted onto nylon or nitrocellulose membranes. *BioTechniques* 8:478–481.

Wahl, G. Rapid detection of DNA and RNA using slot blotting, sequences. *Schleicher and Schuell Appplication Update.*

WHO. 1987. Report of the Meeting of the Study Group on Biologicals. *Acceptability of cell substrates for production of biologicals.* WHO Technical Report Series, No.747.

Zouboulis, C. C., and A. Tavakkol. 1994. Storage phosphor imaging technique improves the accuracy of RNA quantitation using ^{32}P labeled cDNA probes. *Biotechniques* 16:290–294.

APPENDIX 14.1: DNA HYBRIDIZATION ASSAY USING HEAT DEPROTEINIZATION

Reagent solutions

Note: All reagents should be sterile filtered through 0.2 μm nylon filtration unit (Nalgene or equivalent). Store all reagent solutions for up to 3 months at ambient temperature unless otherwise stated. Use MilliQ water or water for irrigation for all solutions.

1. Hybridization buffer (0.25 M Na_2HPO_4/7% SDS/1% Blocking Reagent):

   (Take 17.7 g Na_2HPO_4
   add 35 g SDS
   Water to 500 ml.
   Filter sterilize.
   Add 5 g of Blocking Reagent (Boehringer Mannheim)
   Stir with moderate heating until the Blocking Reagent has dissolved but the solution is still opalescent (about 30–40 min).
   Store at –70°C in 50 ml aliquots for 3 months or at ambient temperature for 2 days.)

2. 0.4 N NaOH:

   (Take 16 g NaOH or 40 ml 10 N NaOH
   Water to 1 liter
   Filter sterilize.)

3. 2 N NaOH:

   (Take 80 g NaOH or 200 ml 10 N NaOH
   Water to 1 liter Filter sterilize.)

4. 20 × SSC (150 mM NaCl/15 mM citric acid)

   (Take 175.53 g NaCl
   88.23 g citric acid
   Water to 1 liter
   Filter sterilize.)

5. 10% SDS:

   (Take 100 g SDS
   Water to 1 liter
   Heat to dissolve.
   Filter sterilize.)

6. 0.5 M EDTA pH 8.0:

   (Take 18.6 g EDTA
   Water to 80 ml
   Adjust pH to 8.0 with 10 N NaOH.
   Adjust volume to 100 ml with water.
   Filter sterilize.)

7. 1 M Tris-HCl pH 8.0:

    (Take 12.1 g Tris
    Water to 80 ml
    Adjust pH to 8.0 with HCl
    Adjust volume to 100 ml with water.
    Filter sterilize.)

8. Washing solution 1 (2 × SSC):

    (Take 50 ml 20 × SSC
    450 ml water.)

9. Washing solution 2 (0.2 × SSC/0.5% SDS):

    (Take 10 ml 20 × SSC
    50 ml 10% SDS
    940 ml water.)

10. Dilution buffer (100 mM Tris-HCl pH 8.0/150 mM NaCl)

    (Take 10 ml 1 M Tris-HCl pH 8.0
    3.33 ml 5 M NaCl
    Water to 100 ml.
    Filter sterilize)

11. 5 M NaCl

    (Take 146 g NaCl
    Water to 500 ml.
    Filter sterilize.)

12. 10 N NaOH

    (Take 200 g NaOH
    Sterile filtered water—400 ml.
    Add NaOH to water in small batches while stirring. Use thick-walled plastic container. Work under the hood. When NaOH is completely dissolved, adjust the volume to 500 ml. Store in a plastic container at ambient temperature.)

Procedure

Notes and Precautions:

Sterile Sarstedt tubes should be used for all procedures. Use gloves, lab coats, protective Plexiglas® shields and lead containers when working with radioactively labeled materials. Discard all radioactive waste into a designated container and dispose of it with appropriate precautions. Use sterile techniques for all procedures.

Preparation of Standard Curve

1. Measure OD of standard DNA stock at 260 nm.

2. Calculate concentration of DNA:

50 μg/ml DNA has $OD_{260} = 1$
X μg/ml DNA has $OD_{260} = A$,
X (μg/ml) = 50 μg/ml $\times A \times 1/1 \times$ dilution factor
where $A = OD_{260}$ of stock DNA

3. Dilute stock solution to 1 pg/μl with water.

4. Prepare standard curve dilutions:

Tube Signed	μl of 1 pg/μl DNA solution	μl water
200	200	300
100	100	400
50	50	450
25	25	475
12	12	500
6	6	500
3	3	500
1	1.5	500
0	0	500

Preparing Protein Samples

1. Use 1 mg of protein per slot for all samples, except for upstream samples containing large amounts of DNA. Calculate the volume (V) of protein solution containing 1 mg of protein:

$$V \text{ (ml)} = \frac{1 \text{ (mg)}}{\text{concentration of protein (mg/ml)}}$$

2. Upstream samples must be diluted 1:1000 (or appropriately):
If the initial 1:1000 dilution proves to be too high or too low, so that the DNA amount could not be accurately estimated, the dilution plan should be changed.
Prepare samples as "sample" 1 in step B3 for 100 μl, 50 μl, 25 μl aliquots of 1:1000 dilution (for example) and as "sample 2" for another 25 μl aliquot.

3. For each protein sample prepare mixtures for "sample" (1), "sample + spike" (2) and "buffer" (3) tubes as follows:

Tube No. (1pg/µl)	µl protein	µl Dilution Buffer	µl DN
1.	X	500 – X (or 2X, whichever is more)	0
2.	X	450 – X (or 2X, whichever is more)	50
3.	0	500	0

X = volume of protein solution calculated in step C1

4. Incubate the obtained mixtures at 95°C in a boiling water bath for 5 min.

5. Spin in a microcentrifuge at 13,000 rpm for 5 min.

6. Transfer supernatant to fresh tubes and use it for blotting.

7. Discard the white precipitate.

Denaturing and Blotting the Samples

1. Add 125 µl of 2 N NaOH to all sample and standard curve tubes.

2. Mix by inverting. Spin the drops down.

3. Incubate at least 7 min at room temperature.

4. Cut Zeta-Probe Membrane (Bio-Rad) to the size of the slot-blot apparatus. Notch the top left corner. Sign the membrane.

5. Wet the membrane in water for about 5 min.

6. Assemble the Minifold II slot-blot apparatus (Schleicher and Schuell), with two pieces of GB002 blotting paper (Schleicher and Schuell) wetted in water and the soaked membrane, in the following order:

    a. top of the slot blot apparatus

    b. membrane

    c. two pieces of blotting paper

    d. midsection of the apparatus

    e. clear plastic spacer

    f. rubber pad

    g. bottom section of the apparatus

7. Secure the assembly with side clamps.

8. Apply vacuum for about 1 min.

9. Arrange samples and standard curve tubes in a rack in the same order as you want them on the membrane.

10. Using a clean pipette tip for each sample, apply each sample to a separate slot.

11. After all the liquid has been pulled through the membrane, wash the wells with 0.5 ml 0.4 N NaOH.

12. After all the liquid has been pulled through the membrane, wash the wells again, this time with 0.5 ml of water.

13. After all the liquid has been pulled through the membrane, disassemble the slot-blot apparatus.

14. Assemble two pieces of blotting paper and place the third one, wetted in water, on top.

15. Remove the membrane from the slot-blot apparatus with forceps and place on the blotting paper assembly.

16. While the membrane is still damp, UV cross-link at 30 mJ for 20 seconds at 254 nm in a Stratalinker (Stratagene) (optional).

Prehybridization and Hybridization

1. Prewarm the hybridization buffer at 65°C. Place 25 ml in a hybridization tray.

2. Place the membrane into the tray, cover with lid, and incubate with shaking for at least 15 min. Longer prehybridization times (1–4 hr) may help reduce background.

3. Calculate the volume of probe solution needed to achieve the final concentration of probe DNA in hybridization buffer of 2–10 ng/ml (excess amounts of probe DNA will lead to high background):

$$\text{Volume (μl)} = \frac{V \times C \times T}{D}$$

where V = volume of hybridization solution (ml)
C = desired concentration of probe DNA (ng/ml)
T = total volume of probe solution (μℓ)
D = amount of DNA taken into labeling reaction (ng)

3. Using precautions for handling ^{32}P, add 1/3 volume of 2 N NaOH to the labeled probe (to final concentration of 0.5 N NaOH), mix by pipetting, and incubate at room temperature for at least 7 min to denature the probe.

4. Working behind the Plexiglas® shield, remove 1 ml of hybridization buffer from the tray and mix with an appropriate aliquot of denatured probe as calculated in step D3 (do not add probe to the tray directly because it may lead to high background).

5. Add mixture back to the tray. Tilt tray back and forth to ensure complete submersion of membrane. Cover the tray with Saran Wrap™ to reduce air volume and minimize evaporation, and then close the lid.

6. Return the tray to the 65°C incubator. Incubate with shaking for at least 2 hr or overnight. Overnight incubation enhances the hybridization signal.

Washing the Filter

1. Prepare a tray with 500 ml of washing solution 1.

2. Operating behind the Plexiglas® shield, remove the filter from the hybridization tray with forceps and place into the tray with washing solution 1. Decant the radioactive hybridization buffer into a liquid radioactive waste container and place the hybridization tray in Rad-Con solution.

3. Cover the lid and incubate the membrane at room temperature with shaking for at least 10 min.

4. Decant washing solution 1 into a designated for radioactive waste sink.

5. Fill the washing tray with washing solution 2 and incubate at 65°C with shaking for at least 30 min.

6. Repeat step 5.

7. Blot the filter on a paper towel and discard the paper towel into radioactive waste container.

8. Check the radioactivity of the filter with Geiger counter. It will help to estimate the exposure time.

9. Wrap the filter in Saran Wrap™ or plastic sheet protector.

10. Place into a film cassette with intensifying screen (for film exposure) or with no intensifying screen for exposure to phosphorimaging plates and tape it in place to prevent slipping.

11. Expose the membrane to Hyperfilm in the darkroom or to phosphoimaging plate in dim light.

12. Place cassette at –70°C for appropriate time for film exposure or keep at ambient temperature for phosphorimaging exposure.

13. Develop exposed film in film processor or process in the phosphorimager.

Analysis of Results

1. Visually determine DNA content in the "protein" (1), "protein + spike" (2) and "buffer" (3) samples by comparing the signal intensity in corresponding slots to the standard curve.

2. Calculate percent spike recovery (% SR):

$$\% \ SR = \text{e.am. of DNA in (2)}/(\text{e.am. of DNA in (1)} + \text{e.am. of DNA in (3)} + \text{e.am. of DNA in spike}) \times 100$$

where e.am. = estimated amount

3. Calculate DNA content of the "protein" sample in pg DNA / mg protein:

$$\text{DNA content} = \text{pg DNA in slot/mg protein applied}$$

4. Report the data on the test request and calculation sheet.

Appendix 14.2: BACTERIOPHAGE ASSAY IN SAMPLES OF MEDIA AND CELLS FROM FERMENTOR

Reagent Solutions

1. 1.5 percent agar:

   15 g Select Agar
   LB to 1 liter
   Autoclave. Store at ambient temperature.

2. Top agar:

   5 g Select Agar
   (Luria-Bertani) LB to 1 liter
   Autoclave. Store at ambient temperature.

3. Plates:

   Melt 1.5 percent LB Agar in a microwave. Cool to about 60°C.
   Add 200 µl kanamycin stock to 500 ml agar. Pour into plates.
   Allow to cool for about 30 min. Tape the lids with parafilm.
   Store inverted at 4°C for up to 3 months.

Procedure

Notes and Precautions:

When doing sample dilutions, use each pipette one time to ensure no phage is introduced into the assay while pipetting. Use sterile techniques for all procedures. Place all waste in autoclavable biohazard bags and make sure they are sent out to autoclave.

Preparation of Indicator Cells for Assay

1. Inoculate 3 ml LB in a plastic tube with 10 µl of indicator cells from MWCB or MCB stock vial. It is also possible to use 50 µl of cultures from expansion flasks stored at 4°C for up to one week.

2. Incubate at 37°C with shaking (200–250 rpm).

3. After 1–2 hr add 2 µl kanamycin stock (or other antibiotic, if needed).

4. Continue incubation overnight.

5. Transfer overnight culture into sterile 125 ml shake flask containing 25 ml LB and 10 µl kanamycin stock.

6. Incubate at 37°C with shaking (200–250 rpm) for 1.5–2 hr (to OD_{600} of about 1).

Preparation of Samples for Use in the Assay

1. Using sterile techniques remove 1 ml from fermentation sample and place in a sterile 2 ml centrifuge tube.

2. Add 10 µl chloroform and shake for about 15 min.

3. Spin at 14,000 rpm in a microcentrifuge for 5 min.

4. Carefully transfer the supernatant into a fresh microcentrifuge tube.

5. Make dilutions of the supernatant as follows:

    For 10^{-2} dilution: take 10 µl undiluted supernatant and add 0.9 ml LB;
    For 10^{-3} dilution: take 0.1 ml of 10^{-2} dilution and add 0.9 ml of LB.

6. Use 0.1 ml of the undiluted sample and of 10^{-2} and 10^{-3} dilutions to perform the test.

Preparing Negative Controls

Take 0.3 ml of indicator cells and 0.1 ml of LB used to dilute the samples as a negative control for the assay.

Assay Procedure

1. Prewarm the plates at 37°C so that the top agar will spread evenly on the surface.

2. Heat top agar to boiling to melt and cool to 47°C in a water bath (it will take about 1 hour).

3. Place 0.3 ml of indicator cells from step A6 in sterile glass culture tubes.

4. Add 0.1 ml of samples prepared in step B. Mix gently. Cap the tubes.

5. Place the mixture and negative controls from Step C on ice for 20 min to allow phages to infect the cells.

6. Add 5 ml of soft agar to each tube, including negative controls, and immediately pour onto marked prewarmed plates.

7. Leave the plates on the bench for 45–60 min to ensure that the top agar has solidified.

8. Place the plates in a 37°C incubator, inverted.

9. Incubate overnight.

10. Score each plate for the number of plaques present.

Appendix 14.3: PLASMID RETENTION ASSAY BY PCR

Reagent Solutions

1. GeneAmp PCR Reagent Kit with Ampli Taq DNA Polymerase (Perkin Elmer, cat #N801-0055)

2. Custom primers

3. Custom competitive template

4. 1.5 percent agarose/1 × TBE with ethidium bromide

5. 1 × TBE

Procedure

Note: Use pipettes with positive displacement (e.g. Gilson's Microman series) for all procedures. Exercise extreme caution to prevent cross-contamination of the reagents and samples with target plasmid.

PCR Reaction

1. Using sterile techniques, remove 1 ml from the sample of cells and media from fermenter.

2. Spin down for 5 min in a microcentrifuge at 14,000 rpm.

3. Transfer supernatant to fresh tube.

4. Dilute custom primers to approximately 50 pmol/μl.

5. Dilute competitive template to approximately 0.5 pg/μl and 0.05 pg/μl (may vary depending on the expected concentration of plasmid in the media).

6. Prepare dNTPs mixture as follows (final concentration 1.25 mM of each):

   take 125 μl of each dNTP,

   add 500 μl of water.

7. Prepare the master mix (final volume depends on the number of samples N):

   16 × N μl dNTP mix

   10 × N μl 10× PCR buffer

   0.5 × N μl of forward primer

   0.5 × N μl of backward primer

   0.5 × N μl of Taq DNA polymerase

   59.5 × N μl water

8. Take an 87 µl aliquot of master mix for no template control (tube #7).

9. Add 3 × (N–1) µl of media sample prepared in step A3 to master mix.

10. Distribute 90 µl aliquots of the master mix to PCR tubes.

11. Number the tubes and add the following volumes of the competitive fragment dilutions prepared in step A5:

Tube #	µl of 0.5 pg/µl	µl of 0.05 pg/µl	µl water
1	10	0	0
2	5	0	5
3	0	10	0
4	0	5	5
5	0	1	9
6	0	0	10

12. Mix the contents of the tubes and spin the drops down in a microcentrifuge.

13. Place the tubes into thermocycler.

14. Run 35 cycles of 1 min at 94°C, 30 sec at 58°C (or appropriate annealing temperature), and 1 min 72°C.

Gel Electrophoresis

1. Pour 1.5 percent agarose gel with ethidium bromide in 1 × TBE. Allow to solidify.

2. Apply 10 µl aliquots of PCR samples and a molecular weight marker to the gel.

3. Run at approximately 8 V/cm for 20–30 min.

4. Photograph the gel.

5. Find the sample where approximately equal amounts of target and competitive templates were amplified (3 µl of media used in reaction contain an approximately equal number of target templates).

6. Calculate plasmid retention as follows:

$$\% \text{ plasmid retention } = \frac{100 - \text{ number of plasmids in media}}{\text{theoretical number of plasmids in cells}} \times 100$$

where, number of plasmids in cells = copy number per cell $\times$ cell concentration and

number of plasmids in media =

$$\frac{\text{concentration of plasmid DNA in media (PCR)}}{\text{molar weight of the amplified fragment}} \times 6.02 \times 10^{23}$$

15

Semiautomated Quantitation of Microbes in Air

Gordon S. Oxborrow

3M Corporation

Microbial air sampling is a process used to detect and evaluate populations of airborne microorganisms that have economic or health consequences. Governmental regulatory bodies, such as the FDA, use air sampling data to evaluate the adequacy of a manufacturing environment for the production of medical supplies and food. Manufacturers of products that are FDA regulated use air sampling data to assure themselves and the regulators that they meet the GMPs. Healthcare providers use air sampling to monitor environments where microbes, both pathogenic and nonpathogenic, may impact the well being of patients or staff members.

Many organizations downplay the importance of air sampling in their operations. It is always to these data, however, that organizations turn to when they discover problems associated with microorganisms. In these instances an organization wants to prove that the environment was adequate for the intended purpose. Consequently, the data must be collected with the best equipment and most appropriate methodologies. This chapter introduces the need for air samplers, provides information concerning the selection and use of air samplers, and describes the elements of an air sampling plan.

AIR SAMPLING: NEED AND DEVELOPMENT?

Why do we need to do air sampling? There are other ways to detect microorganisms in the environment, such as surface sampling or the determination of the number of microorganisms on a product (bioburden). This question is asked continually by administrators and operators where the cleanliness specifications require microbial control. When no direct relationship is observed between viable particles in the air and the bioburden on the product, the answer, of course, is that there are only three basic sources of microbial contamination:

1. Raw or existing materials

2. Personnel

3. The environment

When the environment is controlled, the first two often contribute the largest part of microbial load in the environment; where the environment is not controlled, the environment contributes to the other two. Therefore, to determine and control the contribution from each source, each of the three sources must be monitored.

Early microbiologists recognized that exposure to microbiological growth medium in open containers resulted in microbial growth. This led to the theory of spontaneous generation, which was shown by Pasteur to be false, that life came from life. The practical application was when medical scientists learned that microorganisms could be transmitted from one patient to another through the air, and that the organisms frequently could be detected using open dishes of microbial growth media.

The development of modern air samplers can be attributed to those individuals who were working for governments in the search for biological weapons. It was necessary for them to detect the presence of microorganisms in aerosols, and to determine the types of microorganisms and their distribution. This evolved into the basic designs for microbial air samplers that are used today.

The National Aeronautics and Space Administration (NASA) Planetary Quarantine Group wished to detect and evaluate the microbial content of air in white rooms (clean rooms) where spacecraft were assembled. The spacecraft that were to contact planets other than Earth had to be sterilized in accordance with international agreement. This allowed detection of possible life on other planets.

Oxborrow et al. (1974) evaluated air sampling as an alternative to surface sampling on spacecraft. Air sampling data related mathematically to surface contamination data only in closely controlled and regulated environments; no two environments tested could be described by the same mathematical model. Where small numbers of microorganisms were found per unit volume of air sampled, large volumes of air were required to provide the data necessary for statistical analysis.

Environments in which healthcare products are being manufactured require air sampling techniques that can detect small numbers of viable particles per unit volume of air. Ideally, the sampling procedure is done without disturbing the normal airflow of the room and without adding particles to the area. Where airflow is disturbed, the data collected may not represent the true values of the area being sampled. Modifications to basic air sampling equipment have improved detectability and recoverability of viable particles in environments where low numbers of viable particles are present. For instance, increased flow rates and increased velocities have significantly improved microbial recovery.

In healthcare facilities, businesses, manufacturing areas, and residences, the primary concern may be airborne pathogens or allergens derived from microbial populations. Fungi have been shown to be both pathogens and allergens. Several investigators (Kozak et al. 1980; Samson 1985; Verhoeff et al. 1990; and Su et al. 1992) have reported on the allergenic effect of fungi on people living in areas where high humidity encourages fungal growth. Numerous studies (Artenstein and Miller 1966; Artenstein et al. 1968; Groschel 1980; Laham et al. 1982; Donaldson 1982; Staiab 1985; Kang and Frank 1990; and Morey 1990) have described the relationship between microbial agents in the air and respiratory disease.

AIR SAMPLING DEVICES

Ideally, air sampling devices collect viable microorganisms or viable particles with high efficiency, without damage to the microorganisms. Some samplers are designed to sample particles and some to sample viable microorganisms. In most cases one or more viable microorganisms are attached to, or are transported by, a single particle. Therefore, a system that detects colony forming units (CFUs) per volume of air may seriously underestimate viable microorganisms. The data from one design of sampler should not be expected to yield the same data as a sampler of a different design.

The ideal air sampler should be simple to use, inexpensive, lightweight, easy to move, and should have its own power source. It should sample the air with little or no disruption of the environment and without contributing particles to the sampled area. Unfortunately, no samplers fit all of these criteria.

Table 15.1 lists characteristics of commercial samplers. This information, together with the following brief descriptions, may aid in the selection of a sampler for a particular application. Additional information may be found in *Microorganisms in Cleanrooms* (IES-RP-23.1 [1993]).

Air samplers allow particles to settle onto a surface, or accelerate an airsteam to impact or impinge on surface that may be liquid, solid, or semisolid. Most collection systems depend on Newton's First Law of inertia that states: *"If a body in motion is not acted upon by an external force, its momentum remains constant"* (Law of Conservation of Momentum). Therefore, a particle in straight line motion maintains the same velocity and direction unless it intercepts an immovable plane. For slit, sieve, and impinger samplers, and so on, the speed (particle mass remains constant) of the particle in an airstream is increased such that the particle cannot make a right angle turn at the surface of the collecting medium under a slit, sieve, or critical orifice.

Particles of large mass deposit at lower air velocities than particles with small mass. The speed of the air as it impacts the surface is known as the critical velocity. With centrifugal and cyclone samplers, a particle of a specific size is accelerated in a tangential flow of air until it reaches a critical velocity. At that velocity the particle is thrown out of the airstream and onto the collecting surface.

Table 15.1. Sampler Characteristics

Sampler	Sampling Rate ℓ/min	Maximum CFU/m³ × 10ˣ	Collection Media	Minimum Particle Size Sampled
Large volume (Cassella)	700	3	agar	20
Slit-to-agar	28	3	agar	0.2
Spore trap	10	1	tape	6
Sieve (Andersen)	28	3	agar	1.2
Sieve (SAS)	180	3	agar	10
RCS	40	4	agar	5
Cyclone	75	4	liquid	3
All-glass impinger	125	no limit	liquid	2
Membrane filters	1–30	no limit	filter	0.1
Fallout plates	n/a	no limit	agar	n/a
Nonnutrient fallout plates	n/a	no limit	solid	n/a

The physical characteristic that makes one inertial impaction sampler more efficient than another is the speed of the airstream. A high velocity airstream impacts more and smaller particles onto the collecting surface than a low velocity system.

Membrane filter systems, unlike the inertial impaction samplers, act as true sieves by collecting particles that cannot pass through the pores of the filter. Filters (sieves) are not dependent on a critical velocity to collect particles.

The oldest and most common means of collecting particles is by sedimentation. Viable particles are collected either by direct fallout onto microbiological growth media, or indirectly by fallout onto a nonnutritive material such as stainless steel strips of a given dimension and surface finish. The particles are collected after varying periods of time, and they are dispersed in a liquid menstruum for analysis.

One of the best sources for describing numerous samplers and their characteristics is entitled *Sampling Microbiological Aerosols* (Wolf et al. 1965).

Slit-to-Agar Samplers

Slit samplers draw environmental air through a slit by means of a vacuum. Each sampler type has its own slit size, critical air velocity, and critical distance from the bottom of the slit to the agar surface. The slit width and length, and the volume of air drawn through the slit, determine the critical velocity and the sampler efficiency. Slit sizes range from 0.2 mm to 1.0 mm and air velocities range from 10 m/sec to 50 m/sec, depending on sampler design. Distances from the slit to the agar vary from 5 mm to 25 mm, depending on the velocity of the air through the slit and the dispersion of the air between the slit and the agar. Most samplers draw air through the slit at 1 cfm (28.32 ℓ/min).

Slit samplers usually have a mechanism to rotate or move the collecting growth medium under the slit. The speed of rotation (movement past the slit) allows the user to determine the number of viable particles per unit volume during any time interval. Studies by Favero et al. (1966, 1967, 1968) used slit-to-agar air samplers to show the contribution to the air of microorganisms by personnel in clean rooms and operating rooms over a 12-hour time interval.

Sieve Samplers

A sieve sampler consists of a metal or plastic plate with holes of uniform size and a nutrient collecting medium under the holes. Hole sizes for the most common sieve sampler (Andersen) vary from 1.81 mm for stage one to 0.25 mm for stage six. As air is drawn through the holes in an individual sampler plate, a critical air-flow velocity is reached for particles of a specific size. Those particles impact on the collecting surface. The critical distance from the sieve to the agar surface is controlled by the amount of agar in the collection plate. When plates with different sized holes are sequentially stacked so that air is drawn through the plates with the largest holes first, and finally through the plates with the smallest holes, particles can be separated into a size distribution. This type of information may be useful in determining the source of viable particles. When operations in the area being sampled are known to produce particles of specific sizes, the contribution of each operation to the total microbial load can be determined and subsequently controlled and monitored.

Centrifugal Samplers

Centrifugal samplers operate by accelerating air with entrained particles until a critical velocity is reached for a given particle size. The blades of an impeller force the air onto the agar surface. The particles impact on a nutrient agar surface. The volume of air sampled may be difficult to determine because the air mixes as it enters and exits the sampler. Kaye (1988) proved this for the Reuter Centrifugal Sampler (RCS) manufactured by Biotest, Inc. Newer models of the RCS overcame this deficiency by allowing free-flow of air through the sampler. Sampling times with this type of device are usually short. Large volumes of air moving over the surface of the growth medium cause it to dehydrate.

Cyclone Samplers

The design of the cyclone sampler is similar to that of a standard dust collector. The particles are collected in a liquid menstruum that is injected into the collecting area to form a film on the inner surface of the cyclone. The tangential velocity of the air increases as the diameter of the collector decreases and particles entrained in the air are impinged into the liquid film. The liquid collects at the bottom as the air is ejected through the top of the cyclone. This type of sampler can be used for long sampling times (large volumes of air) since there is a constant makeup of the liquid.

All-Glass Impingers

All-glass impingers (AGIs) draw particle laden air through a critical orifice located at the end of a glass tube that ends just above the liquid in the bottom of a collector vessel. The critical orifice is an opening of a specific size that limits the volume and velocity of the air as it passes through the hole when a vacuum is applied. The collecting vessel has a specific volume of liquid (i.e., phosphate-buffered saline) to collect the particles. The particles impact a thin film of liquid that remains between the orifice and the bottom of the collection vessel. The distance from the orifice to the film of liquid is specific for each sampler. The AGI can be operated for only a short time because of the rapid evaporation of the collecting medium. Particles frequently are carried out of the sampler in air bubbles, or they are impinged on the upper surfaces of the container as bubbles explode upon reaching the surface. The collecting fluid frequently needs an antifoaming agent to keep the foam under control.

Bubblers are a crude type of AGI. They do not lend themselves to quantitative evaluation of the microorganisms in an aerosol. Critical velocities in the airstream are not reached because the hole size of the immersed tube is large and the volume of air sampled is small. This type of sampler does not have a collection surface. Particles remain in the airstream and are released back into the air as the bubbles burst. The collection efficiency is low and inconsistent.

Membrane Filter Samplers

Membrane filters, in contrast to other samplers, are very efficient in the collection of particles. However, fewer viable particles are recovered compared to other sampler types. When Fields et al. (1966) compared the number of viable particles collected per cubic foot of air using membrane filters with those collected using the Andersen Sampler, they found significantly fewer viable particles on the membrane filters. The difference was due to dehydration of the microorganisms on the filter as air was drawn through the filter.

Many types of membrane filter media (cellulose acetate, cellulose nitrate, cellulose mixed esters, alginate, gelatin, or other polymeric types) are used for this type of sampling. Some polymeric filters have etched holes of a specific size and others are depth filters that collect particles on the fibers. The alginate or gelatin filters can be dissolved in a water-based collecting medium for microbial recovery. A wide range of filter pore sizes (from 0.2 mm to 1.0 mm) have been used.

Low air velocities and short sampling times improve the efficiency of collecting viable particles.

Sedimentation Samplers

Sedimentation samplers are usually petri plates of any size that contain a solidified agar growth medium. They are placed on work surfaces with the lids removed. Viable particles that happen to fall onto the surface of the growth medium are collected and allowed to grow. This approach to sampling airborne microorganisms is inefficient; only large particles are collected because the normal air currents in most areas keep smaller particles suspended. Although most people consider this to be a nonquantitative method of sampling, Whyte and Niven (1986) developed a formula to estimate the particles in the air from the data collected on fallout plates. This author does not agree with Whyte's and Niven's approach on a routine basis. The number of particles falling onto stainless steel strips are related to the number of particles in the air only in specific areas under strictly controlled conditions. Many factors preclude the use of fallout plates: particle size, particle distribution, airflow patterns, air velocities, and the many routine activities in the area being sampled.

Nonnutritive sedimentation plates may be made from strips of stainless steel, plastic, glass, or other nonporous materials. Such plates were used extensively to evaluate the clean rooms for spacecraft assembly during the 1960s and 1970s. The usual protocol was to expose trays (containing 20 to 40 strips) to the area of interest, and collect a set number of strips at intervals up to 6 weeks. Viable microorganisms were recovered by washing the strips in a rinse solution and plating out samples with a nutrient agar growth medium. This type of sampler suffers the same type of problems that occur with fallout plates. They are, however, useful in assessing the contribution of airborne particles to the contamination of horizontal surfaces.

SAMPLING PLANS

A sampling plan is a necessary part of any environmental monitoring program. The plan defines the responsibilities, considerations, procedures, the interpretation of data, and the report format.

A proper sampling plan must take the following into consideration:

- The environmental control of the room (temperature, relative humidity, air change rate, air filtration, and air pressure)

- The room geometry and volume

- The room layout (equipment and traffic patterns)

- The type of sampler, the types of microorganisms that may be present

- Where to take the samples

- What effect the sampler will have on the room

- The number of sampling locations

- The volume of air to sample

- The frequency of sampling

- Equipment preparation

Environmental control of the room is necessary because efficacious collection of airborne microorganisms may depend on the humidity and temperature. If, for instance, the temperature is warmer than normal (above 28°C) or if the relative humidity is low (below 30 percent RH), some of the microorganisms may die due to dehydration, or the growth media may dehydrate to the point where microorganisms will not reproduce. Electrostatic attraction increases in dry areas and may influence the collection efficiency of the sampler. In this case a liquid-collecting system may be more appropriate than an agar plate sampler. The types of microorganisms detected are also affected by how the environment is controlled.

The volume of the room may affect the choice of a particular type of sampler. It would not be appropriate to use a high volume sampler in a small area. A high volume sampler would affect the airflow, and would temporarily deplete the microbial population. Sample values would be lower than ordinarily would be expected. For particular microbial types, such as molds, special growth media should be used to collect the greatest number. Rose Bengal Agar is the medium of choice for the collection of fungi (Domeyer 1995).

The selection of sampling locations in a room is critical to the quality of the information obtained. The reason for selecting sampling locations must be clearly stated in order to select the sampling site properly and obtain the desired information. Personnel movement patterns and equipment locations have a large affect on the numbers and types of microorganisms that are collected. To obtain data from a high traffic area, for instance, the sampling plan must state the area to be sampled. If a sampling site is chosen that is not adjacent to the area of concern, the data will not provide a true estimate of the microbial population.

The ideal sampler takes nothing from the area sampled except the desired particles, does not disturb the environment that is being sampled, and leaves nothing that may contaminate the area. Sampler choice also depends on the area being sampled. In an area crowded with equipment or personnel, a small quiet unit, such as the RCS, may be desirable. This type of sampler, however, disturbs the surrounding environment because of the large volumes of air it moves. Consequently, the advantages and disadvantages of each sampler type must be weighed carefully before making a selection.

The volume of air to be sampled must be considered, depending on the sampler type, the number of microorganisms per unit volume, and the size of the area. If the number of microorganisms is excessively high, an accurate count on an agar-collecting surface is difficult to achieve because colonies impinge on or overlap each other. Liquid-collecting systems allow dilution of the collected organisms and impose less restrictions on the volume sampled. Remember, however, that samplers that impact particles onto agar surfaces tend to underestimate CFUs. Samplers that collect particles in liquid are more likely to enumerate viable cells. Liquid samplers generally are higher counts because cells absorbed to a particle have the opportunity to desorb, hence be enumerated as individual units.

The frequency of sampling usually depends on consistency in the number of personnel, the activity of the personnel, and the environmental control in the area being sampled. Consistency is dependent on the control and maintenance of the air handling system, personnel hygiene, attention to housekeeping, and the processes or activities that are being done in the area. When activities and equipment in the area to be sampled are constant, a monthly sampling schedule may be adequate. When activities and/or equipment in an area are highly variable, a weekly or even a daily schedule may be more appropriate. In healthcare facilities, air sampling is done to evaluate a cleaning process or to investigate an outbreak of some pathogenic organism. In a manufacturing operation air samples are collected to determine adherence to proper process conditions. Air sampling is also done in personal residences when allergenic responses are encountered. The acquired data helps distinguish fungal or bacterial sources from other allergens in the environment.

Samplers must be properly cleaned, maintained, and calibrated. Part of the cleaning process is sterilization or disinfection of the parts that come in contact with the collecting medium or affect the recovery. Whenever possible, the collecting head should be sterilized. Samplers that cannot be sterilized may be disinfected with 70 percent isopropyl alcohol or other disinfectants. Contaminated and poorly maintained sampling heads will change the way a sampler operates, resulting in misleading data of both microbial types and numbers.

Air samplers should provide data that is representative of the environment in which they are operated. It is the responsibility of the user to interpret that data. To properly interpret the data, the responsible person must be familiar with microbiology and the potential sources of microorganisms.

The responsible person also should understand the capabilities and function of the sampler. To properly interpret the data, the sampling plan must have a stated objective; the plan must provide guidance for doing the sampling and criteria for the interpretation of the collected data. Proper interpretation of the data requires a knowledge of microbiology and the sources of microbial contamination. Where the number and type of airborne microorganisms is important, cost savings can be achieved in the manufacturing process; the risk to patients and personnel in healthcare facilities can be reduced when the types and numbers of airborne microbes are known.

REFERENCES

Artenstein, M. S., and W. S. Miller. 1966. Air sampling for respiratory disease agents in Army recruits. *Bacteriol. Rev.* 30 (3):571–575.

Artenstein, M. S., W. S. Miller, T. H. Lamson, and B. L. Brandt. 1968. Large-volume air sampling for Meningococci and adenoviruses. *Am. J. Epidemiol.* 87 (3): 567–577.

Domeyer, S. L. 1995. Recovery of fungal organisms from the environment. M.S. Thesis, University of Minnesota School of Public Health.

Donaldson, A. I., N. P. Ferris, and J. Gloster. 1982. Air sampling of pigs infected with foot-and-mouth disease virus: Comparison of Litton and cyclone samplers. *Res. Vet. Sci.* 33 (3):384–385.

Favero, M. S., J. R. Puleo, J. H. Marshall, and G. S. Oxborrow. 1966. Comparative levels and types of microbial contamination detected in industrial clean rooms. *Appl. Microbiol.* 14:539–551.

Favero, M. S., J. R. Puleo, J. H. Marshall, and G. S. Oxborrow. 1968. Comparison of microbial contamination levels among hospital operating rooms and industrial clean rooms. *Appl. Microbiol.* 16:480–486.

Fields, N. D., G. S. Oxborrow, and C. M. Herring. 1974. Evaluation of membrane filter field monitors for microbiological air sampling. *Appl. Microbiol.* 27 (3): 517–520.

Groschel, D. H. 1980. Air sampling in hospitals. *Ann. N. Y. Acad. Sci.* 353:230–240.

IES. 1993. Microorganisms in cleanrooms. RP-23.1. Mt. Prospect, IL: Institute of Environmental Sciences.

Kang, Y. J., and J. F. Frank. 1990. Characteristics of biological aerosols in dairy processing plants. *J. Dairy Sci.* 73 (3):621–626.

Kaye, S. 1988. Efficiency of the "Biotest RCS" as a sampler of airborne bacteria. *Parent. Sci. Tech.* 42 (5):147–152.

Kozak, P. P., J. Gallup, L. H. Cummins, and S. A. Gillman. 1980. Currently available methods for home mold surveys. I. Description of techniques. *Annals of Allergy* 45:85–89.

Laham, M. N., B. Jeffery, and J. L. Carpenter. 1982. Frequency of clinical isolation and winter prevalence of different *Aspergillus* species at a large Southwestern Army medical center. *Annals of Allergy* 48:215-219.

Morey, P. R. 1990. Practical aspects of sampling for organic dusts and microorganisms. *Am. J. Indust. Med.* 18 (3):273–278.

Oxborrow, G. S., A. L. Roark, N. D. Fields, and J. R. Puleo. 1974. Mathematical estimation of the level of microbial contamination on spacecraft surfaces by volumetric air sampling. *Appl. Microbiol.* 27 (4):706–712.

Samson, R. A. 1985. Occurence of moulds in modern living and working environments, *Eur. J. Epidem.* 1 (1):54–61.

Staib, F. 1985. Sampling and isolation of *Cryptococcus neoformans* from indoor air with the aid of the Reuter centrifugal sampler (RCS) and Guizotia abyssinica creatinine agar. *Zentralbl Bakteriol Mikrobiol Hyg.* 180 (5–6):567–575.

Su, H., J. A. Rotnitzky, H. A. Burge, and J. D. Spengler. 1992. Examination of fungi in domestic interiors by using factor analysis: Correlations and associations with home factors. *Appl. Environ. Microbiol.* 58 (1):181–186.

Verhoeff, A. P., J. H. van Wijnen, J. S. M. Boleij, B. Brunekreef, E. S. van Reenen-Hoekstra, and R. A. Samson. 1990. Presence of viable mould propagules in the indoor air of houses. *Toxicol. Indust. Health* 6 (5):133–145.

Whyte, W., and L. Niven. 1986. Airborne bacteria sampling: The effect of dehydration and sampling time, *J. Parent. Sci. Technol.* 40 (5):182–188.

Wolf, H. W., P. Skaliy, L. B. Hall, M. M. Harris, H. M. Becker, L. M. Buchanan, C. M. Dahlgren. 1965. Sampling microbiological aerosols. Public Health Monograph No. 60. Washington, DC: Government Printing Office.

Name Index

Microorganism Index

Subject Index